Geoscientific Investigations From the Indian Antarctic Program

Neloy Khare
Ministry of Earth Sciences, India

A volume in the Advances in Environmental Engineering and Green Technologies (AEEGT) Book Series

IGI Global
PUBLISHER of TIMELY KNOWLEDGE

Published in the United States of America by
IGI Global
Engineering Science Reference (an imprint of IGI Global)
701 E. Chocolate Avenue
Hershey PA, USA 17033
Tel: 717-533-8845
Fax: 717-533-8661
E-mail: cust@igi-global.com
Web site: http://www.igi-global.com

Library of Congress Cataloging-in-Publication Data

Names: Khare, Neloy, editor.
Title: Geoscientific investigations from the Indian Antarctic program / Neloy Khare, editor.
Description: Hershey, PA : Engineering Science Reference, [2022] | Includes bibliographical references and index. | Summary: "This book sheds light on some of the geoscientific aspects of the icy continent, attempting to integrate them into a holistic understanding of Antarctica Geoscience and its trajectory of change"-- Provided by publisher.
Identifiers: LCCN 2021044562 (print) | LCCN 2021044563 (ebook) | ISBN 9781668440780 (hardcover) | ISBN 9781668440797 (paperback) | ISBN 9781668440803 (ebook)
Subjects: LCSH: Geology--Research--Antarctica. | Earth sciences--Research--Antarctica.
Classification: LCC QE48.A6 G46 2022 (print) | LCC QE48.A6 (ebook) | DDC 559.89--dc23/eng/20211006
LC record available at https://lccn.loc.gov/2021044562
LC ebook record available at https://lccn.loc.gov/2021044563

This book is published in the IGI Global book series Advances in Environmental Engineering and Green Technologies (AEEGT) (ISSN: 2326-9162; eISSN: 2326-9170)

British Cataloguing in Publication Data
A Cataloguing in Publication record for this book is available from the British Library.

Advances in Environmental Engineering and Green Technologies (AEEGT) Book Series

ISSN:2326-9162
EISSN:2326-9170

MISSION

Growing awareness and an increased focus on environmental issues such as climate change, energy use, and loss of non-renewable resources have brought about a greater need for research that provides potential solutions to these problems. Research in environmental science and engineering continues to play a vital role in uncovering new opportunities for a "green" future.

The **Advances in Environmental Engineering and Green Technologies (AEEGT)** book series is a mouthpiece for research in all aspects of environmental science, earth science, and green initiatives. This series supports the ongoing research in this field through publishing books that discuss topics within environmental engineering or that deal with the interdisciplinary field of green technologies.

COVERAGE

- Sustainable Communities
- Pollution Management
- Green Transportation
- Industrial Waste Management and Minimization
- Renewable Energy
- Contaminated Site Remediation
- Electric Vehicles
- Water Supply and Treatment
- Radioactive Waste Treatment
- Waste Management

IGI Global is currently accepting manuscripts for publication within this series. To submit a proposal for a volume in this series, please contact our Acquisition Editors at Acquisitions@igi-global.com or visit: http://www.igi-global.com/publish/.

The Advances in Environmental Engineering and Green Technologies (AEEGT) Book Series (ISSN 2326-9162) is published by IGI Global, 701 E. Chocolate Avenue, Hershey, PA 17033-1240, USA, www.igi-global.com. This series is composed of titles available for purchase individually; each title is edited to be contextually exclusive from any other title within the series. For pricing and ordering information please visit http://www.igi-global.com/book-series/advances-environmental-engineering-green-technologies/73679. Postmaster: Send all address changes to above address.

Titles in this Series

For a list of additional titles in this series, please visit:
http://www.igi-global.com/book-series/advances-environmental-engineering-green-technologies/73679

Handbook of Research on Water Sciences and Society
Ashok Vaseashta (International Clean Water Institute, USA & Transylvania University of Brasov, Romania & Academy of Sciences of Moldova, Moldova) Gheorghe Duca (Institute of Chemistry, Moldova State University, Moldova & Academy of Science of Moldova and Romanian Academy, Romania.) and Sergey Travin (Federal Research Center for Chemical Physics and N. N. Semenov of the Russian Academy of Sciences, Russia)
Engineering Science Reference • © 2022 • 728pp • H/C (ISBN: 9781799873563) • US $465.00

Optimization and Decision-Making in the Renewable Energy Industry
Figen Balo (Fırat University, Turkey) Ayşe Topal (Nigde Omer Halisdemir University, Turkey) Ezgi Demir (Sumer Robotics Engineering and Consultancy Ltd., UK) and Alptekin Ulutaş (Sivas Cumhuriyet University, Turkey)
Engineering Science Reference • © 2022 • 304pp • H/C (ISBN: 9781668424728) • US $245.00

Improving Integrated Pest Management for Crop Protection and Food Security
Muhammad Haseeb (Florida A&M University, USA) Lambert H.B. Kanga (Florida A&M University, USA) and Jawwad A. Qureshi (University of Florida, USA)
Engineering Science Reference • © 2022 • 315pp • H/C (ISBN: 9781799841111) • US $195.00

Prevention and Management of Soil Erosion and Torrential Floods
Slobodan Milutinović (University of Niš, Serbia) and Snežana Živković (Faculty of Occupational Safety, University of Niš, Serbia)
Engineering Science Reference • © 2022 • 335pp • H/C (ISBN: 9781799884590) • US $195.00

For an entire list of titles in this series, please visit:
http://www.igi-global.com/book-series/advances-environmental-engineering-green-technologies/73679

IGI Global
PUBLISHER of TIMELY KNOWLEDGE

701 East Chocolate Avenue, Hershey, PA 17033, USA
Tel: 717-533-8845 x100 • Fax: 717-533-8661
E-Mail: cust@igi-global.com • www.igi-global.com

This book is dedicated to the late
Professor Satyendra Kumar Singh
(14.05.1935 - 17.02.2006)

Table of Contents

Foreword xvi

Preface xvii

Acknowledgment xxi

Chapter 1
Glacial Sedimentology of Schirmacher Oasis, East Antarctica: Indian Perspective 1
Ashok Kumar Srivastava, SGB Amravati University, India
Neloy Khare, Ministry of Earth Sciences, India

Chapter 2
Geochemical and Mineralogical Studies of Gneiss-Charnockite Rocks of Schirmacher Region, East Antarctica 32
D. Rameshwar Rao, Wadia Institute of Himalayan Geology, Dehradun, India

Chapter 3
A Comprehensive Study of the Structural Geology of the Schirmacher Hills 68
Sudipta Sengupta, Jadavpur University, India

Chapter 4
Pollen-Spore Transport Into Antarctica and Possible Past Climatic Oscillations 84
Samir Bera, Birbal Sahni Institute of Palaeobotany, India

Chapter 5
Integrated G3 Investigations in Antarctica 97
H. K. Gupta, National Geophysical Research Institute, India
G. S. Mittal, National Geophysical Research Institute, India
S. K. Verma, National Geophysical Research Institute, India
D. V. Divakaria Rao, National Geophysical Research Institute, India
M. S. Joshi, National Geophysical Research Institute, India
R. L. Dhar, National Geophysical Research Institute, India
S. C. Jain, National Geophysical Research Institute, India
A. V. Ramana Rao, National Geophysical Research Institute, India
P. Rama Rao, National Geophysical Research Institute, India
T. Madoom Hussain, National Geophysical Research Institute, India
D. V. Subba Rao, National Geophysical Research Institute, India
S. S. H. Jafri, National Geophysical Research Institute, India
V. N. Sarma, National Geophysical Research Institute, India
B. Vijayagopal, National Geophysical Research Institute, India
K. N. S. Reddy, National Geophysical Research Institute, India
V. M. Tiwari, National Geophysical Research Institute, India
M. B. Vyageshwar Rao, National Geophysical Research Institute, India
Md. Ghous Uzzaman, National Geophysical Research Institute, India

Chapter 6
GPR Surveys in Antarctica to Map Ice Thickness and Rock Topography 121
Rajib Kumar Sinharay, Central Water and Power Research Station, India

Chapter 7
Studies on Seismotectonics and Geodynamical Processes Between India and Antarctica: A Review 138
E. C. Malaimani, CSIR-National Geophysical Research Institute, India
N. Ravikumar, CSIR-National Geophysical Research Institute, India
S. V. R. R. Rao, CSIR-National Geophysical Research Institute, India
S. Padhy, CSIR-National Geophysical Research Institute, India
A. Akilan, CSIR-National Geophysical Research Institute, India
K. Abilash, CSIR-National Geophysical Research Institute, India
L. Prem Kishore, CSIR-National Geophysical Research Institute, India
G. B. Navinchander, CSIR-National Geophysical Research Institute, India
G. S. Srinivas, CSIR-National Geophysical Research Institute, India
U. Gowrishankar, CSIR-National Geophysical Research Institute, India
I. P. Raju, CSIR-National Geophysical Research Institute, India
Y. V. V. B. S. N. Murthy, CSIR-National Geophysical Research Institute, India
M. Venkatrayudu, CSIR-National Geophysical Research Institute, India
A. Nageshwara Rao, CSIR-National Geophysical Research Institute, India

Chapter 8
Hydrographic Surveys in Antarctica 160
Prashant Srivastava, Ministry of Earth Sciences, India

Chapter 9
An Assessment of Geo-Scientific Investigation in the Antarctic Region 194
Rajesh Asthana, Geological Survey of India, India
Amit Dharwadkar, Geological Survey of India, India
Prakash Kumar Shrivastava, Geological Survey of India, India

Chapter 10
An Overview of Survey and Mapping of the Antarctic Region Around Maitri Research Base 215
Nagarajan Balasubramanian, Indian Institute of Technology, Kanpur, India

Chapter 11
Antarctic Laboratory Ground Truth for a Microwave Eye in the Sky 227
Mukund Kurtadikar, Department of Physics, JES College, India

Chapter 12
Ice Core Records for Paleo-Volcanism, Climate, and Snow Accumulation Rates Over the Past 150 Years 240
D. K. Rao, Physical Research Laboratory, Department of Space, India
R. A. Jani, Physical Research Laboratory, Department of Space, India

Chapter 13
India's GeoScience Pursuit in the Antarctica: Initial Attempts 260
Neloy Khare, Ministry of Earth Sciences, India

Compilation of References 269

About the Contributors 297

Index 303

Detailed Table of Contents

Foreword xvi

Preface xvii

Acknowledgment xxi

Chapter 1
Glacial Sedimentology of Schirmacher Oasis, East Antarctica: Indian Perspective 1
Ashok Kumar Srivastava, SGB Amravati University, India
Neloy Khare, Ministry of Earth Sciences, India

Schirmacher Oasis, East Antarctica is under investigation for the last four decades by Indian scientists of various disciplines, of which geological studies constitute a significant aspect. The present attempt is a review of sedimentological studies carried out in the area. A few of the current chapter's elements are sediment characteristics, mode, and pattern of sediment deposition, source rock, provenance, paleoclimate, and weathering condition through clay mineral studies, grain surface studies, etc., with specific aspects more data on the number of samples. Simultaneously, the Oasis also offers good scope for various other significant investigations like paleoclimatic reconstruction, source of glacial sediments, silicates weathering, etc. Globally, they are a front-line work area and recommended.

Chapter 2
Geochemical and Mineralogical Studies of Gneiss-Charnockite Rocks of Schirmacher Region, East Antarctica 32
D. Rameshwar Rao, Wadia Institute of Himalayan Geology, Dehradun, India

The studies of multifaceted problems of gneiss-charnockite rocks in the Schirmacher region of East Antarctica suggest a retrograde clockwise isobaric cooling P-T history of the terrain involving an early granulite phase, a late granulite phase, and a retrograde amphibolite phase metamorphism in the region. Also, a good correlation between fluid and mineral data is observed. The high-density CO2 fluids fall well within the P-T box estimated by mineral thermobarometers, envisaging a pervasive influx of deep-seated CO2 rich fluid from the mantle resulting in the formation of the granulites, followed by decrease CO2 density fluids along with progressive influx of hydrous fluids leading to the generation of retrograde amphibolite facies rocks. The geochemical studies helped trace two-phase evolution of the region (basic magmatism around ~1200 Ma) involving a depleted mantle source implying an accretion of the juvenile crust during the late Mesoproterozoic period and major felsic magmatism around ~880 Ma involving partial melting of mafic-intermediate crust.

Chapter 3
A Comprehensive Study of the Structural Geology of the Schirmacher Hills 68
Sudipta Sengupta, Jadavpur University, India

The Precambrian basement of the Schirmacher Hills records multiple episodes of deformation, metamorphism, migmatization, and emplacement of successive generations of mafic and felsic bodies. The earliest tectono-thermal event (D1/M1), preserved in some mafic and ultramafic enclaves, indicates deformation at great crustal depth. The mineralogical assemblage of these enclaves indicates early high temperature (900o C) and high-pressure (10 Kbar) granulite facies conditions. The second tectono-thermal event also showed deformation under granulite facies metamorphism (D2/M2) under 800-850oC and 8 Kbar. The third group of events (D3/M3) is the most dominant in this region and involved deformation under amphibolite facies conditions with synchronous emplacement of granites and mafic dykes and culminated in regional thrusting, producing a regional inversion where the granulates were emplaced over the amphibolite facies rocks. The later events created upright folds and vertical shear zones under amphibolite facies conditions.

Chapter 4
Pollen-Spore Transport Into Antarctica and Possible Past Climatic Oscillations84
Samir Bera, Birbal Sahni Institute of Palaeobotany, India

During XIX-XX Indian Antarctic Expedition (IAE), two sediment cores were collected. The occurrence of the exotic pollen taxa like Larix, Ulmus, grasses, few herbs, local moss spores, and other cryptogams prove the activity of the palaeowind, which in turn caused transportation microbiota from a long distance. The study of pollen-spores accumulation in polar air (deduced from slide exposures from 40° S) to Antarctic mainland studied for two years (2000 and 2001) as well as surface deposits (45 moss tuft, 15 lichen samples, 10 frozen soils, 10 moraine matrix) is well corroborated with lake sediment study. Ten bulk ice samples from the Antarctic ice cap (5 litres of ice melt) from Schirmacher Oasis were studied to recover trapped palynodebris to understand the depositional pattern of various microbiota in the Antarctic ice sheet.

Chapter 5
Integrated G3 Investigations in Antarctica 97
H. K. Gupta, National Geophysical Research Institute, India
G. S. Mittal, National Geophysical Research Institute, India
S. K. Verma, National Geophysical Research Institute, India
D. V. Divakaria Rao, National Geophysical Research Institute, India
M. S. Joshi, National Geophysical Research Institute, India
R. L. Dhar, National Geophysical Research Institute, India
S. C. Jain, National Geophysical Research Institute, India
A. V. Ramana Rao, National Geophysical Research Institute, India
P. Rama Rao, National Geophysical Research Institute, India
T. Madoom Hussain, National Geophysical Research Institute, India
D. V. Subba Rao, National Geophysical Research Institute, India
S. S. H. Jafri, National Geophysical Research Institute, India
V. N. Sarma, National Geophysical Research Institute, India
B. Vijayagopal, National Geophysical Research Institute, India
K. N. S. Reddy, National Geophysical Research Institute, India
V. M. Tiwari, National Geophysical Research Institute, India
M. B. Vyageshwar Rao, National Geophysical Research Institute, India
Md. Ghous Uzzaman, National Geophysical Research Institute, India

CSIR-NGRI has been carrying out integrated G3 investigations in Antarctica since the second IAE. The geophysical studies of the initial 25 years of IAE included surface and helicopter-borne magnetic, EM, seismic, gravity, and paleomagnetic surveys. A total of 60 line-km magnetic profiles over the ice-shelf revealed the magnetic characteristics of the bedrock beneath the ice cover. Based on these and in conjunction with the early seismic studies, a crustal structural model was evolved. Secondary sulfide mineralisation near a suspected fissure zone in Schirmacher Oasis (SO) was inferred. It was based on the multi-frequency EM and radiometric measurements. The helicopter-borne magnetic N-S profiles over an area of 100x100sq.km with a spacing of about 3.0 km between SO and Wohlthat Mountains (WM) yielded gross features of sub-glacial topography nunataks (exposed peaks of sub-glacial hills).

Chapter 6
GPR Surveys in Antarctica to Map Ice Thickness and Rock Topography 121
Rajib Kumar Sinharay, Central Water and Power Research Station, India

Ground penetrating radar (GPR) surveys have been carried out at Schirmacher Oasis and Dakshin Gangotri located at Queen Maud Land, East Antarctica, during the 22nd Indian Antarctic Summer Expedition, 2002-2003. The present study confirmed the ability of the high-resolution GPR for monitoring the glaciers. It gives information about the health of the glaciers before it collapses. GPR survey over three frozen lakes near the Maitri Station provided the lakes' top ice thickness and bedrock depth. Similarly, the internal layers of the glaciers have been mapped in-situ using GPR. The results can be correlated with the results obtained by ice-core drill wells to understand the different field parameters (i.e., thickness of each layer, dielectric constant, etc.).

Chapter 7
Studies on Seismotectonics and Geodynamical Processes Between India and Antarctica: A Review..138
E. C. Malaimani, CSIR-National Geophysical Research Institute, India
N. Ravikumar, CSIR-National Geophysical Research Institute, India
S. V. R. R. Rao, CSIR-National Geophysical Research Institute, India
S. Padhy, CSIR-National Geophysical Research Institute, India
A. Akilan, CSIR-National Geophysical Research Institute, India
K. Abilash, CSIR-National Geophysical Research Institute, India
L. Prem Kishore, CSIR-National Geophysical Research Institute, India
G. B. Navinchander, CSIR-National Geophysical Research Institute, India
G. S. Srinivas, CSIR-National Geophysical Research Institute, India
U. Gowrishankar, CSIR-National Geophysical Research Institute, India
I. P. Raju, CSIR-National Geophysical Research Institute, India
Y. V. V. B. S. N. Murthy, CSIR-National Geophysical Research Institute, India
M. Venkatrayudu, CSIR-National Geophysical Research Institute, India
A. Nageshwara Rao, CSIR-National Geophysical Research Institute, India

Seismotectonic and the geodynamical processes between India and Antarctica are essential. During the initial 25 years of Indian expeditions, India's National Geophysical Research Institute (NGRI) launched a GPS-Geodesy programme by establishing a permanent GPS tracking station and a state-of-the-art Digital Broad Band Seismological Observatory Maitri during 1997. Both these stations have been continuously operational. It addresses the crustal deformation in the south of the Indian peninsula, the driving mechanisms, and the response of the Indian Ocean lithosphere bringing a deep understanding of the causes of the accumulation processes of strain in the Indian Ocean and the northward movement of the Indian plate.

Chapter 8
Hydrographic Surveys in Antarctica...160
Prashant Srivastava, Ministry of Earth Sciences, India

Hydrographic and physical oceanographic studies could enhance the national data bank. They could also greatly assist the various national agencies in correlating their work with the Antarctic waters' bathymetric and physical properties, leading to an improved understanding of the characteristics of the Antarctica waters. The bathymetric data gathered near the permanent station 'Maitri' could be utilised to exchange data with other nations like Russia and IHO. As per the charting scheme promulgated by the International Hydrographic Organization (IHO), INT Chart Nos. 9050 and 9051 are required to be co-produced by Russia and India. In the INT Chart 9050, vast blank areas exist due to the non-availability of bathymetric data. Thus, concerted efforts are required to fill in these gaps by undertaking systematic hydrographic surveys. These ocean areas are close to the permanent station in Antarctica. It is in the best interest of India to have a well surveyed navigational chart near the permanent station.

Chapter 9

An Assessment of Geo-Scientific Investigation in the Antarctic Region 194

Rajesh Asthana, Geological Survey of India, India
Amit Dharwadkar, Geological Survey of India, India
Prakash Kumar Shrivastava, Geological Survey of India, India

During the initial period of about 25 years (1981-2006), systematic geological mapping covering an area of about 19,000 sq. km, continuous monitoring of Dakshin Gangotri (DG) snout, ice core drilling, and limnological investigations are some of the significant achievements of GSI in the field of Antarctic geoscience. For the first time, ground penetrating radar (GPR) survey for lake bathymetry has been introduced in Schirmacher Oasis. Collaborative work has resulted in quantifying the annual vector movement of the ice sheet in the Schirmacher region.

Chapter 10

An Overview of Survey and Mapping of the Antarctic Region Around Maitri Research Base 215

Nagarajan Balasubramanian, Indian Institute of Technology, Kanpur, India

Survey of India, the premier mapping agency of India, has been associated with Indian Antarctica Research Program since the 10th Indian Scientific Expedition to Antarctica (1991-92). It has been taking part in all the expeditions ever since, except for the 15th expedition. The focus till the 22nd expedition has remained to carry out the surveying and mapping of the entire Antarctica region of Indian interest and provide additional technical support to the various other participating agencies. During the first two expeditions, global positioning system (GPS) and conventional triangulation techniques were used for providing primary control work to facilitate detailed mapping of the area using conventional mapping techniques. In the successive expeditions, initially, mapping was done on the scale of 1:5000 with contour interval 5m and then on a scale of 1:1000 with a contour interval of 1m. 7.8 sq. km of Schirmacher Oasis has been surveyed, and mapping has been undertaken on a large scale. Analogue and digital maps have been prepared for most of the areas for scientific use.

Chapter 11
Antarctic Laboratory Ground Truth for a Microwave Eye in the Sky 227
Mukund Kurtadikar, Department of Physics, JES College, India

A C-band microwave bench setup was carried out during the 22nd Indian Scientific Expedition to Antarctica (December 2002 to March 2003) to measure the dielectric properties of Antarctic geophysical materials like samples of ice, soil, and rocks around Indian Antarctic Station, Maitri. Laboratory validation of these properties of Antarctic frosts and soils was done during the summer period of the expedition, as they are significant for microwave remote sensing applications. Dielectric measurements of nine Antarctic rock samples were made after returning to India, which are substantial from the geological point of view. Here, the authors report the first attempt to measure the dielectric properties of these materials of Antarctica.

Chapter 12
Ice Core Records for Paleo-Volcanism, Climate, and Snow Accumulation Rates Over the Past 150 Years 240
D. K. Rao, Physical Research Laboratory, Department of Space, India
R. A. Jani, Physical Research Laboratory, Department of Space, India

The annual fallout of radionuclides 32Si, 7Be, 210Pb, and 137Cs in the shelf snow samples collected near the Dakshin Gangotri, East Antarctica, has been estimated. The polar fallout of cosmic-ray produced 32Si estimated to be 2.34 X10-5 dpm cm-2 y-1. The fallout of 7Be and 210Pb is estimated to be 4.2 and 1.86 X 10-2 dpm cm-2 y-1. The depth profiles of electrical conductance, 210Pb, δ18O, and cosmogenic radioisotopes 10Be and 36CI have been measured in a 60 m long ice core. Based on 210Pb and δ18O, the mean annual accumulation rates have been calculated. These rates are 0.20 and 0.23 m of ice equivalent per year during the past ~150 years. Based on electrical conductance measurements and using these accumulation rates, a volcanic event, 'Tambora' that occurred in 1815 AD, was identified. δ18O values suggested that the beginning of the 19th century was colder by about 2°C than the recent past and middle of the 18th century. The fallout of 36CI reported here agrees well with the mean global production rate estimated earlier.

Chapter 13
India's GeoScience Pursuit in the Antarctica: Initial Attempts 260
Neloy Khare, Ministry of Earth Sciences, India

India made her maiden entry in the Antarctic realm way back in 1981 by launching her first scientific expedition to the icy continent. Previous chapters dealt with many significant geoscientific studies of the past Indian scientific expeditions to Antarctica. However, few geoscientists participated for a short duration but carried out essential Antarctic studies. This chapter highlights and briefly collates such short geoscientific investigations that are otherwise significant and add to the knowledge base about the icy continent.

Compilation of References 269

About the Contributors 297

Index 303

Foreword

Antarctica is known for its awe-inspiring wilderness, harshest climates, and inhospitable conditions. It gained significance as a sequel to the success of International Geophysical Year (1957-58) when the international scientific community deeply acknowledged the vast potential of this piece of land to unravel the mysteries of the past.

Antarctic ice sheet is a repository of paleoclimatic indicators on earth, extending possibly to past over one million years. International efforts on a multinational collaborative basis have succeeded in raising many deep ice cores while Vostok drilling has gone upt o 3623 m deep, the EPICA dome C records are available up to a depth of 3,139 meter, which accounts for the last 7,40,000 years. The vast dimension of ice mass controls the climate of the Southern hemisphere and acts as a heat sink. Any changes in its mass can alter the global climatic balance and cause sea level perturbations.

It is also a well-recognized fact that the ice-atmosphere-ocean interaction affects the entire global system. Realizing the scientific importance India settled with serious pursuit of science in the Antarctic. Establishment of three Indian permanent research bases in Antarctica strongly demonstrate the significance that India accords to Antarctica. Besides other fields of sciences, Geoscience of this icy continent has been ably investigated by Indian Geologists, Geophysicists, Glaciologists and Paleoclimatologists.

Having completed a long glorious journey of Indian Antarctic Programme, it is most appropriate to review the achievements made in the fields of Antarctic Geology, Geophysics, Glaciology, Limnology and Paleoclimatology in the past.

I am delighted to learn that efforts have been made in this direction and the book on 'Geoscientific Investigations from the Indian Antarctic Program' has been brought out which covers most of the scientific aspects of Antarctic Geosciences. Such a compilation of the work undertaken by India in Antarctic Geosciences, will act as a reference for the future generations.

I compliment the editor and wish him all the very best for his commendable efforts.

Bimlendu B. Bhattacharya
Indian Institute of Technology (ISM), Dhanbad, India

Preface

Polar Regions though remotely situated and hostile in conditions are an integral and highly significant components of the earth's climate system. These regions are the first to suffer from global warming and constitute an archive of vital climatic variations. Besides progressing in other domains of Antarctic Science India has developed significant research programmes, logistics, recourses, and commendable expertise in the field of Antarctic Geo science.

The geological studies that started as reconnoitering appraisal in earlier expeditions eventually gave way to regional systematic mapping. Sustained effort till date has enabled Geological Survey of India (GSI) to complete systematic geological mapping on 1:50,000 scale, in areas exposing outcrops between 15^0 and 4^03' East longitude covering the Wohlthat, Orvin and Muhlig-Hofmann ranges off the Princess Astrid coast in central Dronning Maud Land east Antarctica, covering an area of 19,000 sq.km. The geochemical and petrochemical studies have been carried out in Gruber area exposing a massif anorthosite complex. On the contrary, Geology of Larsemann Hills is especially relevant in Indian context i.e. the Eastern Ghats. The area exposes a veritable sequence of Meta sedimentary units apparently reworked and modified due to the influence of at least two tectono-thermal events. There is a reporting of Ultra High Temperature mineral assemblages which have suffered decompression and are stabilized at lower pressure-temperature. It is believed that the last tectono-thermal event (an early Palaeozoic ~ Pan-African event) imprint indicated by the rocks in Larsemann Hills may be correlated with the Eastern Ghats.

Jadavpur University (India) has completed geological and structural map of the Schirmacher Hills with special emphasis on the tectono-thermal evolution of the area. Five episodes of deformation and three episodes of metamorphism were identified from the Schirmacher Hills. This study has yielded important indicators for the understanding of deformation mechanisms of the constituent minerals at great crustal depth. For the first time such detailed work on shear zones was carried out on Antarctic rocks.

Litho-geological mapping using gamma-ray spectrometric data was carried out over an area of 35 sq.km. Regional geological map on 1:25,000 scale showing distribution of various gneisses/granulites, was prepared. The low abundance of uranium in these rocks is attributed to high grade of metamorphism and variation in U, Th, and K is explained by metamorphic redistribution of a mixed crustal source rock with dominant acid igneous component.

Similarly, the Ice Shelf which is a continuation of polar ice sheet moving away from the continent and floating over the Antarctic seas is being monitored by GSI. SIR-20 dual-channel Ground Penetrating Radar (GPR) with a range of antennas from 15 to 200 MHz has successfully been tested by GSI in Antarctica.

The continuous monitoring of various glaciological parameters has been done as a project of ongoing nature. The current long-term studies include observing accumulation / ablation patterns on the iceshelf, monitoring the recession of snout of DG glacier and the southern ice wall margin in Schirmacher Range and studying the glacial dynamics of the ice sheet, south of Schirmacher Range. The investigations have provided substantial inputs to our knowledge of status of ice shelf conditions and patterns of climatic changes affecting this part of Antarctica.

Besides, the Ice core drilling, using an electromechanical ice core drill machine was initiated in 1992-94 when a borehole of ~60 m was drilled successfully. Subsequently, two more boreholes of ~76 m and ~84 m depths, respectively, were drilled during 15th Expedition. During the 19th Expedition, a surface route for transportation of drill machine and other supplies was established right up to polar plateau, via Humboldt-Somovken glacier region. In the 22nd expedition, an upgraded version of the machine was tested near Tallaksenvarden nunatak and an ice core of ~62 m length was successfully raised. In the 25th expedition, drilling work was taken farther south, deeper into the continental interior, near the mouth of Humboldt Glacier and a core of ~65 m was recovered during the summer season.

Since Maitri is one of the SCAR GPS stations contributing to the SCAR database, Maitri data is included in the International Data Base of the SCAR Epoch GPS campaigns Permanent GPS Station at Maitri, would continue to contribute to SCAR Data Archive so that Maitri remains in the global space geodesy scenario. Additionally, MTS group of National Geophysical Research Institute (NGRI) envisaged to study the deep electrical structure and interpret in terms of geodynamical processes between Antarctica and south India by recording natural electromagnetic signals near Maitri station. Therefore, the continuous recording of signals also helps in studying the co-seismic activity, (if any).

Geodetic and Research Branch of the Survey of India has been associated with the Indian Antarctic Research Program since 10th expedition to provide framework for horizontal and vertical control, which is required for mapping.

The Antarctic lakes provide unique environments for palaeoecological studies. The palaeoclimatic data so far generated from Zub and Long lake goes back to about 8,000 years BP, reflecting arid - warm and humid climatic condition. A multi-institutional and multi-disciplinary approach on many samples, are likely to show great potential to yield more dependable palaeoclimatic data in a chronological sequence relevant to the environmental history of the polar region. Pollen, Diatoms including other algae, and isotope of Carbon and Oxygen are good sources of proxy data for climatic reconstruction. Analysis of these proxy records in the Antarctic glacier region will be of great significance in understanding the climatic changes vis-a-vis glacial fluctuations of this region. These studies would also be useful in establishing the long-term linkage with various aspects of climatic change, glacier fluctuations and their relationship with the EL Nino/Southern Oscillation (ENSO). Glacio-geomorphological evidences suggest that there were several phases of glacier advancements and retreats during Holocene and earlier period in the Antarctic glacier region.

However, for sediment coring in any of the Antarctic lake, proper bathymetry, bottom topography, with a fair idea of sediment distribution, its thickness underlying the lake floor is imperative, as in most cases the lake bottom is uneven, boulderous and could be dangerous at times. Therefore, a pilot project was launched to carryout geophysical survey of various Antarctic lakes.

The Hydrographic Scientific teams have gained valuable experience/expertise in conducting the specialized Hydrographic studies. The opportunities were also effectively utilized by the team to study the effects of environment on equipment/ personal with consequential effects of operation in polar conditions. So far, the National Hydrographic Office (NHO) could undertake meaningful bathymetric survey of over 1000NM.

At this juncture, when country completes a long journey of over four glorified decades of the Indian Antarctic Program, a review of what has been achieved and what more to achieve in the Antarctic Geoscientific research is highly warranted to evaluate the status of Indian geoscientific research in global perspectives.

Accordingly reviews on various aspects of Antarctic Geoscientific research have been incorporated in this book which begins with a chapter on Glacial Sedimentology of Schirmacher Oasis, East Antarctica by Srivastava and Khare. On the other hand, Rameshwar Rao described the details of Geochemical and mineralogical studies of gneiss-charnockite rocks of Schirmacher region, East Antarctica. Similarly, a comprehensive study of the structural Geology of the Schirmacher Hills has been undertaken by Sengupta. In order to illuminate climate oscillations over the Antarctic region Bera used Pollen - spore transport into the icy continent.'

On the contrary, results of the Integrated Geoscientific Investigations in Antarctica have been highlighted by Gupta et al. While GPR Surveys in Antarctica to Map Ice Thickness and Rock Topography have been discussed by Sinharay. Similarly, studies on Seismotectonics and Geodynamical Processes between India and Antarctica have been thoroughly reviewed by Malaimani et al., highlighting the significance of the data obtained from permanent Seismic and GPS observatories at Antarctica.

An overview on the Hydrographic surveys in Antarctica waters has been ably provided by Srivastava. Whereas, an assessment of geo-scientific investigation in the Antarctic region has been made by Asthana et al. Survey and mapping has been a significant activity during various expeditions to icy continent and an overview of Survey and Mapping of the Antarctic Region around Maitri Research Base has been presented by Nagarajan. Interestingly, Kurtadikar highlighted the results of the experiment conducted in the Antarctic laboratory towards ground truthing for a microwave eye in the sky. Rao and Jani detailed Antarctic Ice Core records for paleo–volcanism, climate and snow accumulation rates over the past 150 years. Lastly the initial geoscientific research activities have been collated by Khare which generated significant baseline data in the earlier days of India's journey to the Antarctic region.

It is hoped that our humble effort will facilitate and strengthen mutual exchange of ideas on the Antarctic Geoscience and future endeavors in this field.

January 2022

Neloy Khare
Ministry of Earth Sciences, India

Acknowledgment

I gratefully acknowledge Dr. M. Ravichandran, Secretary to the Government of India, Ministry of Earth Sciences, who has shown keen interest and provided valuable guidance and encouragement for addressing the issues of Antarctic Geoscience in this book.

All contributing authors are deeply acknowledged for their significant scientific inputs to this book. Without their support and cooperation this book would never have materialized in time.

I sincerely acknowledge the valuable technical expertise and guidance provided by various reviewers who have contributed significantly in improving the quality of various papers considered in this book through their critical review on various manuscripts.

The officers and staff of the Ministry of Earth Sciences, Government of India, New Delhi and the Director, of the National Centre for Antarctic and Polar Research, Goa deserve appreciation for their direct or indirect involvement in the preparation of this book.

I deeply admire continuous guidance, inspirations and encouragements received from Dr. Rajiv Nigam, Former Adviser and Head, the National Institute of Oceanography, Goa, Dr K.J. Ramesh, Former Director General, India Meteorological Department, New Delhi, Prof Anil Kumar Gupta, Indian Institute of Technology, Kharagpur, Prof Devesh Kumar Sinha, Delhi University, Dr R.S. Kankara, Adviser, National Centre Of Coastal Research, Chennai and Dr. O.P. Mishra, Director, National Centre of Seismology, New Delhi. Shri Hari Dass Sharma who helped me in various ways in preparation of this book is deeply acknowledged.

I would be failing in my duties if I do not acknowledge my family which rendered all possible support to make this book a realty.

Neloy Khare

Chapter 1
Glacial Sedimentology of Schirmacher Oasis, East Antarctica:
Indian Perspective

Ashok Kumar Srivastava
SGB Amravati University, India

Neloy Khare
Ministry of Earth Sciences, India

ABSTRACT

Schirmacher Oasis, East Antarctica is under investigation for the last four decades by Indian scientists of various disciplines, of which geological studies constitute a significant aspect. The present attempt is a review of sedimentological studies carried out in the area. A few of the current chapter's elements are sediment characteristics, mode, and pattern of sediment deposition, source rock, provenance, paleoclimate, and weathering condition through clay mineral studies, grain surface studies, etc., with specific aspects more data on the number of samples. Simultaneously, the Oasis also offers good scope for various other significant investigations like paleoclimatic reconstruction, source of glacial sediments, silicates weathering, etc. Globally, they are a front-line work area and recommended.

DOI: 10.4018/978-1-6684-4078-0.ch001

INTRODUCTION

The Schirmacher Oasis is a small, ice-free area in perennially glaciated East Antarctica. Hard Precambrian metamorphic rocks primarily represent the Oasis terrain; however, it also offers good scope for studying glacial sediments released and deposited due to the wind, ice, and water. Various erosional and depositional processes of the continental glacier are well evident in the area. There are ample exposures of loose sediments, accumulated due to different geological processes, i.e., i) the scarp face of a thick pile of polar ice covers the northern margin of the Oasis, which serves as a suitable site of deposition for the sediments released due to glacial melt and dropping down with the meltwater, ii) the undulating topography of main rocky land favours small accumulations of the deposits transported by meltwater channels and wind, iii) the inland lake sites, and iv) coastal shelf sediments along with the northern margin grading to shelf region of the Antarctic Ocean. This region also receives the entire area's deposits through meltwater channels. The Oasis's geology, geomorphology, and climatology have been studied during the past annual scientific expeditions. The important geological aspects are general geology (Singh, 1986; Sengupta, 1986, 1988), sedimentology (Lal, 1986; Asthana and Chaturvedi, 1998); P-T-conditions of metamorphism (Kumar, 1986); magnetic characteristic (Gupta and Verma, 1986), etc. In recent years, the Oasis has attracted the attention for its varied aspects e.g., - ecological assessment of freshwater lake (Ingole and Dhargalkar, 1998); petrography, geochemistry, and dating of lamprophyres (Dayal and Hussain, 1997; Hoch and Tobschall, 1988; Hoch, 1999; Hoch *et al.,* 2001); pollen-spores (Sharma *et al.,* 2002; Bera, 2004); the occurrence of quench olivine in basalt dykes (Jafri *et al.,* 2002), structural and thermal studies of graphite (Parthasarathy *et al.,* 2003); petrology and geochemistry of granite, gneiss, and fayalite (Rama Rao, *et al.,* 1995; Hussain and Rao, 1996; Jayapaul *et al.,* 2005); lichens (Nayaka and Upreti, 2005); Arcellaceans (Mathur *et al.,* 2006); bedrock topography and subsurface structure (Sundararajan and Rao, 2005); fungal biodiversity (Singh *et al.,* 2006); Holocene climatic changes (Sharma *et al.,* 2007); diatom flora (Palanisamy, 2007); granulometric analysis (Srivastava, 2007; Srivastava and Khare, 2009, Srivastava *et al.,* 2012); clay mineralogy and weathering pattern (Srivastava, *et al.,* 2011); heavy minerals and sediment sources of various geomorphic units (Srivastava, *et al.,* 2010) and interpretation of paleoclimate (Srivastava, *et al.,* 2018). The documented literature shows that most studies are related to general geology, hard rock petrography, floral and faunal remains, environmental aspects, lakes, etc. However, Oasis's sedimentological characteristics are less explored than many proxies to interpret geological and climatological conditions in the past and present. It is an ongoing site of glacial sedimentation. Hence, it offers an opportunity to work for provenance studies of the glacial sediments, modes, and mediums of sediment transport, weathering pattern, mineralogy and geochemistry, clay mineralogy, etc. In the present chapter, we review the status of work carried out on various aspects of glacial sediments of Schirmacher Oasis, East Antarctica, and the identification of thrust areas for future work.

GEOLOGY AND GEOMORPHOLOGY

The Schirmacher Oasis (Lat. 70° 44' 30" to 70° 46' 30" S and Long. 11° 22' 40" E to 11° 54' 00" E) is an east-west trending narrow strip having a maximum width of about 2.7 km in the central part and covers an area of about 35 sq. km. It has an undulating topography consisting of low elevation hills up to 200 m altitudes and depressions formed due to glacial valleys and lakes.

Surface lithology is mainly represented by the high-grade metamorphic suite of rocks, forming the Precambrian crystalline basement such as:

1. banded gneiss,
2. alaskite, garnet-biotite gneiss,
3. calcgneiss, khondalites, and associated migmatites,
4. Augen gneiss, and
5. streaky gneiss is at places intersected by the dykes of basalt, lamprophyres, pegmatite, dolerite, and apatite (Sengupta, 1986) (Figure 1).

Figure 1. Geological map of the Schirmacher Oasis and sampling sites of glacial sediments

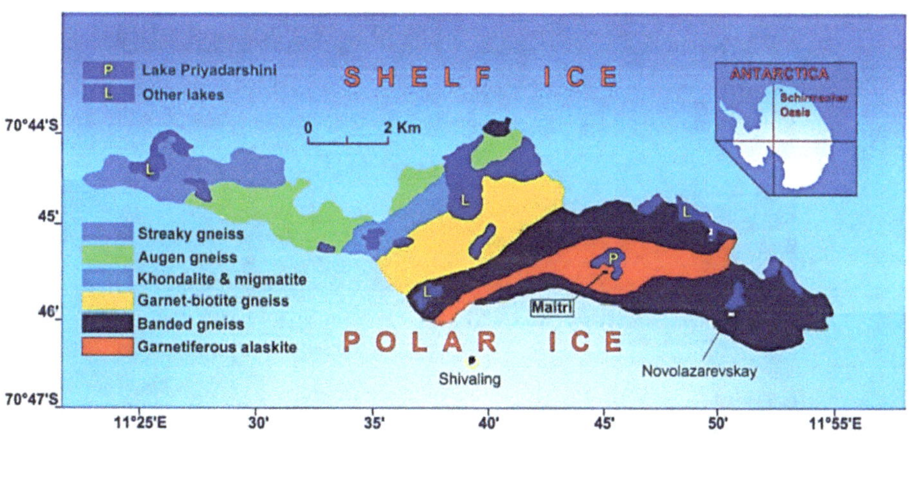

Geomorphologically, the Schirmacher Oasis is distinguishable into three units viz., i) polar ice sheet, ii) Schirmacher mainland including lakes, and iii) coastal shelf area (Figure 2 A-H). All the three units roughly extend in the east-west directions as of the coastline. The first unit, i.e., the polar ice sheet, covers a large area in the south and is impregnated with abundant sand and silt-sized sediments. The second unit, i.e., the main rocky land of Schirmacher, covers an undulating topography due to low altitude hills, depression, and lakes formed of glacier outlets. Its northern periphery shows a general steepness towards the shelf ice with many sub-glacial tills and U-shaped valleys. The third geomorphic unit is the east-west trending shelf region. It is marked by coastal sand, mainly mixed with the mainland's sediments and the polar ice sheet coming through the meltwater channels.

Figure 2. A and B- polar ice sheet exposed south of Maitri; C and D- Main rocky land; E- Lake Priyadarshini; F- An epishelf lake, north of Maitri; G & H: Shelf area

GLACIAL SEDIMENTOLOGY

Glacial sediments are a valuable tool for various interpretations regarding the environment of deposition, nature of deposits, provenance and source rock determination, transportation history, paleoclimate, *etc*. (Reinick and Singh, 1980). The available literature on the sedimentological aspect of the Schirmacher Oasis, i.e., Lal (1986), Asthana and Chaturvedi (1998), Srivastava (2007), Srivastava and Khare (2009), Srivastava, *et al.* (2010, 2011, 2012, 2018) indicates that the area needs further explorations. However, the attempts made are significant mainly due to limitations to access the site, rugged terrain, limited duration of the fieldwork, etc. The studies on sediments are subdivided into five elements viz., i) granulometric studies, ii) heavy mineral studies, iii) clay mineralogy, iv) grain morphoscopy, and v) lake sedimentology.

Granulometric Studies

A preliminary initial report on textural parameters of glacial sediments is provided by Lal (1986), based on the analysis of 12 sediment samples that includes interpretation of specific basic textural parameters, i.e., mean standard deviation, kurtosis, and skewness. He interpreted that fine to medium sand are abundant constituents of the sample, whereas the size varies from coarse-grained sand to clay. Asthana and Chaturvedi (1998) analysed 18 samples, collected from four different glacial units, i.e., along the small channels formed over the polar ice sheet (11 samples), melting spots and depressions in the ice (04 samples) and wind scooped depressions within the sastrugi (03 samples). The report emphasises on basic size parameters of the sediments and their interrelationships. They interpreted that the deposits are coarse to fine-grained in size, moderately to poorly sorted (1.98 to 0.75Φ), positively to very positively skewed, and very platykurtic to very leptokurtic. The binary plots between various parameters indicate a loss of sorting with a decrease in the sediments' mean size. The plots between mean size vs skewness and kurtosis show no relation, whereas skewness and standard deviation show a V-shaped trend. They further concluded the polymodal nature of the sediments, which reflects high fluctuations in transporting medium. They suggested that the deposits are composed of different lithology and provenance. The wind is a significant sediment transport agency in the region, responsible for mixed population sediments. Comparatively better-sorted sediments restricted to wind scooped depressions. Water has a minor role in sorting meltwater channels and melting spots. The saltation transports most of the residues.

The mode and Pattern of deposition of the glacial sediments have recently been based on detailed textural and statistical parameters of glacial sediments accumulated in the polar ice sheet, the main rocky land of Schirmacher, lakes, and shelf ice area.

A total of thirty-seven samples collected from the entire area (Srivastava, 2007; Srivastava and Khare, 2009; Srivastava *et al.*, 2012) have been collected and studied. The locations of the sampling sites are as follows;

Polar Ice Sheet

A-2: Base of polar ice sheet scarp face.

A-3: From the height of about 3m above A-2.

A-4: From the top 2m thick ice of scarp face, approachable in the SW of Maitri.

A-5: From the base of the polar ice sheet, SW of Maitri.

A-6: From the frozen meltwater channel at the top of the polar ice sheet near Shivling, lying at 3km SW of Maitri.

A-32, A-33 and A-34: The ice sheets lying at the base near the scarp margin at three different locations roughly 500m apart, south of Maitri.

Lake

A-7 and A-22: From the margin of the lake located at about 2 km. East of Maitri, on the route to Russian station viz., Novalazarevskay.

A-12 and A-13: From the lake's margin situated adjacent to the previous (sample A-7).

A-18, A-19 and A-20: From the margin of Priyadarshini Lake near Maitri.

A-28 and A-30: From the margin of epishelf lake located in the extreme north.

Main Rocky Land

A-1: From 400 m south of Maitri.

A-11 and A-26: About 4 km NE of Maitri.

A-14: About 2 km NW of Maitri.

A-17 and A-21: From the sand pocket accumulation sites near Maitri at an elevation of 2 m and 4 m, respectively.

A39: Two km east of Maitri.

Coastal Shelf Area

A-8, A-9 and A-10: Shelf approachable 3 km east of Maitri.

A-15 and A-16: From the rocky shelf area about 3 km west of a previous site (sample nos. A-8, A-9 & A-10).

A-23, A-24 and A-25: East of Maitri at about 4 km

A-27, A-35, A-36, A-37 and A-38: From four nearby locations at about 10 km NW of Maitri.

The methodology adopted for grain size analysis is as proposed by Ingram (1971). We calculated all samples' weight percentage and cumulative weight percentage for graphical representations, histograms, and cumulative curves. With the help of these statistical representations, various graphic measures have been computed

(Table-1.1) and interpreted as per the criterion suggested by Folk (1980), Reinick and Singh (1980), Lindholmn (1987), and Sengupta (1996).

Table 1. Graphic measures and statistical parameters of the glacial sediments from different areas (Srivastava et al., 2012)

Stat. Parameter /Sample No.	Φ 95	Φ 84	Φ 75	Φ 50	Φ 25	Φ 16	Φ 5	Φ 1	C in Micron	M in Micron	Mz	σ_I	Sk_I	Kg	SOS	SKS
Polar ice-sheet and meltwater channels																
A-2	4.41	2.98	2.74	1.69	0.36	-0.94	-1.53	36.5	3600	310	1.24	1.88	+0.21	1.10	2.97	2.56
A-3	4.45	2.76	2.26	1.51	0.45	-0.77	-1.40	40.4	3400	350	1.17	1.78	-0.14	1.32	2.925	2.83
A-4	4.42	3.41	3.13	2.66	1.57	0.74	-1.04	18.1	2950	160	2.27	1.48	+0.38	1.43	2.73	0.14
A-5	4.37	3.15	2.81	1.62	0.38	-0.89	-1.44	38.8	3350	320	1.29	1.81	+0.14	0.98	2.905	2.57
A-6	4.32	2.95	2.81	1.97	1.52	1.06	0.47	14.7	1450	260	1.99	1.00	+0.12	1.22	1.925	-0.09
A-32	3.31	2.69	1.95	0.67	-0.73	-1.03	-1.37	56.7	2850	640	0.77	1.63	+0.11	0.71	2.34	3.34
A-33	3.41	2.95	2.76	1.60	-1.14	-1.77	-2.55	44.0	6950	330	0.98	2.08	+0.40	0.62	2.98	2.76
A-34	4.24	2.96	2.61	1.12	-0.85	-1.29	-1.78	48.3	7200	460	0.93	1.98	-0.05	0.71	3.01	3.78
Lake and adjoining area sediments																
A-7	4.25	2.94	2.71	1.49	-0.23	-0.89	-1.43	44.0	2950	360	1.18	1.81	+0.13	0.73	2.84	2.7
A-12	4.71	4.05	3.15	1.67	0.0	-0.49	-1.17	41.7	2550	320	1.64	2.13	-0.006	0.76	2.94	2.54
A-13	3.41	2.44	1.61	0.41	-1.14	-1.43	-1.82	68.5	3700	750	0.47	1.75	+0.09	0.77	2.615	4.41
A-18	4.45	2.83	2.21	1.25	0.36	-0.64	-1.38	45.1	2950	420	1.14	1.74	+0.01	1.29	2.915	3.33
A-19	4.20	3.29	2.93	1.73	-0.66	-1.31	-2.07	40.4	4950	295	1.23	2.19	+0.26	0.71	3.135	2.81
A-20	4.02	3.15	2.84	1.72	-0.87	-1.53	-2.40	40.1	6500	300	1.11	2.04	+0.33	0.70	3.21	2.98
A-22	3.89	3.08	2.80	1.61	-0.31	-1.09	-1.75	41.5	3850	325	1.20	1.89	+0.24	0.74	2.82	2.42
A-28	3.41	2.85	2.59	1.58	0.48	-0.58	-1.45	39.5	3600	340	1.28	1.50	+0.25	0.94	2.43	1.7
A-30	3.94	3.15	2.84	1.51	-1.17	-1.81	-2.57	46.9	6700	350	0.95	2.22	+0.28	0.66	3.255	3.49

Table 2.

Schirmacher main rocky land																			
A-1	4.39	3.38	3.10	2.46	2.02	1.06	0.49	15.0	2250	185	2.3	1.17		+0.10	1.48	1.95	-1.02		
A-11	3.74	2.81	2.20	1.03	-0.49	-1.17	-1.72	49.4	4000	480	0.89	2.25		+0.10	0.83	2.73	3.4		
A-14	4.12	3.22	2.93	2.37	1.62	1.23	-0.79	14.1	3400	192		2.27	1.23		+0.21	1.53	2.455	0.17	
A-17	4.23	3.20	2.93	2.22	1.51	0.70	-1.68	19.7	5100	215	2.07	1.52		+0.25	1.70	2.955	1.47		
A-21	2.93	2.39	1.91	1.50	1.05	0.90	-0.58	35.8	3650	355		1.59	0.89		-0.01	1.12	1.755	0.51	
A-26	4.28	3.44	3.16	2.64	0.75	0.47	-1.38	25.9	3300	160		2.18		1.59		+0.37	0.96	2.83	0.38
A-39	3.45	2.86	2.55	1.22	-0.58	-1.25	-1.96	47.4	4600	430		0.94		1.83		+0.19	0.70	2.705	2.97
Shelf area																			
A-8	4.22	2.92	2.70	1.73	0.66	0.19	-1.42	32.0	3600	295	1.61	1.53	+0.07	1.13	2.82	2.18			
A-9	4.60	3.39	2.95	1.95	0.56	0.31	-1.49	34.2	4800	260	1.88	1.69	+0.15	1.03	3.045	2.19			
A-10	3.09	2.77	2.56	1.85	1.51	1.09	0.47	14.3	2450	280	1.90	0.81	+0.04	1.03	1.31	-1.08			
A-15	4.56	3.39	2.86	1.62	-0.26	-1.06	-1.78	42.0	4100	325	1.31	2.08	-0.06	0.83	3.17	3.1			
A-16	4.30	3.26	2.14	2.19	1.22	0.48	-1.36	23.4	3100	220	1.97	1.55	+0.23	1.35	2.83	1.28			
A-23	4.35	3.42	2.98	2.09	0.69	-0.16	-1.31	28.4	5700	235	1.78	1.75	+0.22	1.05	2.83	1.48			
A-24	3.76	2.87	2.69	1.72	-0.32	-1.26	-2.14	36.3	3200	300	1.11	1.91	+0.37	2.02	2.95	2.46			
A-25	3.71	3.05	2.90	2.62	1.82	1.61	0.54	7.5	3250	165	2.42	0.84	-0.15	1.20	1.585	-2.07			
A-27	4.15	3.05	2.83	1.92	0.07	-1.05	-1.79	37.2	2900	265	1.27	1.92	+0.30	0.88	2.97	2.1			
A-35	3.35	2.86	2.68	1.02	-0.47	-0.97	-1.97	49.4	3820	490	0.97	1.75	+0.08	0.69	2.66	3.28			
A-36	4.05	2.85	2.42	0.93	-0.33	-1.05	-1.77	51.3	5800	520	0.91	1.66	+0.03	0.86	2.91	3.96			
A-37	3.14	2.66	2.00	1.33	0.26	-0.91	-1.37	60.8	0390	395	1.02	1.57	+0.22	1.05	2.255	1.85			
A-38	4.18	3.44	3.19	2.59	0.51	-0.83	-1.52	24.3	3850	168	1.73	1.92	+0.52	0.87	2.85	0.52			

Frequency Curves

The weight percentage histograms of polar ice sheet sediments (08 samples) show the dominance of medium to fine sand to clay, the lake sediments (09 samples) dominantly show the variation of grain size from coarse to fine sand whereas, the main rocky land (07 samples) contains more quantity of medium to very fine sand. The shelf (13 samples) is again represented by coarse to fine sand and coarse silt-sized grains; however, very coarse sand and granules are also present. The comparative study of the four units' cumulative curves shows almost the same trend of grain size class as interpreted through weight percentage histograms. The curves show a general tendency of a low sorting of grains.

Textural Parameters

The average mean size value of all the four geomorphic units predominantly indicates a predominance of medium sand-sized sediments; however, each unit's separate study shows local variation, ranging from coarse to fine sand. In general, all four areas show the dominance of medium to fine grain sediments. The sorting or uniformity of particle size distribution ranges from 1.00 Φ to 2.08Φ (av. 1.70 Φ) in the polar ice sheet, 0.47Φ to 1.64 Φ (av. 1.91 Φ) in the lake sediments, 0.89 Φ to 2.25 Φ (av. 1.49 Φ) in the Mainland and 0.91 Φ to 2.42 Φ (av. 1.61 Φ) in the shelf, indicating overall low sorting of the grains. Though all the four average values indicate low sorting of grains, still, there is an increasing trend of sorting in the standard class of poorly sorted category (1.00 Φ to 2.00 Φ value), i.e., lake < polar ice < shelf < rocky land.

The skewness values derived from the polar ice sheet are dominantly fine skewed (04 samples) and very fine skewed (02 samples). The lake sediments are also dominated by fine skewed category (05 samples) but followed by near-symmetrical (03 samples). The mainland sediments are equally represented by fine skewed and near-symmetrical (03 samples each). Similarly, fine skewed and near-symmetrical categories are equally represented in shelf sediments. In general, the entire area is characterised by near-symmetrical to strongly positive skewed sediments. The kurtosis relates to the peaks of the distribution. The values obtained from polar ice sediments range from 0.6 Φ to 1.43 Φ with an average value of 1.01 Φ, i.e., mesokurtic condition. The lake sediments range from 0.66 Φ to 1.29 Φ, averaging 0.81, a platykurtic condition. The mainland and shelf area ranges from 0.70 Φ to 1.70 Φ and 0.69 Φ to 2.02 Φ respectively, giving the average values of 1.18 Φ and 1.70 Φ respectively. Both the average values denote leptokurtic conditions. The kurtosis values of all the four units indicate the dominance of platykurtic sediments (15 samples) followed by leptokurtic (09 samples) and mesokurtic (08 samples); very leptokurtic (03 samples) and very platykurtic (02 samples) categories are also present.

Bivariate Plots

Bivariate plots between various parameters, such as the C-M and Visher plots, have been constructed to interpret the depositional setup. The bivariate plot of the mean vs standard deviation of polar ice sheet sediments shows sediment clustering near coarse sand's lower and medium sand's upper limits (Figure 3A). It depicts a dominantly poor sorting nature. Lake sediments ranging from 0.95 Φ to 1.28 Φ are poorly to very poorly sorted. The mainland sediments lack any definite pattern, whereas the shelf sediments are also poorly sorted. In general, the sediments from the entire area dominantly fall in carelessly to very poorly sorted categories. A few samples showing moderate sorting with a value of 0.81Φ, 0.84Φ, and 0.89Φ are because of the localised phenomenon, i.e., continuous flow of meltwater causing the removal of fine-grain material. The bivariate scatters of mean size vs skewness show that polar ice and lake sediments tend to have more fine material in medium sand (Figure 3B). The mainland also exhibits the tendency of fine skewness in fine sand fractions. The shelf sediments are dominantly near-symmetrical to fine skewed. The overall assemblage shows a lower bit of coarse sand dominantly represents two clustering groups, i.e., one cluster to an upper fraction of medium sand. In contrast, the other set shows an increasing tendency of fine skewness with the decrease of grain size.

Figure 3. Bivariate plots showing the trends of the scatter; green - polar ice sheet sediments
Violet - mainland; red - lake; blue - shelf

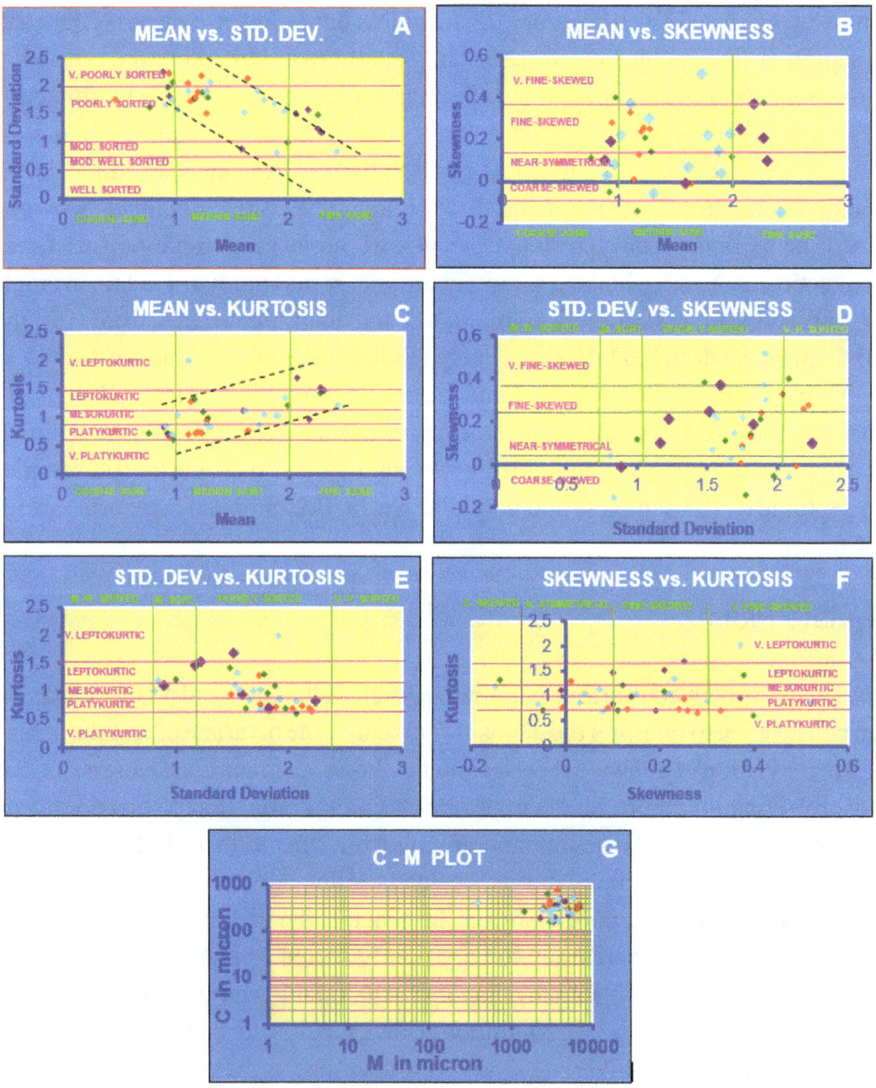

The mean vs kurtosis plot indicates a fragile relationship (Figure 3C). It is difficult for polar ice and lake sediments to establish any relation or control of others' parameters. In the mainland case, the coarse sediments are flat peaked, which tends to leptokurtic with a decrease of grain size. The shelf sediments ranging in medium grain size class fall between excessive peaked to flat-peaked categories. In general, most medium sand-sized sediments are platykurtic to mesokurtic, which tends to be leptokurtic with a decrease in mean grain size.

The standard deviation vs skewness plot of polar ice and mainland sediments lack any definite pattern or clustering (Figure. 1.3D). However, lake sediments show a tendency of increasing fine skewness with a decrease in sorting. The poorly sorted shelf sediments also have a faint tendency of increasing fineness with a reduction of sorting. The entire area analysis represents dominantly a purported category followed by very poorly sorted sediments. There is a tendency of gradational decrease of sorting with the increase of fine-grained sediments. The plot of standard deviation vs kurtosis for the polar ice sheet shows a decrease of kurtosis with a reduction of sorting. The lake sediments are mostly platykurtic near the inferior boundary and very poorly sorted category (Figure. 1.3E). The mainland sediments seem to have no relation between standard deviation and kurtosis. Whereas the shelf sediments also have a faint tendency of proportionate decrease in kurtosis to increasing sorting. A wide scatter plot is observed in skewness vs kurtosis of polar ice and mainland sediments, which lack any defined pattern (Figure. 1.3F). In the lake sediments, the finely skewed platykurtic sediments tend to cluster. In shelf sediments, nearly symmetrical to fine skewed platykurtic to mesokurtic sediments form a narrow zone. The overall picture shows that the near-symmetrical to fine skewed sediments have a gradational occurrence starting with a maximum scatter in the platykurtic category followed by the mesokurtic and leptokurtic categories.

Passega (1957, 1964) and Passega and Byramjee (1969) proposed that the C-M Pattern gives an idea about traction current deposits, turbidities, and the quiet water suspension deposits. The present analysis of thirty-seven samples shows that a total of twenty-four samples have a C-value between 2 to 4 mm, whereas eleven samples are more than 4mm (Figure. 1.3G). The M-value mostly ranges between 150 to 400 mm. accordingly, the plot of C and M values mainly falls in the rolled sediments field as proposed by Passega (1957). Visher (1969) suggested a relationship between grain size and the transport of sediments. All the thirty-seven samples are analysed to compare the sub-population. The general tendency observed in the polar ice sheet is two small sub-populations of traction and suspension, whereas, saltation population dominates and may also show two subpopulations. The lake and mainland sediments show the same trend. The shelf sediment has a minor change, i.e., a noticeable proportion of the traction population. The general slope towards the shelf provides a better chance of adding coarse sediments from adjoining areas. In general, the plots are closer to the beach sands (Visher, 1969), showing low sorting of coarser population. However, the saltation population dominates most of the sediments.

Though the Oasis is distinguishable in the polar ice sheet, including lakes and shelf area, it isn't easy to differentiate based on sediments' textural parameters. Most of the sediments fall into a single domain, i.e., poor to very poor category. If present, the deviations are due to the localised action of meltwater channels or naturally controlled unidirectional wind flow, which occasionally makes a little sorting of sediments per their energy conditions. A clear-cut differential pattern in textural and statistical parameters of the sediments, established by various workers (Folk and Ward, 1957; Friedman, 1961, 1967 and Moiola and Weiser, 1968) for different environments, is challenging to find out in the present case for separate geomorphic and glacial units, i.e., polar ice, mainland, lake, and shelf.

Low to the high-velocity wind is a primary erosional agency in the area, reshuffling the entire area's sediments during four months of the summer when the Oasis is primarily exposed. The high-energy winds act as a powerful agent for mixing sediments belonging to all the three geomorphic units in a very haphazard way. The summer period is also a time of maximum melting of ice, because of which the impregnated sediment gets differential momentum. The larger sediments get to settle in the same place with little or no movement, whereas the fine fraction, apart from localised settlement, also moves with meltwater for a shorter distance. The melting actions of water also add sediments to the entire area irrespective of any definite trend. In the cold months, the rocky surface of Oasis is covered with ice, as well as the perennial ice frozen area, i.e., polar ice and shelf ice, experiences a thicker layer of ice than the summer season. This period also experiences high-velocity winds along with thick snow and sediments. These sediments have no definite pattern for their settlement. Therefore, it is evident that both the geological agents, i.e., the high-velocity winds and meltwater, have a significant role in sediment transport and deposition. The resultant deposit is poorly to very poorly sorted sediments of the entire area, irrespective of any specific control of physical factors on the sediments of polar ice, lake, mainland, and shelf area.

Heavy Mineral Studies

Lal (1986) mentions a few heavy minerals from the sea bed and glacial sediments. However, based on the grains' angular shape, it is commented that the heavy minerals are of the first cycle and have gone less transportation. The work is of a very preliminary type and lacks enough data, proper detailing, referencing, and interpretation. Recently, Srivastava *et al.* (2010) attempted to interpret the source rock and provenance of the glacial sediments based on a detailed study of the heavy minerals from the sediments of various glacial and geomorphological units the area. The study is based on a total of twenty samples representing equally the polar ice-sheet (sample nos. A-2, A-4, A-6, A-32, A-34), lakes (A-7, A-13, A-19, A-22, A-30), main rocky land (A-1, A-11, A-14, A-21, A-39) and coastal-shelf area (A-8, A-15, A-24, A-35, A-38) (Figure-1.1). The minerals are separated by the gravitational settling method using Bromoform (sp. gr. 2.89) proposed by Carver (1971). Minerals' are identified by following Lindholm (1987) and Mange and Maurer (1992). Statistical analysis involves using slides having more than 300 grains. The heavy mineral assemblage consists of ultrastable minerals like zircon, tourmaline, and rutile; moderately stable minerals like garnet, kyanite, sillimanite, enstatite, zoisite, and unstable minerals like hornblende, hypersthene, and andalusite (Figure. 1.4). The stability index of the mineral is as proposed by Pettijohn *et al.* (1973). Chlorite, spinel, topaz, and lawsonite are rare occurrences. However, opaque concentration dominates the entire assemblage.

Figure 4. Heavy minerals identified in various area units; 1A&B. Zircon, 2A&B. Tourmaline, 3A&B. RRutile, 4-C. Garnet, 5. Topaz, 6. Hornblende, 7. Kyanite, 8. chlorite, 9. Enstatite, 10. Sillimanite, 11A&B. Spinel, 12A. Zoisite, 13A&B. Hypersthene, 14A&B. Opaque minerals, 15. Lawsonite, and 16. Anda

Figure 5. A histogram showing a comparative abundance of various heavy minerals

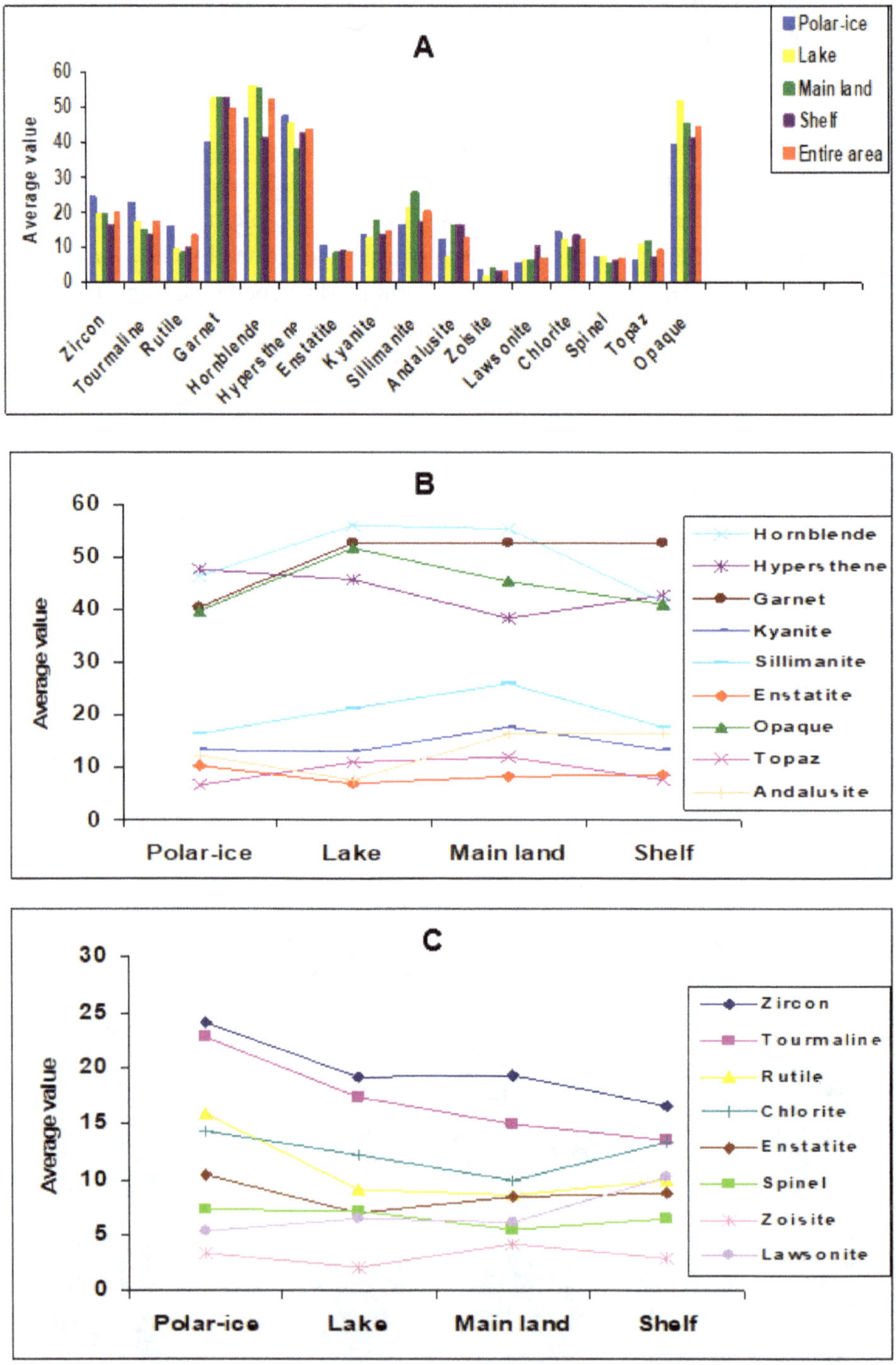

B &C. line diagram showing heavy mineral trends (average concentration)

Barring the opaque, hornblende, hypersthene, and garnet dominate the entire area. Hornblende is comparatively more in the lake and mainland area whereas, garnet has almost equal representation in all the units. Hypersthene shows the maximum average value in the polar ice sheet, followed by the lake and shelf area, whereas the lowest is in the mainland. The higher values of these minerals are indicative of a metamorphic source. Similarly, kyanite, sillimanite, and andalusite have the highest average values in mainland areas. The mainland is a high-grade metamorphic terrain and releases these minerals due to glacial weathering and wind erosion. It is difficult to interpret the controlling factors based on heavy minerals in different sub-environments with a weak trend. All the four sub environments show a distinct trend of sediment accumulations, as well as their release and transport, which may be due to different climatic and weather conditions, transport mechanisms, a reworking of the sediments which also, affect the distribution as well as the stability of grains (Elliot *et al.,* 1992; Ehrmann and Polozek, 1999). In general, the heavy minerals indicate a short distance of sediment transport as the grains' shapes are mostly angular to prismatic or well crystalline grains exhibit subangular preservation of corners, edges, and faces indicating low transportation.

Table 3. Grain counts, frequency numbers, and percentage of heavy minerals in various glacial sub environments of the Oasis (Srivastava et al., 2010)

Heavy Min. /Sample no.	Zircon	Tourmaline	Rutile	Garnet	Hornblende	Hypersthene	Enstatite	Kyanite	Sillimanite	Andalusite	Zoisite	Lawsonite	Chlorite	Spinel	Topaz	Opaque	Total	ZTR Total	ZTR Index
Polar-ice sheet																			
A-2	36 11.07; 5	27 8.30; 5	22 6.76; 5	43 13.2; 5	44 13.5; 6-	43 13.23; 5	08 2.46; 3	14 4.30; 4	11 3.38; 3	06 1.84; 2	05 1.53; 2	02 0.61; 1	21 6.46; 4	05 1.53; 2	06 1.84; 2	31 9.53; 5	325	85	28.33
A-4	35 10.79; 5	36 11.07; 5	20 6.15; 4	30 9.23; 5	47 14.46; 6-	30 9.23; 5	05 1.53; 2	14 4.30; 4	14 4.30; 4	08 2.46; 3	02 0.61; 1	07 2.15; 3	21 6.46; 4	09 2.76; 3	08 2.46; 3	39 12.0; 5	325	91	30.33
A-6	08 2.57; 3	30 9.64; 5	10 3.21; 3	59 18.97; 6	38 12.21; 5	73 23.47; 6+	04 1.28; 2	08 2.57; 3	13 4.18; 4	02 0.64; 1	--	02 0.64; 1	12 3.85; 4	11 3.53; 4	02 0.64; 1	39 12.54; 5	311	48	16
A-32	19 6.98; 5	12 4.41; 4	16 5.88; 4	33 12.13; 5	48 17.64; 6-	39 14.33; 6-	23 8.45; 5	21 7.72; 5	31 11.39; 5	19 6.98; 5	09 3.30; 3	11 4.04; 4	10 3.67; 4	09 2.89; 3	08 2.94; 3	37 13.60; 6-	272	47	15.66
A-34	23 6.99; 5	09 2.73; 3	12 3.64; 4	36 10.94; 5	55 16.71; 6-	54 16.41; 6-	12 3.64; 4	10 3.03; 3	12 3.64; 4	27 8.20; 5	01 0.30; IX	05 1.51; 2	08 2.43; 3	03 0.91; 1	09 2.73; 3	53 16.10; 6-	329	44	14.66
Avg.	**24.2**	**22.8**	**16**	**40.2**	**46.4**	**47.8**	**10.4**	**13.4**	**16.2**	**12.2**	**3.4**	**5.4**	**14.4**	**7.4**	**6.6**	**39.8**	**354.4**	**63.0**	**20.99**
Lakes																			
A-7	05 1.58; 2	20 6.34; 4	08 2.53; 3	62 19.68; 6	70 22.22; 6	59 18.73; 6	02 0.63; 1	05 1.58; 2	07 2.22; 3	03 0.95; 1	--	06 1.90; 2	09 2.85; 3	11 3.49; 3	13 4.12; 4	35 11.11; 5	315	33	11
A-13	18 5.47; 4	20 6.07; 4	07 2.12; 3	70 21.27; 6	56 17.02; 6-	48 14.58; 6-	--	06 1.82; 2	10 3.03; 3	01 0.30; IX	--	01 0.30; IX	15 4.55; 3	06 1.82; 3	05 1.51; 2	66 20.06; 5	329	45	15
A-19	16 4.12; 4	20 5.15; 4	15 3.86; 4	52 13.40; 5	63 16.23; 6-	44 11.34; 5	19 4.89; 4	17 4.38; 4	27 6.95; 5	09 2.31; 3	--	11 2.83; 3	11 2.83; 3	06 1.54; 2	13 3.35; 3	65 16.75; 6-	388	51	17
A-22	23 7.39; 5	09 2.89; 3	11 3.53; 4	34 10.93; 5	42 13.50; 5	52 16.72; 6-	07 2.25; 3	24 7.71; 5	19 6.10; 4	19 6.10; 4	07 2.25; 3	15 4.82; 4	14 4.50; 4	07 2.25; 3	13 4.18; 4	51 16.39; 6-	311	43	14.33
A-30	34 10.62; 5	18 5.62; 4	05 1.56; 2	45 14.06; 6-	49 15.31; 6-	26 8.12; 5	07 2.18; 3	13 4.06; 4	43 13.43; 5	06 1.87; 2	04 1.25; 2	--	12 3.75; 4	06 1.87; 2	11 3.43; 3	41 12.81; 5	320	57	19
Avg.	**19.2**	**17.4**	**9.2**	**52.6**	**56**	**45.8**	**7**	**13**	**21.2**	**7.6**	**2.2**	**6.6**	**12.2**	**7.2**	**11**	**51.6**	**357.6**	**45.8**	**15.26**
Main-land																			
A-1	29 7.79; 5	15 4.03; 4	10 2.68; 3	39 10.48; 5	123 33.06; 7-	31 8.33; 5	06 1.61; 2	12 3.22; 3	12 3.22; 3	38 10.21; 5	--	02 0.53; 1	17 4.56; 4	03 0.80; 1	04 1.07; 2	31 8.33; 5	372	54	18
A-11	24 8.13; 5	20 6.77; 5	05 1.69; 2	61 20.67; 6	29 9.83; 5	13 4.40; 4	03 1.01; 2	33 11.18; 5	29 9.83; 5	11 3.72; 4	11 3.72; 4	07 2.37; 3	07 2.37; 3	12 4.06; 4	22 7.45; 5	19 6.44; 4	295	49	16.33
A-14	16 4.48; 4	18 5.04; 4	08 2.24; 3	67 18.76; 6	43 12.04; 5	50 14; 6-	--	24 6.72; 5	46 12.88; 5	11 3.08; 3	02 0.56; 1	10 2.80; 3	08 2.24; 3	--	12 3.36; 3	42 11.76; 5	357	42	14
A-21	07 2.18; 3	13 4.04; 4	12 3.73; 4	44 13.70; 6-	36 11.21; 5	48 14.95; 6-	15 4.67; 4	13 4.04; 4	23 7.16; 5	16 4.98; 4	05 1.55; 2	08 2.49; 3	08 2.49; 3	09 2.80; 3	13 4.04; 4	51 15.88; 6-	321	32	10.66
A-39	21 6.03; 4	09 2.58; 3	08 2.29; 3	52 14.94; 6-	46 13.21; 5	49 14.08; 6-	18 5.17; 4	07 2.01; 3	20 5.74; 4	05 1.43; 2	03 0.86; 1	04 1.14; 2	10 2.87; 3	04 1.14; 2	09 2.58; 3	84 24.13; 6+	348	38	12.66
Avg.	**19.4**	**15**	**8.6**	**52.6**	**55.4**	**38.2**	**8.4**	**17.8**	**26**	**16.2**	**4.2**	**6.2**	**10**	**5.6**	**12**	**45.4**	**368.6**	**43.0**	**14.30**
Shelf area																			
A-8	33 10.54; 5	24 7.66; 5	09 2.87; 3	86 27.47; 6+	36 11.50; 5	23 7.34; 5	--	06 1.91; 2	27 8.62; 5	02 0.63; 1	--	--	20 6.38; 4	12 3.83; 4	03 0.93; 1	32 10.22; 5	313	66	22
A-15	08 2.46; 3	18 5.53; 4	12 3.69; 4	65 20; 6	53 16.30; 6-	47 14.46; 6-	08 2.46; 3	14 4.30; 4	11 3.38; 4	20 6.15; 4	--	14 4.30; 4	07 2.15; 3	04 1.23; 2	14 4.30; 4	30 9.23; 5	325	38	12.66
A-24	10 3.15; 4	12 3.78; 4	16 5.04; 4	25 7.88; 5	43 13.56; 6-	49 15.45; 6-	15 4.73; 4	14 4.41; 4	20 6.30; 4	10 3.15; 4	03 0.94; 1	11 3.47; 4	15 4.73; 4	09 2.83; 3	06 1.89; 4	59 18.61; 6	317	38	12.66
A-35	25 7.26; 5	08 2.32; 3	08 2.32; 3	40 11.62; 5	59 17.15; 6-	46 13.37; 5	14 4.06; 4	12 3.48; 3	13 3.77; 4	31 9.01; 5	04 1.16; 2	16 4.65; 4	16 4.65; 4	05 1.45; 2	05 1.45; 2	42 12.20; 5	344	41	13.66
A-38	07 2.16; 3	06 1.85; 2	05 1.54; 2	48 14.81; 6-	64 19.75; 6	49 15.12; 6-	07 2.16; 3	21 6.48; 4	17 5.24; 4	18 5.55; 4	08 2.46; 3	10 3.08; 3	09 2.77; 4	03 0.92; 1	10 3.08; 3	42 12.96; 5	324	18	06
Avg.	**16.6**	**13.6**	**10**	**52.8**	**51**	**42.8**	**8.8**	**13.4**	**17.6**	**16.2**	**3**	**10.2**	**13.4**	**6.6**	**7.6**	**41**	358	40.2	19.90
Entire area avg.	**19.8**	**17.2**	**13.4**	**49.5**	**52.2**	**43.6**	**8.55**	**14.4**	**20.25**	**13.05**	**3.2**	**7.1**	**12.5**	**6.7**	**9.3**	**44.5**	359	**48.0**	17.61

The high-grade metamorphic terrain is the primary source of the sediments is well evident by the dominance of specific minerals viz., hornblende, hypersthene, garnet, kyanite, sillimanite, andalusite, lawsonite, and spinel. (Mange and Maurer, 1992). Similarly, glacial-marine surface sediments of the Weddell Sea, Antarctica, show the dominance of mineralogically low mature heavy minerals, i.e., garnet, green hornblende, clinopyroxene (Diekmann and Kuhn, 1999). The common source of rutile is the high-grade metamorphic rock (Force, 1980). The ZTR index is low and ranges from 14.3 to 20.99. The low index shows sediment immaturity due to rapid erosion, short transportation, and rapid deposition (Hubert, 1962). Garnet, a dominant metamorphic constituent, is occurring in lakes and mainland areas. It appears together with kyanite, sillimanite, and andalusite. It is comparable to the Palaeocene sediments of Seymour island of Larsen ice-shelf, northern Antarctica (Elliot *et al.*, 1992). Apart from metamorphic terrain, other sediment supply sources are indicated by rounded, sub-rounded to sub-angular grains of zircon and tourmaline showing transportation, usually considered the derivatives of non-metamorphic to low-grade metamorphic terrains (Deer *et al.,* 1992). Pelitic and psammitic rocks are familiar sources of tourmalines with a widespread occurrence in the assemblage (Henry and Guidottic, 1985; Mange and Maurer, 1992).

Clay Mineralogy

Clay mineralogy is an almost unexplored aspect of the area; however, the presence of illite as a dominant constituent and subordinate chlorite in the sea bed, sediment, and lake terrace have been reported based on XRD analysis of a single sample each from the respective area (Lal, 1986).

The author's group has also carried out basic studies on the clays separated from the Oasis's sediments of different glacial units (Srivastava *et al.*, 2011). These clay minerals also serve as an effective tool for interpreting paleoclimate (Srivastava *et al.*, 2018). Total of sixteen samples representing equally the polar ice-sheet, i.e., sample nos. C-1 (A-2), C-2 (A-5), C-3 & C-4 (A-33 & 34); lakes C-5 (A-7), C-6 (A-13), C-7 (A-19), C-8 (A-28); main rocky land C-9 (A-11), C-10 (A-14), C-11 (A-17), C-12 (A-39) and coastal shelf area C-13 ((A-8), C-14 (A-23), C-15 (A-27) and C-16 (A-36) have been selected for the analysis. These sediment samples consist of a loose admixture of sand-silt-clay dried and separated into the fine fraction (<0.0625 mm, i.e., silt and clay) through the sieving. The fine clay from these sediments was fractioned according to the procedure as proposed by Jack-son (1979). The fine clay, so obtained, was saturated with Ca or K for mineralogical analysis. Diffractograms were taken on parallel oriented mounts of the saturated samples at different temperatures. The fine clay fractions so obtained were subjected for mineral identification by X^{-rd} analysis of oriented mounts saturated with either Ca or glycolate. K-treated samples were analysed at 250C, 1000C, 3000C, and 5500C. The instrument used is Phillips Analytical X'Pert, Ni-filtered, Cu Ka radiation with a scanning speed of 202Ø min-1. The remaining fraction of the same fine clay initially separated for XRD analysis was subjected to thermal analysis. It is carried out in an air atmosphere, up to the temperature of 10000C, with a scan rate of 100C/min for thermo-gravimetric (TG) and differential thermal (DT) analysis on a computerised TA instrument, model Perkin Almer Diamond TGDTA.

Thermograms

Both TGA and DTA curves of all the samples have similar trends showing the uniformity of clay mineral contents in the admixture (Figures. 1.6 &1.7). The thermogravimetric analysis shows a minor loss or gain in weight, likely due to the loss of bound water in clays (Liptey, 1973) (Table-1.3). All the samples' DTA curves show a small notch of endothermic peak around 585⁰C. It is a usual position of kaolinite (Figure. 1.7, Table-1.3) due to the removal of structural water from the aluminium atoms (O' Gorman and Walker, 1973; Kotoky *et al.,* 2006).

Table 4. DTA and TGA values of the samples

Samples		DTA				TGA	
		Onset	Peak	Peak height	Area	Wt. Gain	Wt. loss
Polar ice -sheet	C1	582.55⁰C	585.51⁰C	-0.616 µV	23.710 µV x sec	0.459%	--
	C2	582.49⁰C	585.75⁰C	-0.674 µV	29.314 µV x sec	0.718%	--
	C3	581.50⁰C	585.46⁰C	-0.902 µV	45.289 µV x sec	0.460%	0.225%
	C4	582.70⁰C	586.15⁰C	-0.772 µV	33.186 µV x sec	0.450%	--
Lakes	C5	583.24⁰C	586.13⁰C	-0.410 µV	14.546 µV x sec	0.509%	0.744%
	C6	582.70⁰C	585.94⁰C	-0.670 µV	28.741 µV x sec	0.515%	0.149%
	C7	582.84⁰C	586.22⁰C	-0.605 µV	23.574 µV x sec	0.462%	0.114%
	C8	583.57⁰C	586.02⁰C	-0.269 µV	6.780 µV x sec	1.032%	0.172%
Main rocky land	C9	583.24⁰C	585.67⁰C	-0.324 µV	10.549 µV x sec	0.711%	0.067%
	C10	583.25⁰C	585.99⁰C	-0.374 µV	12.666 µV x sec	0.574%	--
	C11	582.68⁰C	585.93⁰C	-0.538 µV	21.092 µV x sec	0.457%	0.158%
	C12	583.00⁰C	586.09⁰C	-0.506 µV	17.019 µV x sec	0.629%	--
Coastal shelf	C13	582.68⁰C	586.17⁰C	-0.476 µV	16.849 µV x sec	0.525%	0.124%
	C14	582.79⁰C	586.14⁰C	-0.435 µV	16.427 µV x sec	0.648%	0.172%
	C15	582.47⁰C	585.59⁰C	-0.380 µV	17.130 µV x sec	0.988%	--
	C16	583.26⁰C	585.75⁰C	-0.387 µV	21.171 µV x sec	0.813%	0.201%

Table 5.

Minerals identified	Polar ice-sheet				Lake s				Main rocky land				Coastal shelf			
	C1	C2	C3	C4	C5	C6	C7	C8	C9	C10	C11	C12	C13	C14	C15	C16
Smectite	+	+	+	+	+	+	+	+	+	+	+	+	+	+	+	+
Vermiculite	-	+	+	+	+	+	+	-	-	+	+	+	+	+	+	+
Chlorite	-	-	-	-	-	-	-	+	+	-	-	-	-	-	-	-
Illite	+	+	+	+	+	+	+	+	+	+	+	+	+	+	+	+
Kaolinite	+	+	+	+	+	+	+	+	+	+	+	+	+	+	+	+
Quartz	+	+	+	+	+	+	+	+	+	+	+	+	+	+	+	+
Feldspar (K, Ca, Na)	+	+	+	+	+	+	+	+	+	+	+	+	+	+	+	+
Mixed layer	-	-	+ (CaEg)	-	-	+ (Ca)	+ (Ca)	+ (Ca, CaEg, K25)	+ (K110)	-	+	-	+ (Ca, K25)	+ (Ca, K350)	- (Ca)	-
Amphibole	+	+	+	+	+	+	+	+	+	+	+	+	+	+	+	+
Vm+Sm	+	-	-	-	-	-	-			-	-	-	-	-	-	-
Vm+Cl	-	-	-	-	-	-	-			-	-	-	-	-	+	+

Figure 6. TGA curves showing an almost similar trend

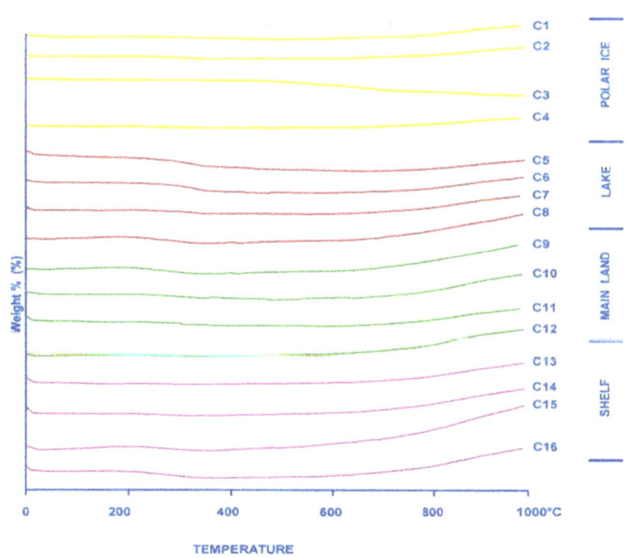

Figure 7. DTA curves showing a small notch at 585⁰C

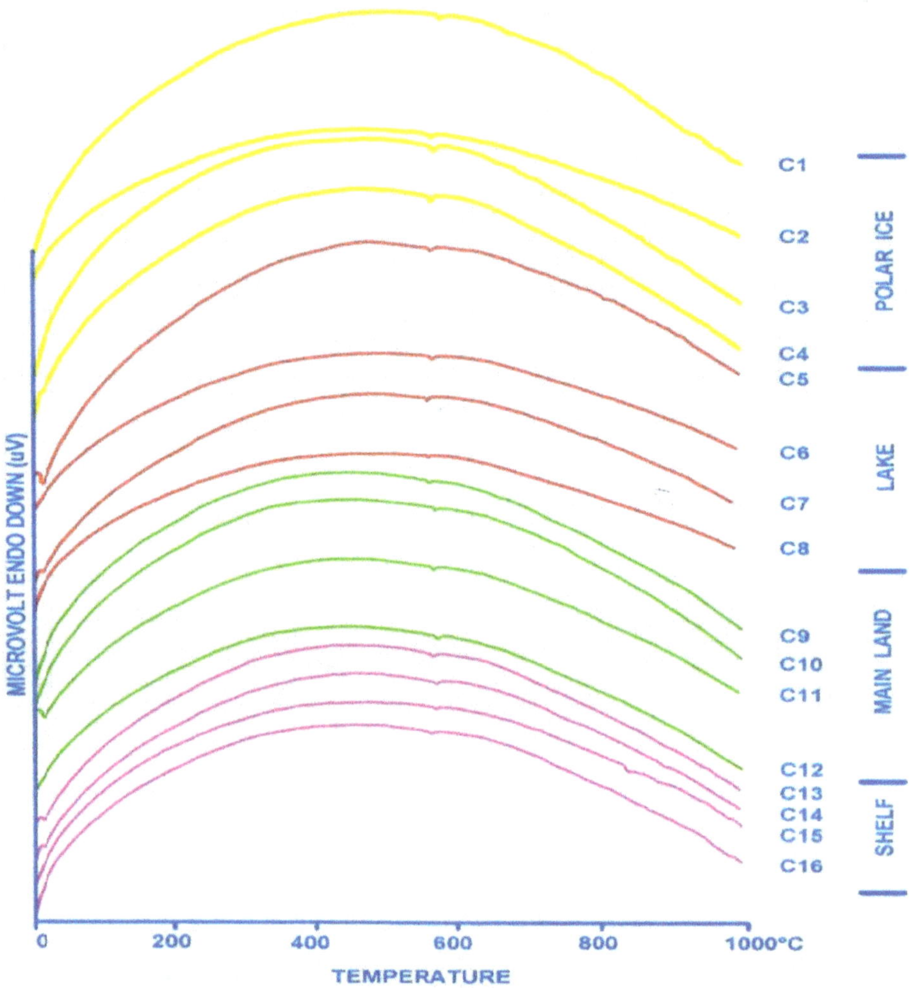

X-Ray Diffract Grams

All the four diffractograms of each sample, i.e., saturated with Ca (Figure. 1.8A), glycolated (Figure. 1.8B), were treated with K and heated at 25⁰C and 550⁰C (Figures. 1.8C & 1.8D) and analysed for clay mineral identification (Table 1.4).

Figure 8. A. X-RD spectra of samples, B. X-RD spectra of samples (CaEg), C. X-RD spectra of K-saturated samples and heated at 25 degrees Celsius, D. X-RD spectra of K-saturated sample and heated at 550 degrees Celsius

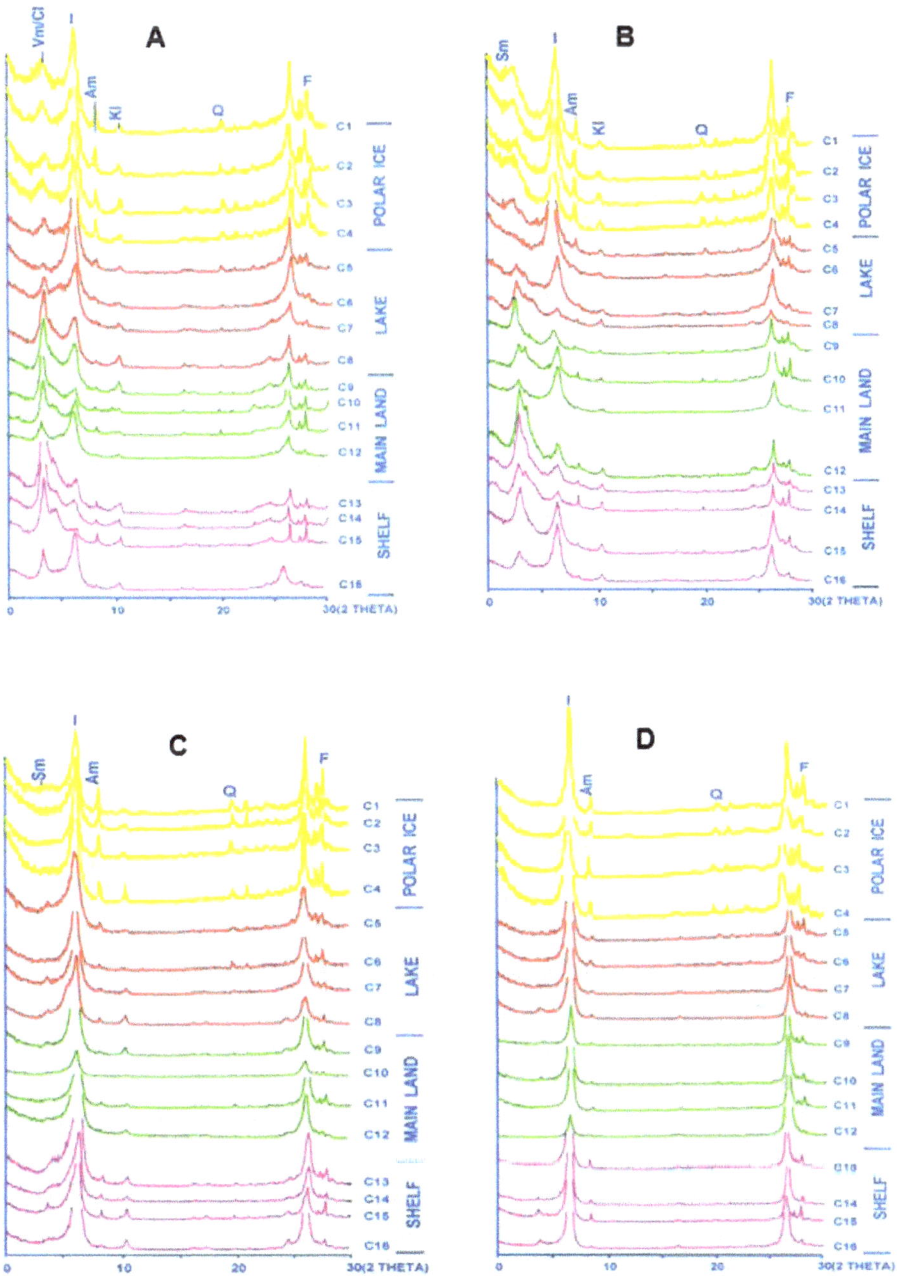

The peak values, as proposed by Jackson (1979), has been the criteria for identification. The X-ray patterns indicate a well-defined and robust illite peak in all the samples, apart from kaolinite, chlorite, smectite, and vermiculite. The peaks of primary minerals like quartz, feldspar, mica, and amphibole are also represented. Illite is a common mineral identified in all the samples. The mineral is considered a detrital clay derivative of acidic crystalline rock (Biscaye, 1965; Griffin *et al.,* 1968; Windom, 1976). In the current setup, illite is a characteristic mineral of high latitudes (Windom, 1976) and a valuable tool to interpret the weathering pattern of older sediments (Holmes, 2000). The mineral is widespread along the Antarctic continental margin and East Antarctic Craton (Tingey, 1991; Petschick *et al.,* 1996; Hillenbrand and Ehrmann, 2001, 2005). The source of illite in East Antarctica may be the biotite-bearing highly metamorphosed rocks (Setti *et al.,* 2001; Ehrmann *et al.,* 2005). Ehrmann *et al.* (2003) suggested that any metamorphic rock of low to a high grade as felsic to basic plutons and dykes can produce illite in the glacial regime. Kaolinite is also present or in all the samples. It is mainly considered a weathering product under tropical conditions (Biscaye, 1965; Singer, 1984; Keller, 1970) and widespread in the Antarctic Peninsula. It is a prominent mineral of the Cenozoic Battye Glacier Formation, North Prince Charle Mountain, East Antarctica (Ehrmann *et al.,* 2003). However, a low percentage in Cenozoic sediments, McMurdo Sound, Antarctica (Ehrman *et al.,* 2005). Robert and Maillot (1990) doubtfully interpreted both the source and origin of kaolinites. Since the kaoline cannot form in polar conditions, Ehrmann et *al.* (2003) suggested that the same from Battye Glacier Formation might have been produced in an interval of warmer and wetter conditions d chemical weathering of orthogneiss, mafic granulites, quartzite, etc. In the study area, feldspar is one of the Archaean gneiss's major constituents forming the basement; however, the prevailing cold climate is not favourable for kaolinite formation. The temperature variation and humidity might have played a significant role in kaolinite formation through chemical weathering. The drill-core sediments of the Paleocene-Eocene boundary at Maud Rise, Weddell Sea region, East Antarctica (Robert and Kennett, 1994), Jeong and Yoon (2001) report the mineral as a chemical weathering product. Or from the sea-floor sediments of South Shetland Island and soil of King George Island, West Antarctica, and interpreted that the kaolinites from both the sea are not the product of chemical weathering.

The chlorite peaks are recorded in one sample of the lake and main rocky land, mixed with vermiculite in two samples of the coastal-shelf area. Apart from tropical conditions (Biscaye, 1965), frequent reporting is from the cold climate of Polar Regions, *i.e.,* Greenland ice-sheet (Drab *et al.,* 2002); Antarctica Peninsula (Tingey, 1991) including Victoria Land Basin (Setti *et al.,* 2001), Pagodroma Group (Ehrmann *et al.,* 2003); McMurdo Sound (Kristoffersen *et al.,* 2000; Ehrmann *et al.,* 2003); ODP sites in Antarctic Sea (Hillenbrand and Ehrmann, 2001); etc. Windom

(1976), Setti *et al.* (2001), and Hillenbrand and Ehrmann (2001, 2003) interpreted that the physical weathering of basement metamorphics and igneous rocks supplies the chlorite of the southern polar areas; and, probably the altered calc-alkaline volcanic was the primary source.

Smectite, a silicate, is also recorded in all the sediment. The origin of smectite differs in different climates (Güven, 1988). The marine setup can be of authigenic or detrital origin (Chamley, 1989). In Antarctica, smectite is well reported from Cenozoic and Quaternary sediments, including drill core, and widely interpreted for palaeoclimatic reconstruction (Campbell and Claridge, 1987; Robert and Maillot, 1990; Ehrmann and Mackensen, 1992). Ehrmann (2001) and Setti *et al.* (2001) studied the abundance, concentration, and crystallinities of smectite from the drill core sediments of the continental shelf of McMurdo Sound in the Ross Sea and Victoria Land basin, Antarctica. Both studies conclude that the smectite formation in the recent Antarctic environment is only a subordinate process and is due mainly to the authigenic approach favoured by the weathering of basaltic rocks. In the present area of Schirmacher Oasis, basaltic rocks are well represented in the form of intrusions, cross-cutting in highly metamorphosed terrain.

Vermiculite is a product of pedogenic or diagenetic origin (Chamley, 1989). Its report from the surface sediments and drill cores are comparatively less from Antarctic Peninsula as well as Antarctic Ocean, however, widely reported from the soils, i.e., Lassiter Coast, south of Antarctic Peninsula (Boyer, 1975); Prince Charles Mountain (Bardin *et al.,* 1979; Bardin, 1982). However, most of the vermiculite is pedogenic (Claridge and Champbell, 1989). Kristofferson *et al.* (2000) reported its occurrence from shallow drill core of continental shelf, Weddell Sea, Antarctica, and considered its origin from the primary mica due to hydration.

Grain Morphoscopy

Grain surface study of glacial sediments through SEM is one of the significant\ aspects which provide important clues about the environment, weathering pattern, diagenetic Pattern, *etc.* Lal (1986) reported sub-parallel steps, striations, conchoidal fractures, and V-shaped markings on quartz grains and interpreted the sediments' subaqueous transport. Asthana and Chaturvedi (1998) made a detailed SEM study of the quartz grains and reported various features like a sharp-edged highly angular outline of quartz grain, high relief, fractured blocks, arcuate fracture pattern, parallel and semi-parallel steps, and uneven striation (first-order mechanical texture); subdued to well-developed fractures, randomly oriented scratches and grooves, upturn plates, meandering ridges, concavities, collision pits, etc. (second-order mechanical texture) and silica precipitation, dissolution pits, and hollows, secondary overgrowth, solution channels, silica tubes, surface etchings, *etc.* (texture of chemical origin). The high energy of mechanical abrasion, the chemical in the channel's sediments, and melting spots. Alkalinity-acidity, pH condition, and temperature have a significant role in producing chemical textures. The sediments of wind scooped depressions characterise the mechanical texture.

Lake Sedimentology

Schirmacher Oasis marks numerous lakes; however, it lacks heavily for their sedimentological studies. Preliminary investigations on the Priyadarshini Lake, adjacent to the Indian station, i.e., Maitri, give a brief idea about grain size, mineralogy, organic matter in the sediment core of Priyadarshini Lake, Sinha and Chatterji (2000). Based on Priyadarshini Lake's bathymetric survey, a 3-D model of a part of the lake is prepared (Khare *et al.*, 2008). However, there is a lack of a detailed approach for lake sedimentology despite having a broad scope.

FUTURE PROSPECT

The sedimentological work on the glacial sediments of Schirmacher Oasis for grain size analysis, heavy minerals, clay mineralogy, surface texture analysis, lake sedimentology, *etc.*, requires more data to authenticate the results drawn. Some of the crucial aspects, like geochemistry, weathering pattern, the chemical framework of the sediments, tectonic setup, etc., are entirely lacking. There is clear-cut differentiation of various geomorphological and glacial units in the area, which differs for their sediment characteristics, like - polar ice sheets, rocky mainland, lake, coastal shelf area, meltwater channels, etc. The sedimentological details of these units, mainly the polar ice sheet and lakes in stratigraphic order, coupled with time marker evidence, can provide an excellent geologic history, paleoclimate, and paleoenvironment of the region. The work done concludes that there is ample scope for sedimentological studies in the area.

ACKNOWLEDGMENT

Ministry of earth science, government of India, and the national Centre for Antarctic & Ocean research GOA are deeply acknowledged for their support in varied ways. Financial assistance in the form of a significant research project awarded to AKS (No. 24/287/06EMR-II) by CSIR, New Delhi, is acknowledged.

REFERENCES

Asthana, R., & Chaturvedi, A. (1998). The grain size behaviour and morphoscopy of supraglacial sediments, south of Schirmacher Oasis, E. Antarctica. *Journal of the Geological Society of India*, *52*, 557–568.

Bardin, V. I. (1982). Composition of East Antarctic moraines and some problems of Cenozoic history. In C. Craddock (Ed.), Antarctic Geoscience (pp. 1069-1076). Univ. Wisconsin Press.

Bardin, V. I., Bubnova, M. I., & Gerasimova, V. M. (1979). Clay minerals in unconsolidated deposits of the Prince Charles Mountains. *Inf. Bull. Sov. Antarct. Exped.*, *71*, 120–128.

Bera, S. K. (2004). Late Holocene palaeo-winds and climatic changes in Eastern Antarctica as indicated by long-distance transported pollen-spores and local microbiota in polar lake core sediments. *Current Science*, *86*(11), 1485–1488.

Biscaye, P. E. (1965). Mineralogy and sedimentation of recent deep-sea clay in the Atlantic Ocean and adjacent seas and oceans. *Bulletin of the Geological Society of America*, *76*(7), 803–832. doi:10.1130/0016-7606(1965)76[803:MASORD]2.0.CO;2

Boyer, S. J. (1975). Chemical weathering of the rocks on the Lassiter Coast, Antarctic Peninsula, Antarctica. *New Zealand Journal of Geology and Geophysics*, *18*(4), 623–628. doi:10.1080/00288306.1975.10421561

Campbell, I. B., & Claridge, G. G. C. (1987). *Antarctica: Soil, Weathering Processes, and Environment*. Elsevier.

Carver, R. E. (1971). *Procedures in Sedimentary Petrology*. Wiley.

Chamley, H. (1989). *Clay Sedimentology*. Springer. doi:10.1007/978-3-642-85916-8

Claridge, G. G. C., & Campbell, J. B. (1989). Clay Mineralogy. In P. J. Barrett (Ed.), Antarctic Cenozoic History from the CIROS-1 Drillhole, McMurdo Sound (pp. 185-193). DSIR Bulletin.

Dayal, A. M., & Hussain, T. S. M. (1997). Rb-Sr ages of lamprophyre dykes from Schirmacher Oasis, Queen Mound Land, East Antarctica. *Journal of the Geological Society of India*, *50*, 457–460.

Deer, W. A., Howie, R. A., & Zussman, J. (1992). *An Introduction to the Rock-Forming Minerals* (2nd ed.). Longman Sc. Technic.

Ehrmann, W., & Mackensen, A. (1992). Sedimentological evidence from the formation of an East Antarctic ice sheet in Eocene/Oligocene time. *Palaeogeography, Palaeoclimatology, Palaeoecology*, *93*(1-2), 85–112. doi:10.1016/0031-0182(92)90185-8

Ehrmann, W., & Polozek, K. (1999). The heavy mineral record in the Pliocene to Quaternary sediments of the CIROS-2 drill core, McMurdo Sound, Antarctica. *Sedimentary Geology*, *128*(3-4), 223–244. doi:10.1016/S0037-0738(99)00071-8

Ehrmann, W., Bloemendal, J., Hambrey, J. M., Mckelvey, B., & Whitehead, J. (2003). Variations in the composition of the clay fraction of the Cenozoic Pagodrama Group, East Antarctica: Implications for determining provenance. *Sedimentary Geology*, *161*(1-2), 131–152. doi:10.1016/S0037-0738(03)00069-1

Ehrmann, W., Setti, M., & Marinoni, L. (2005). Clay minerals in Cenozoic sediments of Cape Roberts (McMurdo Sound, Antarctica) reveal palaeoclimatic history. *Palaeogeography, Palaeoclimatology, Palaeoecology*, *229*(3), 187–211. doi:10.1016/j.palaeo.2005.06.022

Elliot, D. H., Hoffman, S. M., & Rieske, D. E. (1992). The provenance of Paleocene strata, Seymour Island. In Recent Progress in Antarctic Earth Science (pp. 347-355). Terra Sci. Pub. Comp. (TERRAPUB).

Folk, R. L., & Ward, W. (1957). Brazos river bar: A study in the significance of grains size parameters. *Journal of Sedimentary Petrology*, *27*(1), 3–26. doi:10.1306/74D70646-2B21-11D7-8648000102C1865D

Folk, R. L. (1980). *Petrology of Sedimentary Rocks*. Hemphill Austin.

Force, E. R. (1980). The provenance of rutile. *Sedimentary Petrology*, *50*, 485–488.

Friedman, G. M. (1961). The distinction between dune, beach, and river sands from their textural characteristics. *Journal of Sedimentary Petrology*, *31*(4), 514–529.

Friedman, G. M. (1967). Dynamic processes and statistical parameters compared to the size-frequency distribution of beach and river sands. *Journal of Sedimentary Petrology*, *37*(2), 327–354.

Griffin, J. J., Windon, H., & Goldberg, E. D. (1968). The distribution of clay minerals in the World Ocean. *Deep-Sea Research*, *15*, 433–459.

Gupta, H. K., & Verma, A. K. (1986). Magnetic characteristics in Antarctica over geological contacts in Schirmacher Hill Region and Ice Shelf near Dakshin Gangotri (70° 05' 37" S, 12° 00' 00" E). In *Antarctica. Sci. Rep. 3rd Ind. Sci. Exp. Antarctica*. DOD, Govt. of Ind. Pub. Technical Publication.

Güven, N. (1988). Smectites. In S. W. Bailey (Ed.), *Hydrous Phyllosilicates* (pp. 497–552). Mineralogical Society of America.

Henry, D. J., & Guidottic, C. V. (1985). Tourmaline: As a petrogenetic indicator in abyssal and Alpine-type peridotites and spatially associated lavas. *Contributions to Mineralogy and Petrology*, *86*, 54–76.

Hillenbrand, C. D., & Ehrmann, W. (2001). Distribution of clay minerals in drift sediments on the continental rise west of Antarctic Peninsula, ODP leg 178, sites 1095 and 1096. *Proc. ODP, Science., Results.*, *178*, 1-29. 10.2973/odp.proc.sr.178.224.2001

Hillenbrand, C. D., & Ehrmann, W. (2003). Palaeoenvironmental implications of Tertiary sediments from Kainan Maru Seamount and northern Gunnerous Ridge. *Antarctic Science*, *15*(4), 522–536. doi:10.1017/S0954102003001640

Hillenbrand, C. D., & Ehrmann, W. (2005). Late Neogene to Quaternary environmental changes in the Antarctic Peninsula region: Evidence from drift sediments. *Global and Planetary Change*, *45*(1-3), 165–191. doi:10.1016/j.gloplacha.2004.09.006

Hoch, M. (1999). Geochemistry and petrology of ultramafic lamprophyres from Schirmacher Oasis, East Antarctica. *Mineralogy and Petrology*, *65*(1-2), 51–67. doi:10.1007/BF01161576

Hoch, M., & Tobschall, H. J. (1998). Minettes from Schirmacher Oasis, East Antarctica - indicators of an enriched mantle source. *Antarctic Science*, *10*(4), 476–486. doi:10.1017/S0954102098000571

Hoch, M., Rehkamper, M., & Tobschall, H. J. (2001). Sr, Nd, Pb, and O Isotopes of minettes from Schirmacher Oasis, East Antarctica: A case of mantle metasomatism involving subducted continental material. *Journal of Petrology*, *42*(7), 1387–1400. doi:10.1093/petrology/42.7.1387

Holmes, M. A. (2000). Clay mineral composition of glacial erratics, McMurdo Sound. In J. D. Stilwell & R. M. Feldmann (Eds.), *Palaeobiology and Palaeoenvironments of Eocene Rocks, McMurdo Sound East Antarctic, Antarctic. Research*. doi:10.1029/AR076p0063

Hubert, J. F. (1962). A zircon-tourmaline-rutile maturity index and the interdependence of the composition of heavy mineral assemblages with the gross composition and texture of sandstones. *Sedimentary Petrology*, *32*, 440–450.

Hussain, T. S. M., & Diwakar, Rao, V. (1996). Geochemistry of polyphase gneisses from Schirmacher Oasis, East Antarctica. *Journal of the Geological Society of India*, *47*, 303–312.

Ingole, B., & Dhargalkar, V. (1998). Eco-biological assessment of a freshwater lake at Schirmacher Oasis, East Antarctica, with reference to human activities. *Current Science*, *74*(6), 529–534.

Ingram, R. L. (1971). Sieve analysis. In R. E. Carver (Ed.), *Procedures in Sedimentary Petrology* (pp. 49–68). Wilson Interscience.

Jafri, S. H., Moeen, S., Charan, S. N., Narayan, B. L., & Diwakar Rao, V. (2002). The occurrence of Quench olivine crystals in a basaltic dyke from Schirmacher Oasis, Queen Mound Land, East Antarctica. *Journal of the Geological Society of India*, *60*, 67–74.

Jackson, M. L. (1979). *Soil Chemical Analysis-Advance Course* (2nd ed.). Department of Soil Science, Wisconsin University.

Jayapaul, D., Kundu, A., Gaur, M. P., & Krishnamurthy, K. V. (2005). Petrology of A-Type hornblende granite and fayalite granite enclaves from cDML, East Antarctica: Some constraints on their origin. *Geological Society of India*, *65*, 562–576.

Jeong, G. Y., & Yeon, H. I. (2001). The origin of clay minerals in soils of King George Island, South Shetland Islands, West Antarctica, and its implication for the clay mineral composition of marine sediments. *Journal of Sedimentary Petrology*, *71*(5), 833–847. doi:10.1306/2DC4096C-0E47-11D7-8643000102C1865D

Keller, W. D. (1970). Environmental aspects of clay minerals. *Journal of Sedimentary Petrology*, *40*, 788–813. doi:10.1306/74D720A4-2B21-11D7-8648000102C1865D

Khare, N., Chaturvedi, S. K., Saraswat, R., Srivastava, R., Raina, R., & Wanganeo, A. (2008). Some morphometric characteristics of Priyadarshini water body at Schirmacher Oasis, Central Dronning Maud Land with special reference to its bathymetry. *Indian Journal of Geo-Marine Sciences*, *37*(4), 435–438.

Kotoky, P., Bezbarruah, D., Baruah, J., Borah, G. C., & Sharma, J. N. (2006). Characterisation of clay minerals in the Brahmaputra river sediments, Assam, India. *Current Science*, *91*(9), 1247–1250.

Kristoffersen, Y., Strand, K., Vorren, T., Harwood, D., & Webb, P. (2000). Pilot shallow drilling on the continental shelf, Dronning Maud Land, Antarctica. *Journal of Antarctic Science*, *4*(4), 463–470. doi:10.1017/S0954102000000547

Kumar, R. (1986). Note on the P-T Conditions of metamorphism of Schirmacher Range, East Antarctica. Sci. Rep. In *3rd Ind. Sci. Exp. Antarctica*. DOD, Govt. of Ind. Pub. Technical Publication.

Lal, M. (1986). Sedimentology of the glacial sand and lake terraces sediments from Schirmacher Oasis and sea bed sediment of Princess Astrid Coast, Queen Maud land Antarctica. Sci. Rep. In *3rd Ind. Sci. Exp. Antarctica*. DOD, Govt. of India Publ. Technical Publication.

Lindholm, R. C. (1987). *A Practical Approach to Sedimentology*. Allen and Unwin Publ. doi:10.1007/978-94-011-7683-5

Liptey, G. (1973). *Atlas of Thermoanalytical Curves*. Akademiai Kiado.

Mange, M. A., & Maurer, H. F. W. (1992). *Heavy Minerals in Colour*. Chapman and Hall. doi:10.1007/978-94-011-2308-2

Mathur, A. K., Asthana, R., & Ravindra, R. (2006). Arcellaceans (thecamoebians) from core sediments of Priyadarshini Lake, Schirmacher Oasis, East Antarctica. *Current Science*, *90*(12), 1603–1605.

Moiola, R. J., & Weiser, D. (1968). Textural parameters: An evaluation. *Journal of Sedimentary Petrology*, *38*(1), 45–53.

Nayaka, S., & Upreti, D. K. (2005). Schirmacher Oasis, East Antarctica, a lichenologically interesting region. *Current Science*, *89*(7), 1069–1071.

O' Gorman, J. V., & Walker, P. L. Jr. (1973). The thermal behaviour of mineral fractions separated from selected American coals. *Fuel*, *52*(1), 71–79. doi:10.1016/0016-2361(73)90016-1

Palanisamy, M. (2007). Snedra ulna (Nitzsch) Ehrenberg: A new generic record in Schirmacher Oasis, East Antarctica. *Current Science*, *92*(2), 179–181.

Parthasarathy, G., Sharma, S. R., Ravindran, T. R., Arora, A. K., & Hussain, S. M. (2003). Structural and thermal studies of graphite from East Antarctica. *Geological Society of India*, *61*, 335–343.

Passega, R. (1957). Textures as a characteristic of clastic deposition. *Am. Assoc. Petrol. Geol.*, *41*, 1952–1984.

Passega, R. (1964). Grain size representation by C-M Pattern as a geological tool. *Jour. Sed. Pet.*, *34*(4), 830–847. doi:10.1306/74D711A4-2B21-11D7-8648000102C1865D

Passega, R., & Byramjee, R. (1969). Grain size image of clastic deposits. *Sedimentology*, 13.

Petschick, R., Khun, G., & Gingele, F. (1996). Clay minerals distribution in surface sediments of the South Atlantic: Sources, transport, and relation to oceanography. *Marine Geology*, *130*, 203–229.

Pettijohn, F. J., Potter, P. E., & Siever, R. (1973). *Sand and Sandstone*. Springer-Verlag. doi:10.1007/978-1-4615-9974-6

Rama Rao, P., Ramana Rao, A. V., & Poornachandra Rao, G. V. S. (1995). Geological, geochemical, geochronological, and palaeomagnetic studies on the rocks from parts of East Antarctica. In *Sci. Rep., 8th Ind. Sci. Exp.* Antarctica. DOD, Govt. of India Publ. Technical Publication.

Reinick, H. E., & Singh, I. B. (1980). *Depositional Sedimentary Environments*. Springer-Verlag. doi:10.1007/978-3-642-81498-3

Robert, C., & Kennett, J. P. (1994). An Antarctic humid episode of Palaeocene- Eocene boundary: Clay mineral evidence. *Geology*, *22*(3), 211–214. doi:10.1130/0091-7613(1994)022<0211:ASHEAT>2.3.CO;2

Robert, C., & Maillot, H. (1990). Palaeoenvironments in the Weddel Sea area and Antarctic climates, as deduced from clay mineral associations and geochemical data ODP Leg 11. Proceedings. ODP, Science Research, 113, 51-70.

Sengupta, S. (1986). Geology of Schirmacher range (Dakshin Gangotri), East Antarctica. Sci. Rep. 3rd Ind. In *Sci. Exp. Antarctica*. DOD, Govt. of Ind. Pub. Technical Publication.

Sengupta, S. (1988). History of successive deformation in relation to metamorphism-migmatitic events in the Schirmacher Hills, Queen Maud Land, East Antarctica. *Journal of the Geological Society of India*, *32*, 295–31.

Sengupta, S. M. (1996). Introduction to Sedimentology. Oxford & IBH Publishing Co-Pvt. Ltd.

Setti, M., Marinoni, L., & López-Galindo, A. (2001). Crystal chemistry of smectite in sediments of CRP-3 drill core (Victoria Land basin, Antarctica): Preliminary results. *Terra Antarctica*, *8*(4), 543–550.

Sharma, C., Bera, S. K., & Upreti, D. K. (2002). Modern pollen-spore rain in Schirmacher Oasis, East Antarctica. *Current Science*, *82*(1), 88–91.

Sharma, C., Chauhan, M. S., & Sionha, R. (2007). Studies on Holocene climatic changes from Priyadarshini Lake sediments, Schirmacher Oasis East Antarctica: The Palynological Evidence. *Geological Society of India*, *69*, 92–96.

Sinha, R., & Chatterji, A. (2000). Thermal structure, sedimentology, and hydrogeochemistry of lake Priyadarshini, Schirmacher Oasis, Antarctica. In *Sci. Rep., Ind. Sci. Exp. Antarctica*. DOD, Govt. of India Publ., Technical Publication.

Singer, A. (1984). The palaeoclimatic interpretation of clay minerals in sediments: A review. *Earth-Science Reviews*, *21*(4), 251–293. doi:10.1016/0012-8252(84)90055-2

Singh, R. K. (1986). Geology of Dakshin Gangotri hill range, Antarctica. Sci. Rep. In *3rd Ind. Sci. Exp. Antarctica*. DOD, Govt. of Ind. Publ. Technical Publication.

Singh, S. M., Puja, G., & Bhat, D. J. (2006). Psychrophilic fungi from Schirmacher Oasis, East Antarctica. *Current Science*, *90*(10), 1388–1392.

Srivastava, A. K. (2007). Detailed sedimentological studies of the glaciogenic sediments of Schirmacher Oasis, Antarctica. In *Sci. Rep., 25th Ind. Sci. Exp.* Antarctica. DOD, Govt. of India Publ., Technical Publication.

Srivastava, A. K., & Khare, N. (2009). Granulometric analysis of glacial sediments, Schirmacher Oasis, East Antarctica. *Journal of the Geological Society of India*, *73*(5), 609–620. doi:10.100712594-009-0047-3

Srivastava, A. K., Ingle, P. S., & Khare, N. (2010). Textural characteristics, distribution pattern, and provenance of heavy minerals in glacial sediments, Schirmacher Oasis, East Antarctica. *Journal of the Geological Society of India*, *73*, 393–402. doi:10.100712594-009-0019-7

Srivastava, A. K., Ingle, P. S., & Khare, N. (2018). Controlling factor for nature, Pattern, and accumulation of the glacial sediments of Schirmacher Oasis, East Antarctica: Comments on paleoclimatic condition. *Polar Science*, *18*, 113–122. doi:10.1016/j.polar.2018.05.004

Srivastava, A. K., Ingle, P. S., Lunge, H. S., & Khare, N. (2012). Grain-size characteristics of deposits derived from different glacigenic environments of the Schirmacher Oasis, East Antarctica. *Geologos*, *18*(4), 251–266. doi:10.2478/v10118-012-0014-0

Srivastava, A. K., Khare, N., & Ingle, P. S. (2011). Characterisation of clay minerals in the sediments of Schirmacher Oasis, East Antarctica: Their origin and climatological implications. *Current Science*, *100*(3), 363–372.

Sundararajan, N., & Rao, M. (2005). A note on the petrophysical properties and geological interpretation in Schirmacher Oasis East Antarctica. *Geological. Society of India*, *65*, 497–503.

Tingey, R. J. (1991). The Geology of Antarctica. Clarendon Press.

Visher, G. S. (1969). Grain size distributions and depositional processes. *Journal of Sedimentary Petrology*, *39*(3), 1074–1106.

Windom, H. L. (1976). Lithogenous material in marine sediments. In Chemical Oceanography (pp. 103-113). Academic Press.

Chapter 2
Geochemical and Mineralogical Studies of Gneiss-Charnockite Rocks of Schirmacher Region, East Antarctica

D. Rameshwar Rao
Wadia Institute of Himalayan Geology, Dehradun, India

ABSTRACT

The studies of multifaceted problems of gneiss-charnockite rocks in the Schirmacher region of East Antarctica suggest a retrograde clockwise isobaric cooling P-T history of the terrain involving an early granulite phase, a late granulite phase, and a retrograde amphibolite phase metamorphism in the region. Also, a good correlation between fluid and mineral data is observed. The high-density CO2 fluids fall well within the P-T box estimated by mineral thermobarometers, envisaging a pervasive influx of deep-seated CO2 rich fluid from the mantle resulting in the formation of the granulites, followed by decrease CO2 density fluids along with progressive influx of hydrous fluids leading to the generation of retrograde amphibolite facies rocks. The geochemical studies helped trace two-phase evolution of the region (basic magmatism around ~1200 Ma) involving a depleted mantle source implying an accretion of the juvenile crust during the late Mesoproterozoic period and major felsic magmatism around ~880 Ma involving partial melting of mafic-intermediate crust.

DOI: 10.4018/978-1-6684-4078-0.ch002

INTRODUCTION

One of the most challenging modern geology problems is understanding the nature and causes of deep crustal metamorphism, particularly the granulite facies metamorphism. Mineral pressure-temperature (P-T) estimates and CO_2-rich fluid inclusions will help understand the role of the physical conditions of metamorphism and how the granulite facies assemblage develops at deep crustal levels. In particular, the mineral zoning studies have yielded different P-T paths in amphibolite-granulite terranes. Some workers have inferred essentially isothermal paths during the initial stages of retrogression, whereas others have found nearly isobaric paths. Harley (1989) demonstrated that both near-isothermal decompression and near-isobaric cooling P-T paths are important, and the fluid-absent conditions are typical of most granulites at or near the time of their recorded thermal maxima. The granulites also preserve valuable information about the lower crust. The relationships of magma sources and magma generation processes for these lower crustal rocks can be constrained using the combined isotope and trace elemental studies.

The present work aims to study the metamorphism and geochemistry of the rocks from the Schirmacher region of East Antarctica. The Schirmacher region of east Antarctica (Figure-2.1) forms a part of the widespread ‘Mesoproterozoic Maud Orogenic Belts’ (Groenewald et al., 1995) of intensely deformed granulite and amphibolite facies rocks. In this paper, the Proterozoic metamorphic evolution of the Schirmacher region has been attempted by studying the mineral assemblage developed in gneiss-charnockite rocks using the conventional geothermobarometric formulations and fluid inclusion studies. Further, the petrogenesis of the rocks has also been attempted through geochemical and isotopic analyses.

GEOLOGY OF THE AREA

The geological, structural, and metamorphic framework of the Schirmacher region of East Antarctica has been described by Rameshwar Rao et al. (1996, 1997); Ravikant (2005); Ravikant and Kundu (1998); Sengupta (1988; 1993); Stackebrandt et al. (1988). The felsic rocks form the dominant lithology of the Schirmacher region, constituting more than 85% of the exposed bedrocks (Figure-2.1; cf. Sengupta, 1988). They are light-coloured, medium to coarse-grained, and show gneissic foliation. Based on mineralogy, they are described as garnet-biotite gneiss, augen-gneiss, biotite-gneiss, and leucogneiss. The leucogneiss is seen injected into all other rocks as bands and lenses.

Figure 1. Geological map of the Schirmacher region, East Antarctica (after Sengupta, 1988). The inset shows the location map of the region

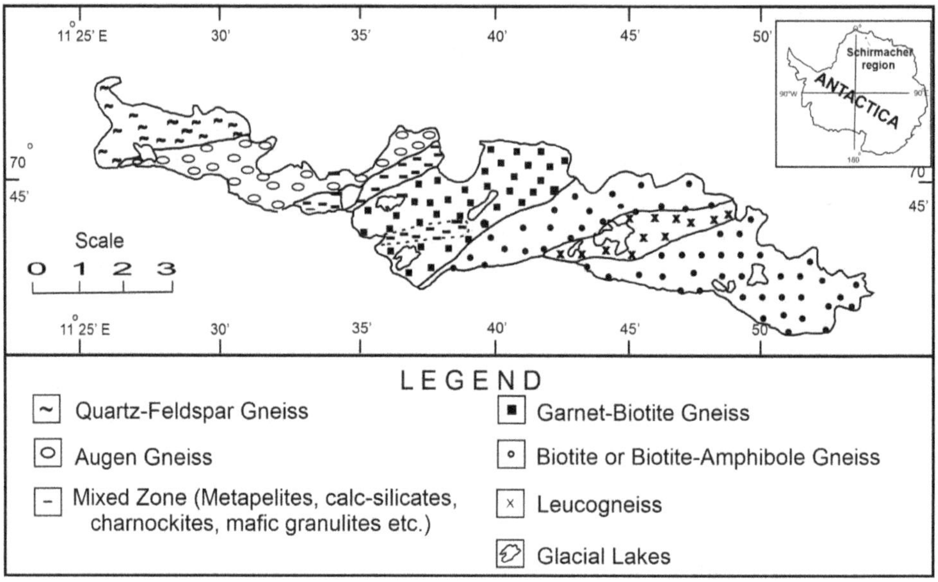

The felsic rocks further contain enclaves of charnockites, mafic granulites, calc-silicates, and metapelites. The mafic granulites are more commonly seen in the central and northeastern parts of the region. They occur as concordant sills, lenses, or bondage structures, intercalated with the charnockites, metapelites, and the dominant felsic gneiss of the region. They are characterised by black to greyish colour and are generally medium-grained. Foliation is usually not distinct. Layered mafic-felsic gneisses traversed by pegmatites and distinctive bands of mafic granulites folded along with metapelites and calc-silicate rocks can also be observed. Mafic granulites occurring as sills within gneisses have undergone intense recrystallisation and deformation during later amphibolite facies metamorphism. The amphibolites, the other conspicuous rock type of the region, mainly represent retrograded pyroxene granulites. Amphibolitization of pyroxene granulites is generally seen where the enclaves are either penetrated by veins of granite pegmatite or fringed by a diffuse zone of pegmatite, and the least altered enclaves of pyroxene granulites mainly occur within the host charnockites.

PETROGRAPHY

The quartzo-feldspathic gneisses of the Schirmacher region were grouped into garnet-biotite gneiss, biotite-gneiss, augen gneiss, streaky gneiss, banded gneiss, and garnetiferous alaskite by Sengupta (1988). They, however, show similar characters under the microscope (e.g., Chakraborty et al., 1985; Kaul et al., 1985; Rameshwar Rao et al., 1996), and in general, show inequigranular texture and consist of K-feldspar (orthoclase/microcline), plagioclase, quartz, biotite, hornblende & garnet, and with opaques, zircon, sphene, allanite, apatite, and epidote as accessory minerals. Leucogneiss also has a minor amount of muscovite, which occurs inter-layered with biotite and at a place as a secondary mineral after plagioclase.

Quartz grains in felsic rocks exhibit extensive undulose extinction, development of sub-grains, and recrystallised grains, indicating dynamic recrystallisation via sub-grain rotation and grain boundary migration. Quartz, except in high strain zones, does not show much flattening. The highly elongated quartz ribbons with a large aspect ratio in some quartzo-feldspathic mylonites indicate metadynamic recrystallisation at high rates of hot deformation (Sengupta, 1988). K-feldspar is mainly microcline and microcline microperthite as well as early untwined microperthite. The early K-feldspar is medium grained porphyroclasts containing plagioclase and quartz, formed as megacrysts during granite crystallisation. The predominant microcline microperthite and microcline with quartz and fine-grained plagioclase must have crystallised during the late stages. The K-feldspar invariably shows uneven undulose extinction and minor development of sub-grains and crystallised grains. They replace the early formed K-feldspars and plagioclases. The serrated boundaries and recrystallised grains indicate that they were deformed by recrystallisation accommodated dislocation creep. The gneisses also show a fair amount of perthites with or without microcline cross-hatched twinning. Perthite development can be related to deformation. Even though these gneisses show intense high-temperature ductile deformation, there is no definite microstructural evidence of different perthite types related to deformation.

Plagioclase in felsic rocks occurs in three varieties: as inclusions in early K-feldspar and rarely in garnet, as a medium to coarse-grained porphyroclasts with serrated boundaries myrmykitic plagioclase around the K-feldspar margins. In general, plagioclase shows sericitisation. Apart from polysynthetic growth twins, several plagioclase grains show thin twins that end abruptly within the grain. They are commonly bent and appear to be mechanically produced. Some of the plagioclase grains are divided into three or more grains with misorientation and deformational twinning. In contrast, others with undulose solid extinctions and bent twin lamellae show sub-grains and recrystallised grains. Even though the feldspars show minor intragranular cracking without displacement, most cracks are transgranular, which could be produced much later.

Garnet is present in some of the quartzo-feldspathic gneisses, primarily associated with biotite. Small biotite flakes surrounding the garnet, replacing the garnet along the margins and along with the fractures in the garnet. In some gneiss, the garnet clots are sporadically distributed and impart foliation. In some cases, the replacement is advanced, producing atoll garnets. Some of the small garnets having quartz inclusions at the margins in symplectic intergrowth indicate that these garnets appear to have developed or grown over the biotite. This texture shows that garnet was stable or generated during the high-temperature ductile deformation. The garnets having early quartz grains as inclusions and surrounded by well-developed recrystallised quartz grains indicate that these garnets have been developed by nucleation and crystal growth at the curved boundaries of quartz grains, which produce hollow monocrystal of quartz. These textures and other textural characters indicate that garnet grew in the granite melt, and some of the new garnets were stable during deformation.

Biotite in felsic rocks occurs mainly with garnet in clusters with different orientations and isolated laths defining foliation. In general, the biotites were stable during deformation. However, minor, rare patches of chlorite occur in biotites due to auto-metamorphic alteration by late-stage fluids. In this case, the biotite alteration is due to late-stage melt crystallising quartz and K-feldspar. Biotite at places also shows pleochroic haloes around allanite and zircon. Muscovite, a minor mineral in these rocks, occurred interlayered with biotite and was used as textural evidence for considering muscovite as primary, while it appears to be secondary. Hornblende forms sub-idioblastic short prismatic and poikiloblastic grains in these rocks. Simultaneously, the accessory minerals like allanite, epidote sphene, apatite, zircon, and opaques occur as idioblastic to sub-idioblastic grains. Zircons also occur as six-sided wedge-shaped grains, while apatite occurs as euhedral to subhedral grains, and opaques as octahedral or small fine grains. These accessory minerals occur as inclusions either in biotite, hornblende, feldspars, or quartz.

The mafic granulite, the other necessary rock type of the region, is not prominent enough to be shown on the geological map of a 1:50,000 scale, nor can they be differentiated in the field. The mineral assemblage of the mafic granulites shows variation from garnet-bearing two-pyroxene granulites and garnet free pyroxene granulites to transitional amphibolite-pyroxene granulites. The rocks under microscope mainly comprise orthopyroxene, clinopyroxene, plagioclase, opaques, garnet, with minor amounts of quartz and accessories of apatite, sphene, zircon, and rare allanite. The clinopyroxene is a primary constituent in these rocks and is in textural equilibrium with orthopyroxene and garnet (Rameshwar Rao et al., 1997). They also contain substantial amounts of brownish or greenish-brown hornblende, ranging from 20 to 40 modal percentages. The hornblende replaces the pyroxenes to different degrees. Brown to reddish-brown biotite also returns pyroxene via amphibole in these rocks. The mafic granulites are mostly equigranular, medium to coarse-grained, showing xenoblastic granular texture. Some of the mafic granulites which occur as boudin preserve primary granulite facies assemblage. The texture is equigranular and polygonal with straight boundaries, and the grains meet a triple point. Generally, the mafic granulites are non-foliated. However, the sheared varieties impart a weak foliation due to the alignment of amphiboles and biotite and elongation of pyroxenes and plagioclase. The disequilibrium assemblage of the mafic granulites indicates a superimposition of amphibolite facies metamorphism over the earlier granulite facies metamorphism. The post-granulite deformation has extensively affected the mafic granulites producing varieties of deformation microstructures and the development of amphiboles and biotites. The increasing intensity of deformation can be correlated with the high modal percentage of amphiboles and biotites. The deformation microstructures indicate a high-temperature environment of deformation.

Orthopyroxene in mafic granulites is mainly hypersthene, which occurs as even-grained xenoblastic to granular aggregate with a pale green to pink pleochroism. In some sections, they meet at triple point junctions indicating the recrystallisation nature. In a few cases, orthopyroxene, along with clinopyroxene, occurs as coarse-grained porphyroblasts, which suggest that they belong to the primary mineral assemblage equilibrated under granulite facies conditions. Some of the orthopyroxene show very fine exsolution lamellae of clinopyroxene, and these grains do not become extinct altogether. The contacts between orthopyroxene and clinopyroxene and plagioclase are primarily straight and, in some cases, having xenoblastic grain boundaries. In general, the orthopyroxene shows undulose extinction and minor recrystallisation. The clinopyroxene is a primary constituent in textural equilibrium with orthopyroxene. It occurs as well developed to xenoblastic grain boundaries and often contains fine exsolution lamellae. Like orthopyroxene, the clinopyroxene contains small garnet grains (one or two), probably due to exsolution.

Plagioclase in mafic granulites occurs as even grains with xenoclastic to well defined straight grain boundaries. It is twinned but does not show any optical zoning. Most of the grains show undulose extinction indicating a superimposed moderate to intense deformation. The plagioclase-plagioclase contact is often polygonal, and the grains meet a triple point junction indicating recrystallisation under higher temperatures. Their connection with pyroxenes is sharp with concave or convex grain boundaries. The opaques are of two varieties. One is the primary with well-bounded crystal faces. The other is secondary with anhedral grains formed during conversion (hydration) of pyroxenes to amphiboles/ biotites under lowering temperature conditions.

The garnet in mafic granulites is present as coronas around pyroxenes or at the contact between pyroxene and plagioclase. However, they persisted during the decompression event, contrary to widespread evidence of its breakdown under similar conditions elsewhere (Harley, 1989). In some cases, garnets occur as anhedral poikiloblasts containing numerous inclusions of plagioclase and pyroxenes. The coronatic garnets, in some instances, merge to form coarse grain garnets. These poikiloblasts and other coarse garnets seem to have attained granoblastic shapes subsequently. In one case, the garnet lamellae on orthopyroxene-plagioclase contact continue to develop across an orthopyroxene grain, probably along the kink band boundary of orthopyroxene. The textural criteria indicate that the garnet has formed by the reaction: opx + pl = grt + qtz.

The charnockites, on the other hand, are medium to coarse-grained and show equigranular texture. They are mainly composed of plagioclase, quartz, orthopyroxene, opaques ±garnet. Plagioclase in these rocks occurs as subhedral grains and shows undulose extinction. They show twinning but do not show optical zoning. The orthopyroxene occurs as even-grained xenoblastic to granoblastic aggregates leading to pleochroism from nearly colourless, pale green to brownish pink, with little compositional zoning. The re-equilibration of pyroxenes produced garnet probably during the conversion of the primary igneous assemblage to charnockites. In some instances, the garnets have merged to form large porphyroblasts. The samples show sub-idioblastic garnet-quartz symplectite along with the interface of orthopyroxene and plagioclase. Compositionally the garnets are almandine than the garnets from the associated felsic gneisses. The rocks show conversion of pyroxenes and garnet to amphiboles (cpx + grt = hbl + pl). The garnet and orthopyroxene altering to hornblende suggest the partial hydration of early granulite facies assemblages. Apart from amphibole, brown to reddish-brown biotite is also found replacing orthopyroxene. These minerals, the hornblende, and biotite define weak foliation in these rocks.

GEOCHEMICAL STUDIES

The major trace and REE data of the rocks from the Schirmacher region were analysed using the XRF instrument at the Wadia Institute of Himalayan, Dehradun, and ICP-MS of the National Geophysical Research Institute, Hyderabad. The geochemical compositions are used to characterise the gneiss-granulite rocks and their evolution. The mean, minimum and maximum values of geochemical data of mafic granulites, charnockites, biotite gneiss, garnet-biotite gneiss, augen gneiss, and leucogneiss are given in Table-2.1.

Table 1. Geochemical data of mafic granulites, charnockites, biotite-gneiss, garnet-biotite gneiss, augen gneiss, and leucogneiss from Schirmacher region of East Antarctica

	Mafic Granulites		Charnockites		Biotite Gneiss		Garnet-Biotite Gneiss		Augen Gneiss		Leucogneiss	
#		19		3		4		13		7		8
	Min-Max	Mean	Min-Max	Mean	Min-Max	Mean	Min-Max	Mean	Min-Max	Mean	Min-Max	Mean
SiO_2	37.17-50.17	46.04	57.27-67.19	62.82	68.87-71.18	69.64	56.36-70.68	64.45	57.03-69.97	63.59	68.36-72.14	69.98
Al_2O_3	9.13-17.23	13.56	0.76-1.09	0.94	0.49-0.64	0.56	0.50-1.73	1.00	0.41-1.52	0.85	0.01-0.68	0.22
$Fe_2O_3^T$	10.99-23.26	15.04	15.98-18.30	17.33	14.64-15.31	14.97	13.61-17.64	15.74	15.03-18.98	16.47	14.03-17.39	15.60
MgO	6.85-19.71	10.09	5.22-6.85	6.00	2.96-4.48	3.97	2.45-11.50	7.13	2.41-9.51	5.88	0.79-4.74	2.32
CaO	3.92-13.69	10.17	0.10-0.12	0.11	0.05-0.08	0.07	0.03-0.19	0.11	0.03-0.14	0.09	0.02-0.09	0.04
Na_2O	2.19-3.24	2.69	0.07-4.40	1.83	0.37-0.88	0.62	0.42-5.25	1.52	0.22-2.40	1.44	0.01-0.84	0.44
K_2O	0.24-4.50	1.27	3.44-5.06	4.15	1.91-2.59	2.28	1.48-4.14	3.13	2.08-3.85	2.88	0.94-2.16	1.37
TiO_2	0.20-4.66	1.56	3.69-3.87	3.79	3.70-3.92	3.80	1.80-4.78	3.53	2.36-5.48	3.88	3.76-5.33	4.74
MnO	0.14-0.32	0.20	1.90-2.27	2.07	3.54-5.12	4.36	1.90-5.33	3.15	2.49-5.18	4.10	3.17-5.53	4.65
P_2O_5	0.12-0.32	0.20	0.28-0.91	0.55	0.06-0.19	0.14	0.08-0.41	0.24	0.07-1.28	0.40	0.04-0.47	0.15
Ga	18-19	19	16-18	17	15-17	16	14-19	17	15-19	17	14-18	15
Pb	8-16	11	6-31	17	3-14	9	4-52	21	8-40	20	2-16	9
Rb	5-309	42	34-111	69	107-153	134	50-119	84	75-168	110	78-147	123
Sr	5-538	149	24-171	108	78-127	112	112-166	138	93-210	145	15-256	118
Y	7-75	28	5-40	23	20-32	24	12-64	40	18-34	25	10-65	26
Zr	6-318	111	187-290	248	126-219	164	148-521	287	142-845	374	31-192	90
Nb	2-19	7	2-8	4	3-5	4	2-15	8	1-8	3	2-11	5
REE*		6		1		3		3		2		2

Continued on following page

Table 1. Continued

	Mafic Granulites		Charnockites		Biotite Gneiss		Garnet-Biotite Gneiss		Augen Gneiss		Leucogneiss	
#		19		3		4		13		7		8
La	1.25-11.30	5.23	-	22.71	25.13-38.61	32.33	31.50-53.64	40.69	23.79-26.09	24.94	8.35-31.10	19.73
Ce	6.20-60.42	24.68	-	32.36	56.73-83.32	70.70	72.24-116.41	88.59	50.62-57.83	54.23	17.76-75.15	46.46
Nd	6.22-44.92	16.64	-	9.84	26.25-36.58	32.30	34.56-52.80	41.67	27.31-27.89	27.60	8.72-34.71	21.72
Sm	1.79-6.54	3.31	-	2.04	8.26-10.60	9.24	10.03-14.56	11.88	7.79-9.22	8.51	2.58-11.08	6.83
Eu	0.94-2.99	1.63	-	1.78	1.12-1.32	1.19	1.18-2.28	1.62	1.80-2.49	2.15	0.40-0.47	0.44
Gd	3.09-13.56	5.70	-	1.83	4.69-5.66	5.20	8.06-11.50	9.72	6.88-7.37	7.13	3.09-8.57	5.83
Dy	2.31-4.38	3.32	-	1.64	2.85-4.02	3.47	8.04-12.03	10.13	6.49-7.83	7.16	5.07-15.00	10.04
Er	1.83-8.74	3.82	-	0.93	0.67-1.57	1.10	5.01-6.57	5.85	3.02-4.36	3.69	3.38-10.25	6.82
Yb	1.97-6.90	3.52	-	-	-	-	-	-	-	-	-	-
Lu	0.14-0.53	0.31	-	0.15	0.10-0.15	0.12	0.65-1.32	1.03	0.47-0.85	0.66	0.50-1.86	1.18

Note:- # represents the number of samples for major and trace elements data. REE* represents the number of samples for REE data. The complete data is available in Rameshwar Rao (2006)

The geochemical data of felsic rocks (Table-1) show that they have a range of SiO_2 from 56-73%, Al_2O_3 from 13-19%, and Na_2O+K_2O from 5-11%. They also generally show Na_2O >3.2% and have relatively low K_2O/Na_2O values than S-type granitoid at equivalent SiO_2 levels. Further, they offer wide compositional variation from granodiorite to adamellite (Figure-2.2). They are metaluminous to peraluminous (Figure-2.3), with the alumina saturation index (ASI) being less than 1.1 (mean value of ~1.03). They show calc-alkaline differentiation trends on the AFM diagram (Figure-2.4) and log $CaO/(Na_2O+K_2O)$ vs SiO_2 diagrams. The felsic rocks and charnockite samples show a generally coherent (enriched or depleted) variation on Harker's plots. They offer a negative correlation in CaO, FeO^T, MgO, MnO, Al_2O_3, P_2O_5, TiO_2, and Zr. However, elements like Na_2O, K_2O, and Rb show a positive correlation but display some scattering degree, resulting from element mobility during late- to post-crystallisation deformation.

Figure 2. Na_2O vs K_2O classification diagram for felsic rocks, showing the adamellite-granodiorite composition of rocks under study

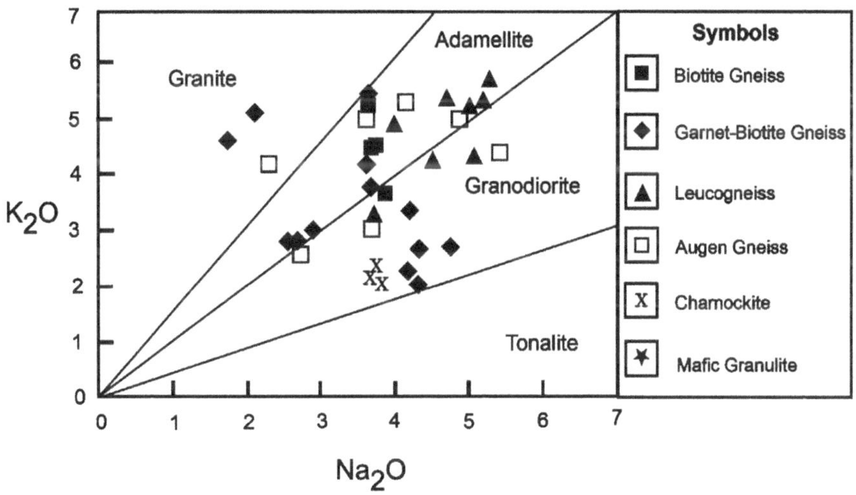

Figure 3. Plotting of mol. A/CNK against mol. A/NK of felsic rocks shows their peraluminous to meta-aluminous nature. Symbols as in Figure-2.2

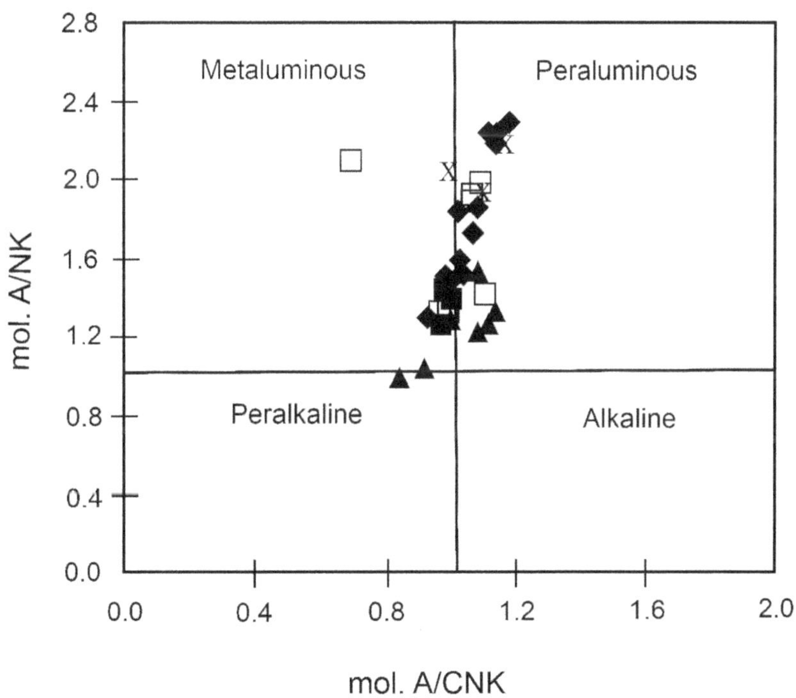

Figure 4. AFM ternary plot showing a calc-alkaline trend for the rocks understudy. Symbols as in Figure 2

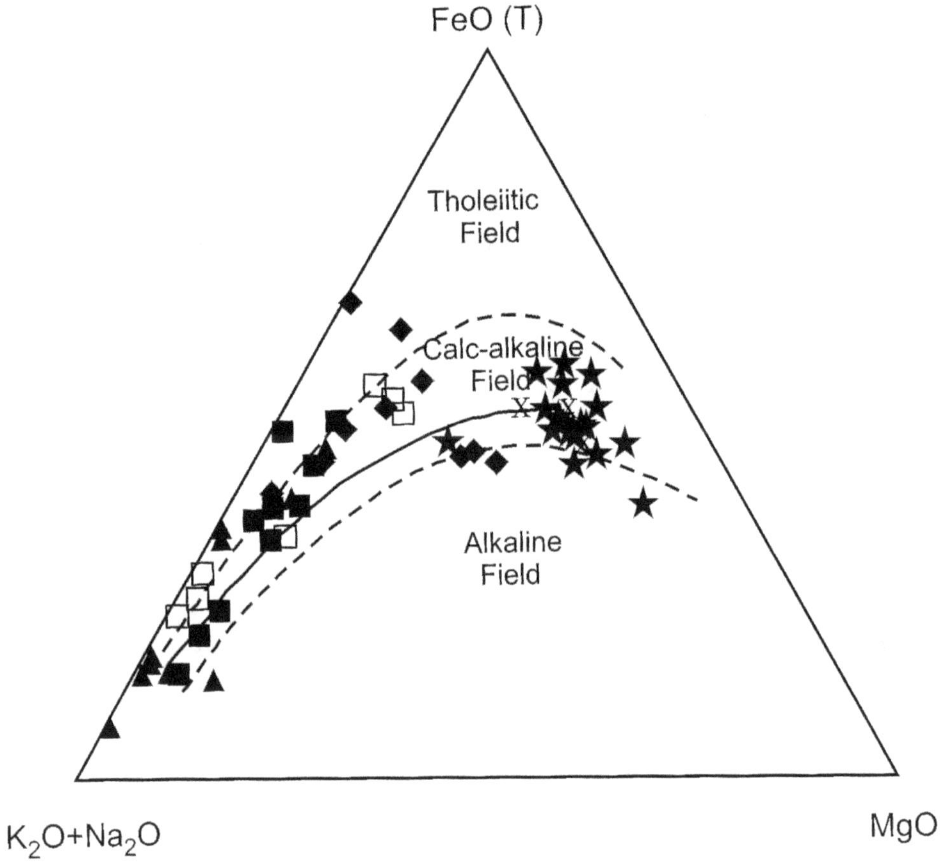

The chondrite normalised REE pattern of garnet-biotite gneiss, augen gneiss, and leucogneiss show a general enrichment of LREE and moderate fractionation of HREE. In contrast, biotite gneiss has a depleted HREE pattern. A negative Eu anomaly characterises all samples. In contrast, the analysed charnockite sample shows a positive Eu anomaly, and its REE pattern resembles that expected from partial melting of an igneous rather than the sedimentary rocks. The evolution of REE patterns is also best explained with fractional crystallisation with LREE depletion and HREE enrichment and leucogneiss marked by 'V' or wing-shaped REE distribution patterns. Further, the geochemical data used for tectonic discrimination of the Schirmacher region indicate that the felsic rocks have volcanic arc affinity. For example, the felsic rocks and charnockite samples plot in the Group III field on SiO_2 vs Rb/Zr plot (Harris et al., 1986) and also in volcanic arc granite field on Y+Nb vs

Rb diagram (Figure-2.5) after Pearce et al. (1984) and a ternary plot of Rb/10-Hf-Ta*3, after Harris et al. (1986). Further, in Zr/2-Y*2-Sr ternary diagram, most of the analysed samples concentrate in the area defined for volcanic arc granite/ collision granite fields, and on Hf/3-Th-Ta geotectonic diagrams, the position of the samples near the apex indicating an evolved island arc character (cf. Cabanis et al., 1987).

Figure 5. Y+Nb ppm vs Rb ppm tectonic variation diagram showing the rocks plotting in volcanic arc granite fields defined after Pearce et al. (1984). symbols as in figure-2.2

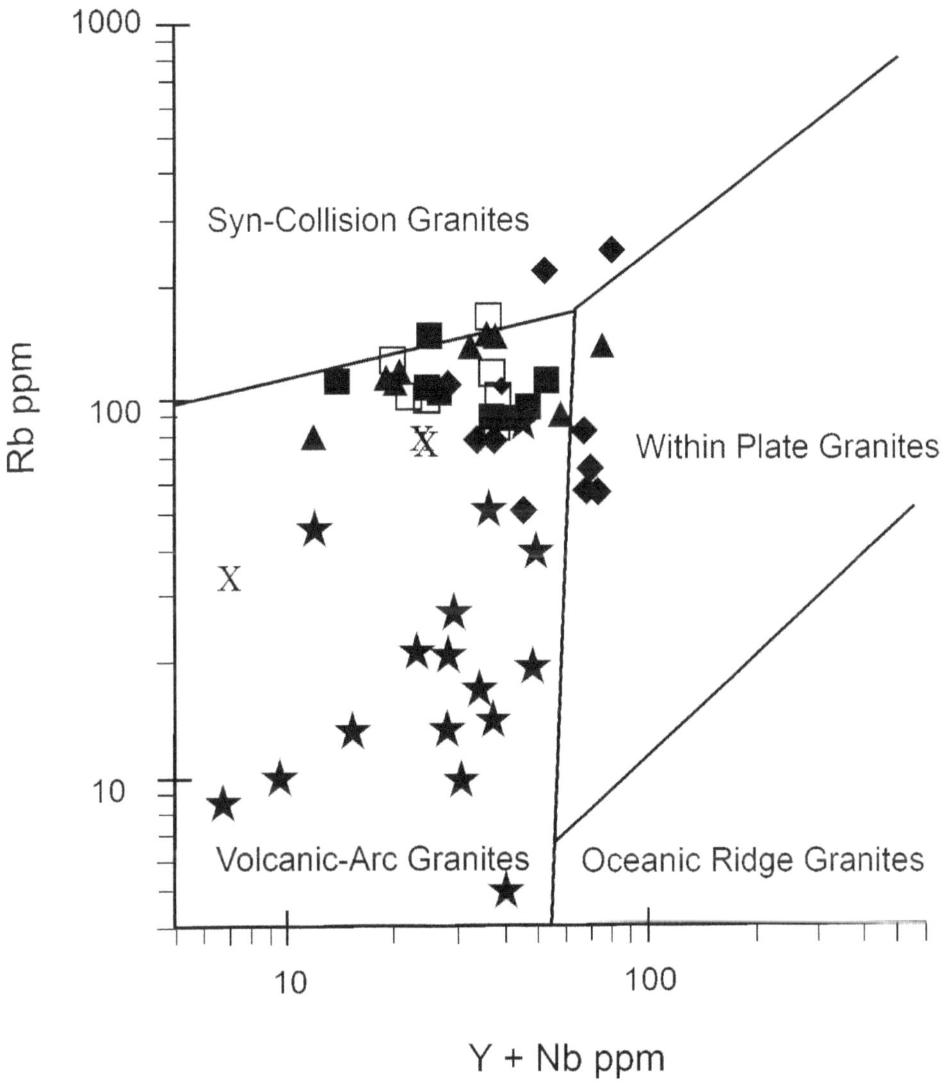

On the other hand, the geochemical studies of mafic granulites of the region show that the majority of the analysed samples plot in the tholeiitic field or show tholeiitic trends on Jensen ternary diagram (Fe+Ti)-Al-Mg, AFM, and FeO*/MgO vs SiO_2 suggesting their metatholeiitic nature (Figure-2.6) as well as general preservation of original composition. They also represent Mg-basalts (classified after Kersting & Arculus, 1994) with MgO>7% and Al_2O_3<16%. They are characterised by mg# in the range of 45 to 77 (mean value of ~60), indicating that the mafic rocks cropping out in the Schirmacher region are not evolved. Their CIPW normative parameters reveal igneous affinity, characterised by the relatively high abundance of normative pyroxenes and olivine (cf. Borsi et al., 1995). The rocks are notable for the wide range of their LIL elemental ratios of large-ion lithophile elements against high-field strength elements. The heavy rare-earth elements show that their ratios are higher than N-type MORB or the primitive mantle. Variation diagram of selected trace elements plotted against Zr, an element essentially incompatible in mafic systems and assumed to be relatively insensitive to amphibolite facies metamorphism and fluid circulation, shows that the incompatible element abundances and REE tend to increase with increasing Zr. These rocks' REE patterns show variable LREE content and more or less flat HREE pattern, with moderate but variable LREE enrichment than MORB (Figure-2.7). The primordial mantle-normalised diagram shows a distinctive negative Nb anomaly compared to OIB, N-type, and E-type MORB. Although, in broad terms, the metatholeiites have an underlying primitive MORB-like HFSE composition coupled with a low concentration of REE and low Sr and Ba concentrations, other processes have influenced their unique characteristics of LILE, Ta, and Nb enrichment relative to MORB.

Figure 6. SiO_2 vs FeO/MgO classification of mafic granulites, wherein the samples are seen plotting in the tholeiitic field. Symbols as in Figure-2.2*

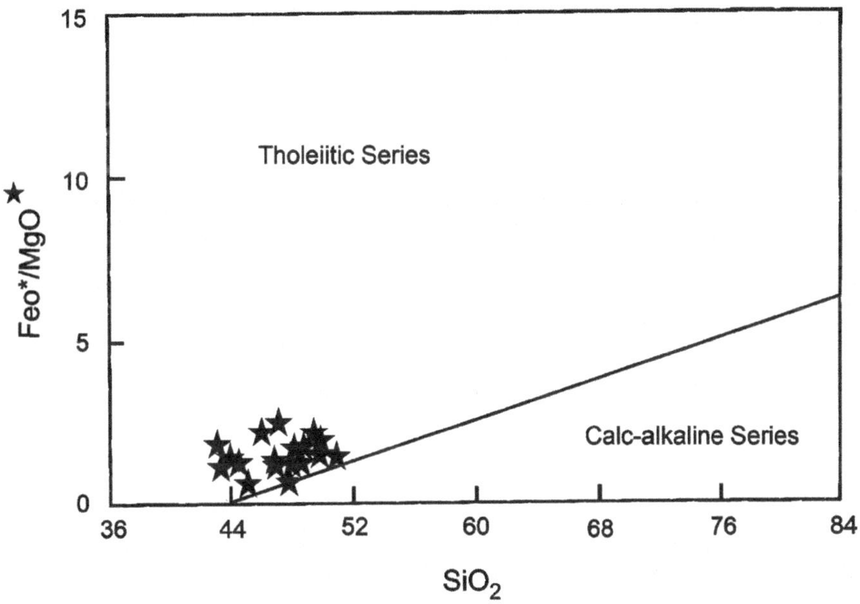

Figure 7. Normalized REE plot of mafic granulites, charnockite, and felsic rocks in comparison with those of Oceanic Island Basalts, N-type MORB and E-type MORB (values after Sun & Mc Donough, 1989)

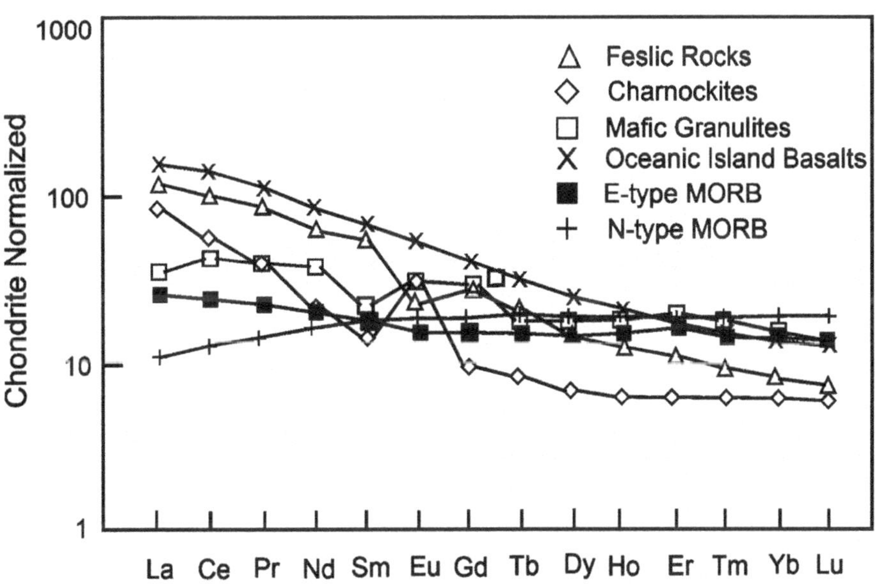

GEOCHRONOLOGICAL STUDIES

The whole-rock Nd and Sr isotopic compositions of the mafic granulites of the Schirmacher region, collected from different sill bands spreading in an area of 2 sq. km, were determined using the VG-354 Mass Spectrometer facility at NGRI, Hyderabad. The strontium isotope analyses of felsic rocks were carried out on Thermal Ionization Mass Spectrometer (VG-354) at KDM Institute of Petroleum Exploration, Dehradun. The analytical procedure and precision of the analyses carried out by NGRI are described in Kumar et al. (1998). The age calculations are done using the Provost (1990) program. The age of mafic granulites is not well constrained: Sm-Nd dating and Rb-Sr dating giving a poor fit for the analysed samples (Figure-2.8). The observed scatter in the Nd-Sr isotopic systematic about the regression exceeds expected for the experimental error associated with each data point. The uncertainty on the ages can be related to the restricted range of $^{87}Sr/^{86}Sr$ ratio (0.11 to 1.34) and $^{147}Sm/^{144}Nd$ ratio (0.135-0.162) or the small population of the samples or to partial isotopic resetting, especially Rb-Sr systematic during subsequent geological activity or probably due to sills representing different levels. The Rb and Sr isotopic studies on garnet-biotite gneiss and biotite gneiss samples produce a better-defined isochron age of 884±40 Ma with a Sr_i ratio of 0.7109±20 (Figure-2.8; Rameshwar Rao et al., 1999). Rama Rao et al. (1995), however, have also reported ages of 853±51 Ma with a Sr_i ratio of 0.7085±18 for garnet-biotite gneiss and 773±26 Ma with a Sr_i ratio of 0.7079±26 for leucogneiss.

The Sm-Nd dating of six analysed samples corresponds to an 'age' of 968±249 Ma, with the Nd_i ratio of 0.51158±24 and MSWD of 13.02. Three out of the six analysed samples (RA-54, RA-62, and RA-112 also gave a similar age with a lower MSWD of 5.47 (age: 961±73 Ma, with a Nd_i ratio of 0.51156±18). Similarly, the Rb-Sr dating of seven analysed samples corresponds to an 'age' of 886±73 Ma with a Sr_i ratio of 0.70282±64 and a high MSWD of ~126. Four out of the seven analysed samples (RA-56, RA-71, RA-87, and RA-112) gave a similar age with a much lower MSWD value of ~36 (age: 893±51 Ma, with a Sr_i ratio of 0.70215±45). The initial Sr ratio (Sr_i) calculated for the samples ranges between 0.702 to 0.705 and are like most basalts, usually less than 0.7055 (Rameshwar Rao et al., 2000).

Figure 8. Sm-Nd and Rb-Sr isochron diagrams of mafic granulites, and Rb-Sr isochron diagram of felsic rocks (cf. Rameshwar Rao et al., 2000)

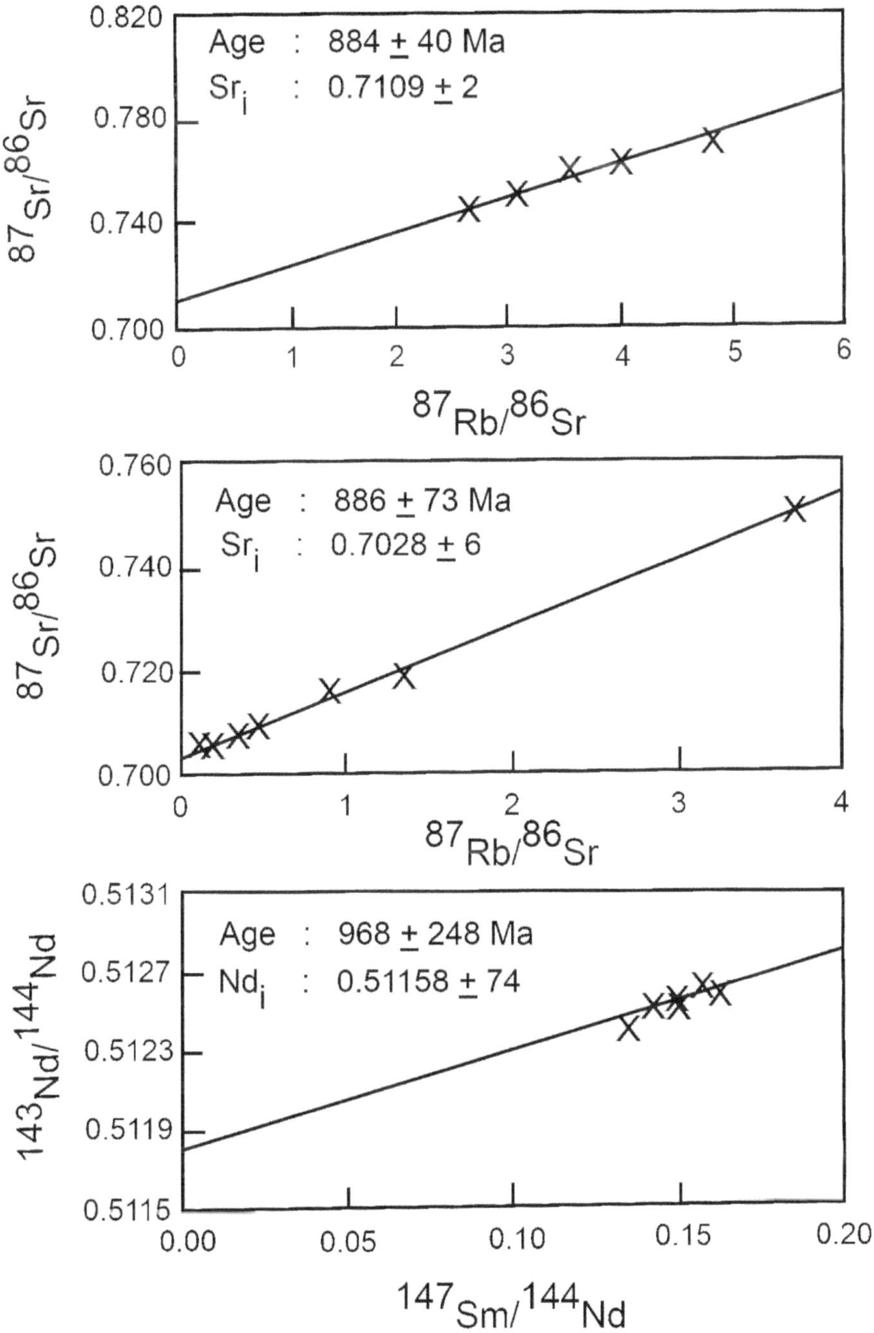

Depleted mantle model ages (T_{DM}) calculated after De Paolo (1981) for the mafic granulites show that they have a relatively restricted range of Nd model ages of 1120 to 1357 Ma (with a mean value of ~1200 Ma), in marked contrast to the wide range shown by T_{UR}^{Sr} ages. The small range defined by the model ages reflects rare-earth elements' relative immobility compared with Rb and Sr during tectono-thermal events. The $\sum Nd_i$ calculated for the samples under study at 1200 Ma range from +4.22 to +6.07. The small range of ε_{Nd} values can be produced from melting a single mantle source and constitute evidence against mixed sources (Farmer and De Paolo, 1983). These rocks' positive $\sum Nd_i$ values also suggest the absence of participation of older crustal material in the genesis of these rocks and offer the addition of juvenile crustal material during Mesoproterozoic times.

FLUID INCLUSION STUDIES

Fluid inclusion studies at Wadia Institute of Himalayan Geology, Dehradun, India, were conducted on doubly polished thin rock wafers (about 0.5 mm). The microthermometric runs were performed on SGE gas flow heating-freezing system.

The stage was calibrated using pure CO_2 inclusions and the Gonzales fluorite. The quoted values have a precision of +0.2°C. The densities and isochors are calculated using Flincor Programme after Brown (1989).

Fluid Inclusion Studies of Felsic Rocks

The felsic rocks of the Schirmacher region show three types of inclusions: (i) Monophase Carbonic inclusions. These inclusions, in general, are of 10±5-micron size. They occur in various disposition patterns like rarely isolated inclusions found in garnet and commonly in small groups and planar arrays wherein they coexist with aqueous carbonic inclusions. (ii) Carbonic aqueous inclusions these inclusions are the most abundant of all the inclusion types. They varied in the inclusions from <5 microns to about 20 or even up to 25 microns. The CO_2 proportion of these inclusions varies from more than 80 vol % to about 30 vol %. (iii) Aqueous inclusions are found in trails and healed cracks, often associated with late H_2O-CO_2 inclusions. Most of these inclusions are <20 microns in size and are subregular in shape. These inclusions are commonly found in quartz grains. Some of them show decrepitation features and other leakages. The deformed isolated aqueous inclusions represent the earliest stage, followed by the entrapment of carbonic inclusions, aqueous carbonic inclusions, and the aqueous inclusions.

The microthermometric estimates of monophase inclusions show that the melting temperatures are in the range of -56.2 to -57.8°C, which nearly coincides with the melting temperatures of pure CO_2 compositions (-56.6°C). Homogenisation of monophase CO_2 inclusions varied from -12.5 to +16.2°C with peaks around -6 and +8°C. Those found in trails homogenised between +3.8 to +16.2°C. This homogenisation data of isolated CO_2 inclusions are consistent with a CO_2 density range of 0.996 to 0.933 g/cm^3. The densities estimated for the CO_2 inclusions in trails are relatively low, varying between 0.905 and 0.811 g/cm^3.

The microthermometric measurements of aqueous carbonic inclusions agree with the H_2O-CO_2 composition of the trapped fluid. The carbonic phase in these inclusions shows melting at around -57.5°C and homogenisation between +15.6 to +29.2°C with a peak around +18°C. These temperatures agree with a CO_2 density of 0.816 to 0.624 g/cm^3. The total homogenisation of aqueous carbonic inclusions into the gas phase took place at a temperature of 190 to 360°C. The inclusions showing the lowest CO2 density of 0.624 g/cm^3, perhaps indicates the waning stage of CO_2 fluid with an average composition of X_{CO2}=0.225, X_{H2O}=0.767, and X_{NaCl}=0.070 during retrograde conditions. The observations show a gradual decrease in CO_2 density at successive stages of metamorphic evolution after primary melt crystallisation.

In the aqueous inclusions, the first melting occurred at a temperature around -22°C, close to H_2O-NaCl-KCl eutectic. The final ice-melting temperatures ranged from -3.5 to -5.8°C, indicating the low saline aqueous nature of the inclusions with nearly 5 to 9 wt % NaCl equisalts. These H_2O-NaCl inclusions homogenised between 210 and 300°C with maxima around 260°C (Rameshwar Rao et al., 1996; Rameshwar Rao et al., 1998; Sharma & Rameshwar Rao, 2006).

Fluid Inclusion Studies of Mafic Granulites

The principal fluid phases observed in mafic granulites include: (i) CO_2-rich inclusions. These are primary monophase inclusions identified as isolated scattered inclusions and in planar arrays. They are mostly <20 µm in size and are subrounded, polygonal, or subregular in shape. (ii) Carbonic-aqueous inclusions, these inclusions commonly occur in trails and groups, locally associated with monophase carbonic inclusions. Their population densities are relatively low and are generally 5-15 µm in size and sub-rounded, irregular, or elongated in shape. (iii) Aqueous fluid inclusions are <10 µm two-phase inclusions and are sub-regular in shape. They are distributed along late trails, denoting a predominance of water-rich fluid at a late stage. The inclusions' petrography suggests that isolated monophase CO_2 inclusions were trapped earliest, followed by CO_2 and CO_2-H_2O inclusions in planar arrays, indicated by the coexistence of the latter two types. Further, the aqueous fluids represent a late entrapment, as evident from their occurrence in secondary trails.

Melting of solid CO_2 in isolated CO_2 inclusions occurred between -57.8 and -56.5°C and trail bounded CO_2 inclusions between -57.4 and -56.5°C, suggesting nearly pure CO_2 in these inclusions. The overall homogenisation temperature range in these rocks is from +13.5 to -25.4°C. Within this range, the isolated primary inclusions were homogenised between -25.4 and -10.5°C, with a maximum at around -21°C. In contrast, the homogenisation temperature values for trail-bounded CO_2 inclusions vary between +4.6 and +13.5°C, showing a peak at about +11°C. The density of isolated CO_2 inclusion ranges from 0.986 to 1.057 g/cm^3 with a histogram peak at 1.037 g/cm^3. The density has also been estimated for the trail bounded CO_2 inclusions in which H_2O was optically undetectable. These values vary from 0.899 to 0.835 g/cm^3. The histogram peak is at a density of 0.854 g/cm^3 (Rameshwar Rao et al., 1997; Sharma & Rameshwar Rao, 2006).

METAMORPHIC STUDIES

Selective samples were analysed for their mineral chemistry using JEOL-JCXA-8600 Super Probe at Roorkee to quantify the metamorphic conditions of gneiss-charnockite rocks of the Schirmacher region. The probe was run under comparable operating conditions of 15 keV accelerating voltage, 20 nA probe current, and 1 to 2 μm beam diameter. ZAF corrections after Philibert (1963) were applied for the correction of analytical data. Multiple analyses of core and rim compositions of minerals in mutual contact or near were done to evaluate the P-T conditions of different metamorphism episodes and are depicted in Figure-2.9.

Figure 9. Schirmacher rocks showing retrograde P-T path. The boxes represent P-T conditions of (i) charnockites occurring in enclaves (box A), (ii) charnockites inter-bedded with felsic gneisses (box B), (iii) felsic gneisses (box C), (iv) mafic granulites (boxes D and E), and (v) leucogneiss (box F). The triple point of Al-silicates is after Holdaway (1971), and the metamorphic reaction of hornblende breakdown is after Binns (1968)

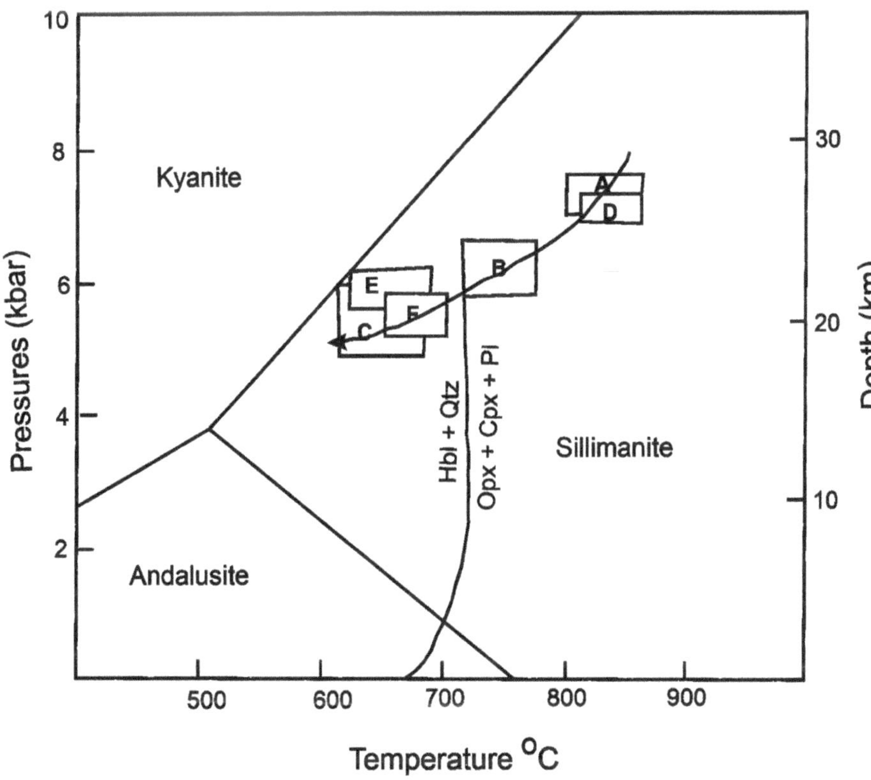

Temperatures of metamorphism are calculated using Fe-Mg exchange reactions between coexisting phases (e.g., grt-cpx, grt-opx, grt-hbl, grt-bt) or Ca-Mg transfer between coexisting pyroxenes. For clinopyroxene-garnet geothermometer, methods like Ellis and Green (1979), Ganguly (1979), Dahl (1980), Krogg (1988), Pattison and Newton (1989), and Sengupta et al. (1989) were used. For orthopyroxene-garnet geothermometer, methods like Harley (1984), Sen and Bhattacharya (1984), Lee and Ganguly (1988), and Carswell and Harley (1990) were used. For garnet-hornblende geothermometer, Graham Powell (1984) method is used. For the garnet-biotite geothermometer, methods like Thompson (1976), Ferry and Spear (1978), Hodges and Spear (1982), Ganguly and Saxena (1984), Perchuk and Lavrent'eva (1983), Indares and Martignole (1985), Bhattacharya et al. (1992), Gessmann et al. (1997), and Holdaway et al. (1997) were used. The feldspar geothermometry was carried out using Stormer (1975), Whitney and Stormer (1977), Powell and Powell (1977), Haselton et al. (1983), Price (1985), Perchuk et al. (1991) methods. The temperature estimates were made using software programs after Rameshwar Rao and Subba Rao (1994), Rameshwar Rao (1995), Rameshwar Rao et al. (1996), and Holdaway et al. (1997).

Geobarometric calibrations based on direct experimental data within the chemical system of interest or reversals of end-member equilibria coupled with activity-composition relationships for the relevant solid solutions are used. For clinopyroxene bearing granulites, pressures have been evaluated using the assemblage grt-an-di-qtz geobarometers. For orthopyroxene bearing granulites, pressures have been assessed for the grt-an-en-qtz and grt-an-fs-qtz geobarometers. The methods applied include Wells (1979), Perkins and Newton (1981), Newton and Perkins (1982), Bohlen et al. (1983), Perkins and Chipera (1985), Eckert et al. (1991), Bhattacharya et al. (1991), and Lal (1993). For the assemblage grt-bt-ms-pl-hbl-qtz methods like Ghent and Stout (1981), Hodges and Crowley (1985), Hoisch (1990), and Kohn and Spear (1990) were used.

The garnet and biotite matrix assemblage of the Schirmacher region's felsic rocks gave temperatures in the range of 648 to 728°C. As all the thermometers produce geologically reasonable results, the mean temperatures were calculated around 654±27°C. The temperature obtained using two-feldspar thermometers ranged from 434 to 456°C and gave a mean value of 444°C with a standard deviation of 9, suggesting the re-equilibrium of feldspars lower temperatures. Further, it also coincides with a compositional system change to H_2O dominating fluids (350 to 480°C and 1.5 to 2.4 kbar), resulting in most hydrous minerals in the rocks (cf. Rameshwar Rao et al., 1998). The pressure estimates for a grt-ms-bt-pl-qtz assemblage of felsic rocks almost show consistent results for the methods used, ranging from 4.8 to 5.8 kbar with a mean value and standard deviation of 5.4±0.4 kbar (Rameshwar Rao et al., 1998; Rameshwar Rao, 2000).

The calculated P-T estimates on the charnockite rocks occurring as enclaves within the felsic gneisses (M1 phase metamorphism of Sengupta, 1993) show mean temperature and pressure estimates of 827±29°C and 7.3±0.3 kbar, respectively, for core compositions. However, the rim compositions only show a slight drop in pressure (6.8±0.3 kbar) without much temperature variation. On the other hand, the core compositions of the charnockite rocks occurring inter-bedded with gneisses (RA-78) show temperature and pressure estimates of 752±22°C and 6.2±0.4 kbar, respectively (Rameshwar Rao, 2000). This event of granulite facies metamorphism probably represents the M2 phase of metamorphism as reported by Sengupta (1993) but needs further confirmation from more data.

There are two groups of temperature estimates for mafic granulites of the Schirmacher region, one falling around 837±26°C, shown by primary mineral assemblages like opx-cpx, and the other falling around 652±33°C, shown by reset and secondary minerals. Similarly, we see two groups of pressure estimates for the mafic granulites of the Schirmacher region. The pressure estimates from the core compositions range between 6.6 to 7.6 kbar with a mean value of 7.1±0.2 kbar, and for the rim, compositions range between 5.1 to 6.0 kbar with a mean value of 5.9±0.3 kbar (Rameshwar Rao et al., 1997).

MAIN SCIENTIFIC FINDINGS

Careful documentation of the geochemical and P-T estimates in the gneiss-charnockite rocks combined with the characteristics of the fluids and their tectonic implication is discussed in the following paras to understand the evolutionary history of the Schirmacher terrane, East Antarctica.

The P-T conditions of charnockites, mafic granulites, and felsic gneisses have shown three phases of metamorphism. An early granulite facies metamorphism involving charnockitization (7.3±0.3 kbar, 827±29°C) as observed in enclave charnockites, and a last amphibolite facies metamorphism was resulting in granitisation (5.4±0.4 kbar, 654±27°C). While, the P-T conditions for core compositions of mafic granulites (7.1±0.2 kbar, 837±26°C) and rim compositions (5.9±0.3 kbar, 652±33°C) represent the late granulite facies metamorphism and retrograde conditions, respectively. A P-T path from nearly 810±50°C; 7.1±0.3 kbar to 660±35°C; 5.7±0.3 kbar for the granulite-amphibolite terrain of the Schirmacher region is observed (Figure-2.8). The proposed values also compare favourably with the estimates based on fluid inclusion studies for these rocks (Rameshwar Rao et al., 1997). However, the minor differences observed are invisible water in CO_2 fluid inclusions (Brown & Lamb, 1986; Lamb et al., 1991). It could be due to inclusion modification during uplift and changed P-T conditions (Sterner & Bodnar, 1989).

To understand further, the estimates based on fluid inclusion studies for peak densities match with the proposed P-T conditions of granulite rocks. The change in grade of early granulite facies metamorphism is indicated by the increasing dominance of CO_2 over H_2O in the early fluid phases. The isochors for early CO_2 inclusions coincide with the P-T boxes of early metamorphic conditions deduced from the mineral composition (1.057 to 0.986 g/cm^3). The low density of the trail-bounded CO_2 inclusions and the entrapment of mixed H_2O-CO_2 inclusions (0.996 to 0.933 g/cm^3), post-dating the peak granulite metamorphism, are attributed to a regional retrogression and hydration event, which has also been inferred through the mineral phase thermobarometry. Further, the mineralogical composition and the mineral chemical data show that the felsic rocks were formed during amphibolite facies metamorphism, changing the fluid regime from CO_2 to mixed CO_2-H_2O. Furthermore, the retrogressive P-T conditions shown by granulite rocks match the P-T conditions estimated for the region's felsic rocks and other leucogneiss. The consistency is also observed for the densities of late-stage fluid.

The continuum in the fluid composition from initial deformed CO_2 rich to later CO_2-H_2O then H_2O-NaCl imply that the early deformed aqueous inclusions were trapped during the initial crystallisation of the partitioned H_2O rich melt at the lower crustal level. Later, the fluids show the evolution from carbonic aqueous to H_2O-NaCl composition consistent with the amphibolite facies metamorphism and further retrogression. These successive lowering of P-T conditions and the observed densities and compositional change of the fluid argues for retrograde effects. Santosh (1986) and Andersen et al. (1991) have also inferred retrogression in the presence of carbonic and saline aqueous fluid in other areas.

As a word of caution, it is often interpreted that the occurrence of high-density CO_2-rich fluid inclusions in granulite facies rocks is strong evidence for the infiltration of a massive amount of CO_2 during the peak metamorphism. The presence of scapolite with significant quartz and calcite in the Schirmacher region calc-silicate rocks suggests a CO_2 present regime during the granulite stage. However, as Harley (1989) suggested, in the absence of definite criteria for the timing of fluid inclusion entrapment in these complex rocks, it is dangerous to correlate the high-density fluid inclusions with the peak metamorphism because of their potential for entrapment at other conditions along P-T path. Thus, with the present level of information, it is suggested that CO_2-rich fluids were present near the granulite stage. However, abundant CO_2 flushing at peak P-T conditions needs further confirmation.

The range of P-T conditions shown by gneiss-charnockite rocks implies tectonic burial of 19 to 24 km and geothermal gradients of around 35°C/km. The pressures recorded in felsic gneisses and charnockites correspond to the crustal depths of 19-24 km. Considering the present-day crustal thickness as ~35 km and presuming that no significant addition to the lower crust has occurred since 1.2 Ga ago, a 54-60 km thick crust could be inferred for the Schirmacher region of East Antarctica. Further, the estimated P-T conditions of the rocks imply a clockwise P-T-t path with a gradual decrease in temperature of around 250°C and a drop in pressure of about 1700 bars suggesting a dP/dT gradient of ~7±1 bar/°C. The proposed dP/dT gradient argues for an isobaric cooling history of the terrane, thereby indicating that, following the maximum P-T conditions, the rocks nearly cooled with a minimal uplift, in contrast to other decompressional terranes which rapidly uplifted and exhumed after the thermal peak.

Origin of Tholeiitic Magmas

The geochemical characteristics of the region's mafic granulites indicate that they are Mg-metatholeiites with low Nb and TiO_2 and high LILE/HFSE and LILE/HREE ratios. The rocks' LILE concentrations are enriched relative to the mantle, while their HFSE concentrations are comparable to those of depleted MORB. The chondrite-normalised REE patterns of these rocks also indicate their LREE enrichment with flat HREE concerning N-type MORB. Despite LILE and LREE enrichment, the individual Nd and Sr initial isotopic composition and HFSE and HREE compositions are consistent with the mantle source. The average $\sum Nd_i$ of +5.26 is a distinctly mantle-type value and, along with high mg# values (~60), indicate that a large part of the region is isotopically "unevolved." The shallow gradient of the combined Nd-Sr plot has no equivalents in oceanic volcanic. It requires the decoupling of the isotopic systems in the mantle, which could be due to mantle metasomatism. Alternatively, it could be due to variable and small degrees of partial melting of mantle source resulting in a greater range of Rb/Sr ratios and restricted Sm/Nd ratios or contamination of basaltic magma by crustal material during subduction processes. The geochemical and isotopic data of mafic granulites gets support for the latter possibility, wherein arc magmatism resulting from partial dehydration melting of down-going oceanic crust is involved.

Thomas (1989), Jacobs and Thomas (1994), Groenewald et al. (1995) have reported that a significant portion of the Maud Belt initially consists of volcanic arcs, a feature of other Proterozoic accretionary provinces. The supporting evidence in the region for this is the metatholeiites showing volcanic arc affinity, having a high Th/Ta ratio (>4) which can be related to processes involving subduction (cf. Cocherie et al., 1994), showing Nb trough on the normalised mantle diagram characterising the subduction zone magmatism (e.g., Saunders et al., 1980; Holm, 1985; Ohta et al., 1996), and showing a low La/Nb ratio of ~0.7 relative to chondrite ratio of 0.95 reflecting the mantle source characteristics of a recycled oceanic crust, which indicate dehydration/partial melting of oceanic crust in a subduction zone. The LILE enrichment can also explain the presence of fluids during dehydration of subducting oceanic lithosphere, high LILE/HFSE and LILE/REE (e.g. Ba/La ~16) ratios of metatholeiites (Saunders et al., 1980; Gill, 1981; Ewart, 1982; Thorpe, 1982; Morra et al., 1997). As a result, most 'primitive' tholeiitic magmas are formed, where olivine plays a dominating fractionating role (Ringwood, 1974; Green, 1980). This crust formation event goes well with what had been given by Rogers (1996), i.e., after the initial formation of Ur, Artic, Atlantic grew at various times during their histories by accretion of juvenile (subduction-related) material and cratonic fragments from other continents. These magmas form a modified thickened crust, which represents the present-day granulite rocks of the region and explains the involvement of isobaric cooling path in the region, wherein the early granulite conditions were attained in a crust of near-normal thickness; and owe its origin to the intrusion and underplating of large amounts of magma that not only provide heat for the granulite facies metamorphism. It also adds a large volume of primarily mafic material to the base of the crust (Bohlen, 1987; Ellis, 1987; Bohlen & Mezger, 1989).

Origin of Felsic Magmas

Geochemical and petrographic data suggest that the felsic rocks of the Schirmacher region can be classified as I-type granitoids and support an origin by partial melting of lower crustal igneous material rather than metapelites source as suggested by Hussain and Divakara Rao (1996). The rocks are characterised by a wide SiO_2 range of 56-73% and Na_2O+K_2O range of 5-11%, have a mean mol A/CNK ratio of ~1.03 and define a calc-alkaline trend on the AFM diagram. The presence of biotite and hornblende as principal primary minerals and the conspicuous absence of aluminosilicate, cordierite, or white-mica minerals in these rocks support the I-type nature of these rocks. The other petrographic evidence, like the accessory minerals occurring as idioblastic to subidioblastic grains, presence of sphene which is listed by Chappell and White (1974) as a characteristic mineral of I-type, accessory zircon occurring as tiny crystals and unlike zircon in S-type rocks, they are scattered through the rocks rather than concentrated in biotite also support the I-type nature of these felsic rocks. Further, the presence of numerous enclaves of granulite rocks in the felsic rocks, and on many of the interelement diagrams, the charnockite rocks lying in line with the felsic rocks suggest a mafic-intermediate source (cf. Rameshwar Rao et al., 1999). The main mafic phases in the felsic rocks are biotite and hornblende, the common constituents of a primary source. These, along with the REE pattern of the charnockites with distinctive positive Eu anomaly, also suggest that the early formed charnockites and mafic rocks are sources of felsic melts. Thus, it is proposed that the partial melting of the earlier created mafic-intermediate crustal material has resulted in the average chemical composition of felsic igneous rocks. However, it cannot be ruled out that at a mature stage where the thickened crust is involved, there is also an increased likelihood of sediments being incorporated into the melting processes and imposing characteristic chemistry and mineralogy in the derived magmas (Green, 1976). An origin for the extensive I-type granites, which principally involved a partial melting in the crust of older igneous rocks which themselves have been derived from the mantle, was advocated by Chappell and Stephen (1988).

The felsic melts so generated underwent fractional crystallisation and diffusive differentiation. The hot, more basic quartz monozodiorite/quartz monozonite melts settle to the base, while those of adamellite to granodiorite compositions move to the top. The parental magma in the N.W. region of Schirmacher represents the melts at the bottom, resulting in mostly the porphyritic rocks (the present-day Augen gneiss) into which the differentiated melts of biotite rich rocks and garnet-biotite-rocks intrude. The leucogranites are formed in the second stage with less melting of the same source, which gets support from the field relations of the leucogneiss with other rocks, from their REE patterns and their younger age (773±26 Ma; cf. Rama Rao et al., 1995) than other felsic rocks of the region (further details are available in Rameshwar Rao et al., 1999).

CONCLUSION

In summary, the metamorphic studies suggest three phases of metamorphism in the Schirmacher region of East Antarctica, an early granulite facies metamorphism, a late granulite phase of metamorphism, and the third phase of retrograde amphibolite facies metamorphism. The mineralogical geothermobarometry indicates that such metamorphism has occurred at a 19-24 km depth, ca. 35°C/km geothermal gradient, suggesting syn-metamorphic thickness of around 54-60 km. The estimated P-T conditions of the rocks imply a clockwise P-T-t path with a gradual decrease in temperature of about 250°C and a reduction in pressure around 1700 bars. They have a dP/dT gradient of ~7±1 bar/°C, arguing for an isobaric cooling history of the terrane.

A good correlation between fluid and mineral data is also observed. The isochors are typical for high-density CO_2 fluids to fall well within the P-T box estimated by mineral thermobarometers. Based on fluid inclusions, a pervasive influx of deep-seated CO_2 rich fluid from the mantle is envisaged in the formation of the granulites. The subsequent stages showed a decrease in CO_2 density and a progressive influx of hydrous fluids, and the generation of retrograde amphibolite facies metamorphism in the area.

The geochemical and isotopic studies of gneiss-charnockite rocks suggest a two-stage crystallisation history of the origin of tholeiitic and felsic rocks of the Schirmacher region. In the first stage, early ‘tholeiitic’ magmas are formed through arc-magmatism involving dehydration of the subducting slab and large-scale mantle melting. In the second stage, the generation of felsic melts through partial melting of early developed mafic-intermediate crustal material.

ACKNOWLEDGMENT

The author is grateful to the Director, Wadia Institute of Himalayan Geology, Dehradun, for providing facilities and permission to publish this work. Drs. Rajesh Sharma and N.S. Gururajan deserves a special appreciation for their contribution to fluid inclusion and petrographic studies. He is thankful to Drs. T.V. Sivaraman, V. Balaram, G.V.C. Pantulu, N.K. Saini and Mr Tamal Ghosh for analytical support. The author is also grateful to the Department of Ocean Development, Government of India, for logistic support during the Twelfth Indian Scientific Expedition to Antarctica. Special thanks to anonymous reviewers whose suggestions have immensely helped in improving the manuscript.

REFERENCES

Andersen, T., Austrheim, H., & Buke, E. A. J. (1991). Fluid induced retrogression of granulites in the Bergen Arcs, Caledonides of W Norway: Fluid inclusion evidence from amphibolites facies shear zones. *Lithosphere*, *27*, 29–42.

Bhattacharya, A., Krishna Kumar, K. R., Raith, M., & Sen, S. K. (1991). An improved set of a-X parameters for Fe-Mg-Ca garnets and refinements of the orthopyroxene-garnet thermometer and orthopyroxene-garnet-plagioclase-quartz barometer. *Journal of Petrology*, *32*(3), 629–656. doi:10.1093/petrology/32.3.629

Bhattacharya, A., Mohanty, L., Maji, A., Sen, S. K., & Raith, M. (1992). Non-ideal mixing in the phlogophite-annite binary: Constraints from experimental data on Mg-Fe partitioning and a reformulation of the biotite-garnet geothermometers. *Contributions to Mineralogy and Petrology*, *111*(1), 87–93. doi:10.1007/BF00296580

Binns, R. A. (1968). Hydrothermal investigation of the amphibolite-granulite facies boundary. Special Publication in. *Geological Society of America*, *2*, 341–344.

Bohlen, S. R. (1987). Pressure-temperature-time paths and a tectonic model for the evolution of granulites. *The Journal of Geology*, *95*(5), 617–632. doi:10.1086/629159

Bohlen, S. R., & Mezger, K. (1989). Origin of granulite terranes and the formation of the lowermost continental crust. *Science*, *244*(4902), 326–329. doi:10.1126cience.244.4902.326 PMID:17738304

Bohlen, S. R., Wall, V. J., & Boettcher, A. L. (1983). Experimental investigation and application to garnet granulite equilibria. *Contributions to Mineralogy and Petrology*, *83*(1-2), 52–56. doi:10.1007/BF00373079

Borsi, L., Petrini, R., Talarica, R. F., & Palmeri, R. (1995). Geochemistry and Sr-Nd isotopes of amphibolite dykes of northern Vitoria Land, Antarctica. *Lithosphere, 35*, 245–260.

Brown, P. E. (1989). Flincor: A microcomputer programme for the reduction and investigation of fluid inclusion data. *American Journal of Mineralogy, 74*, 1390–1393.

Brown, P. E., & Lamb, W. M. (1986). Mixing of H_2O-CO_2 in fluid inclusions: Geobarometry and Archean gold deposits. *Geochimica et Cosmochimica Acta, 50*(5), 847–852. doi:10.1016/0016-7037(86)90360-1

Cabanis, B., Chantraine, J., & Rabu, D. (1987). Geochemistry study of the Brioverian (late Proterozoic) volcanic rocks in the northern Armorican Massif (France), Implications for geodynamic evolution during the Cadomian. In T. C. Beckinsale & D. Rickard (Eds.), Geological Society of London Special Publications: Vol. 33. Geochemistry and Mineralisation of Proterozoic Volcanic Suites' (pp. 525–539). Academic Press.

Carswell, D. A., & Harley, S. L. (1990). Mineral barometry and thermometry. In Eclogite Facies Rocks. Blackie Press.

Chakraborty, S. K., Kaul, M. K., & Raina, V. K. (1985). Mineral variation in the gneisses from Dakshin Gangotri, Antarctica. *Department of Ocean Development. Technical Publication*, (2), 23–28.

Chappell, B. W., & Stephens, W. E. (1988). Origin of infracrustal (I-type) granite magmas. *Transactions of the Royal Society of Edinburgh. Earth Sciences, 79*(2-3), 71–86. doi:10.1017/S0263593300014139

Chappell, B. W., & White, A. J. R. (1974). Two contrasting granite types. *Pacific Geology, 8*, 173–174.

Cocherie, A., Rossi, P. L., Fouillac, A. M., & Vidal, Ph. (1994). Crust and mantle contribution to granite genesis - an example from the Variscan batholith of Corsica, France, studied by trace element and Nd-Sr-O-isotope systematics. *Chemical Geology, 115*(3-4), 173–211. doi:10.1016/0009-2541(94)90186-4

Dahl, P. S. (1980). The thermal-compositional dependence of Fe^{+2}-Mg^{+2} distribution between coexisting garnet and pyroxene: Application to geothermometry. *American Journal of Mineralogy, 65*, 852–866.

De Paolo, D. J. (1981). Neodymium isotopes in the Colorado Front Range and crust-mantle evolution in the Proterozoic. *Nature, 291*(5812), 193–196. doi:10.1038/291193a0

Eckert, J.O. Jr, Newton, R.C., & Klepp, A, O.J. (1991). The ΔH of reaction and recalibration of garnet-pyroxene-plagioclase-quartz geobarometer in CMAS system in solution calorimetry. *American Journal of Mineralogy*, *76*, 148–160.

Ellis, D. J. (1987). Origin and evolution of granulites in the normal and thickened crust. *Geology*, *15*(2), 167–170. doi:10.1130/0091-7613(1987)15<167:OAEOGI >2.0.CO;2

Ellis, D. J., & Green, D. (1979). An experimental study of the effect of Ca upon garnet-clinopyroxene Fe-Mg exchange equilibria. *Contributions to Mineralogy and Petrology*, *71*(1), 13–32. doi:10.1007/BF00371878

Ewart, A. (1982). The mineralogy and petrology of Tertiary-Recent orogenic volcanic rocks with special reference to the andesite-basaltic compositional range. In R. S. Thorpe (Ed.), *Andesite: Orogenic andesites and related rocks* (pp. 25–95). Wiley.

Farmer, G. L., & De Paolo, D. J. (1983). Origin of Mesozoic and Tertiary granite in the Western United States and implications for pre-Mesozoic crustal structure. Nd and Sr isotopic studies in the Geoclue of the northern Great Basin. *Journal of Geophysical Research*, *88*(B4), 3379–3401. doi:10.1029/JB088iB04p03379

Ferry, J. M., & Spear, F. S. (1978). Experimental calibration of the partitioning of Fe and Mg between biotite and garnet. *Contributions to Mineralogy and Petrology*, *66*(2), 113–117. doi:10.1007/BF00372150

Ganguly, J. (1979). Garnet and clinopyroxene solid solutions, and geothermometry based on Fe-Mg distribution coefficient. *Geochimica et Cosmochimica Acta*, *43*(7), 1021–1029. doi:10.1016/0016-7037(79)90091-7

Ganguly, J., & Saxena, S. K. (1984). Mixing properties of aluminosilicate garnets: Constraints from natural and experimental data, and applications to geothermobarometry. *American Journal of Mineralogy*, *69*, 88–97.

Gessmann, C. K., Spiering, B., & Raith, M. (1997). Experimental study of the Fe-Mg exchange between garnet and biotite: Constraints on the mixing behaviour and analysis of the cation exchange mechanism. *American Journal of Mineralogy*, *82*(11-12), 1225–1240. doi:10.2138/am-1997-11-1218

Ghent, E. D., & Stout, M. Z. (1981). Geobarometry and geothermometry of plagioclase-biotite-garnet-muscovite assemblages. *Contributions to Mineralogy and Petrology*, *76*(1), 92–97. doi:10.1007/BF00373688

Gill, J. B. (1981). *Orogenic andesites and plate tectonics*. Springer. doi:10.1007/978-3-642-68012-0

Graham, C. M., & Powell, R. (1984). A garnet-hornblende geothermometer: Calibration testing and application to the Pelona Schist, Southern California. *Journal of Meteorology and Geology*, *2*(1), 13–31. doi:10.1111/j.1525-1314.1984.tb00282.x

Green, T. H. (1976). Experimental generation of cordierite - or garnet-bearing granitic liquids from a pelitic composition. *Geology*, *4*(2), 85–88. doi:10.1130/0091-7613(1976)4<85:EGOCGG>2.0.CO;2

Green, T. H. (1980). Island arc and continent-building magmatism - a review of petrogenic models based on experimental petrology and geochemistry. *Tectonophysics*, *63*(1-4), 367–385. doi:10.1016/0040-1951(80)90121-3

Groenewald, P. B., Moyes, A. B., Grantham, G. H., & Krynauw, J. R. (1995). East Antarctic crustal evolution: Geological constraints and modelling in Western Dronning Maud Land. *Precambrian Research*, *75*(3-4), 231–250. doi:10.1016/0301-9268(95)80008-6

Harley, S. L. (1984). An experimental study of the partitioning of Fe and Mg between garnet and orthopyroxene. *Contributions to Mineralogy and Petrology*, *86*(4), 359–373. doi:10.1007/BF01187140

Harley, S. L. (1989). The origin of granulites: A metamorphic perspective. *Geological Magazine*, *126*(3), 215–247. doi:10.1017/S0016756800022330

Harris, N. B. W., Pearce, J. A., & Tindle, A. G. (1986). Geochemical characteristics of collision zone magmatism. In M. P. Coward & A. C. Ries (Eds.), Geological Society of London Special Publication: Vol. 19. Collision Tectonics (pp. 67–81). Academic Press.

Haselton, H. T. Jr, Hovis, G. L., Hemingway, B. S., & Robie, R. A. (1983). Calorimetric investigation of the excess entropy of mixing in an albite-sanidine solid solution: Lack of evidence for Na, K short-range order and implications for two-feldspar thermometry. *American Journal of Mineralogy*, *68*, 398–413.

Hodges, K. V., & Crowley, P. D. (1985). Error estimation and empirical geothermobarometry for pelitic systems. *American Journal of Mineralogy*, *70*, 702–709.

Hodges, K. V., & Spear, F. S. (1982). Geothermometry, geobarometry and the Al_2SiO_5 triple point at Mt. Moosilauke, New Hampshire. *American Journal of Mineralogy*, *67*, 1118–1134.

Hoisch, T. D. (1990). Empirical calibration of six geobarometers for the mineral assemblage quartz+muscovite+biotite+plagioclase+garnet. *Contributions to Mineralogy and Petrology*, *104*(2), 225–234. doi:10.1007/BF00306445

Holdaway, M. J. (1971). Stability of andalusite and the aluminous silicate phase diagram. *American Journal of Science*, *271*(2), 97–131. doi:10.2475/ajs.271.2.97

Holdaway, M. J., Mukhopadhyay, B., Dyar, M. D., Guidotti, C. V., & Dutrow, B. L. (1997). Garnet-biotite geothermometry revised: New margules parameters and a natural specimen data set from Maine. *American Journal of Mineralogy*, *82*(5-6), 582–595. doi:10.2138/am-1997-5-618

Holm, P. E. (1985). The geochemical fingerprints of different tectonomagnetic environments using hygromagmatophile element abundances of tholeiitic basalts and basaltic andesites. *Chemical Geology*, *51*(3-4), 303–323. doi:10.1016/0009-2541(85)90139-1

Hussain, S. M., & Divakara Rao, V. (1996). Geochemistry of polyphase gneisses from Schirmacher Oasis, East Antarctica. *Journal of the Geological Society of India*, *47*, 303–312.

Indares, A., & Martignole, J. (1985). Biotite-garnet thermometry in the granulite facies: The influence of Ti and Al in biotite. *American Journal of Mineralogy*, *70*, 272–278.

Jacobs, J., & Thomas, R. J. (1994). Oblique collision at about 1.1 Ga along the southern margin of the Kaapval continent, south-east Africa. *Geologische Rundschau*, *83*, 322–333.

Kaul, M. K., Chakraborty, S. K., & Raina, V. K. (1985). Petrography of the biotite-hornblende bearing quartzo-feldspathic gneisses from Dakshin Gangotri, Schirmacher Hill, Antarctica. *Scientific Report, Department of Ocean Development. Technical Publication*, *2*, 29–38.

Kersting, A. B., & Arculus, R. J. (1994). Klyuchevskoy volcano, Kamchatka, Russia: The role of high-flux recharged, tapped, and fractionated magma chamber(s) in the genesis of high Al_2O_3 from high-MgO basalt. *Journal of Petrology*, *35*(1), 1–41. doi:10.1093/petrology/35.1.1

Kohn, M. J., & Spear, F. S. (1990). Two new geobarometers for garnet amphibolites with application to southeastern Vermont. *American Journal of Mineralogy*, *75*, 89–96.

Krogg, E. J. (1988). The garnet-clinopyroxene Fe-Mg geothermometer-reinterpretation of existing experimental data. *Contributions to Mineralogy and Petrology*, *99*(1), 44–48. doi:10.1007/BF00399364

Kumar, A., Charan, S. N., Gopalan, K., & Mac Dougal, J. D. (1998). A long-lived enriched mantle source for two Proterozoic carbonatite complexes from Tamil Nadu, southern India. *Geochimica et Cosmochimica Acta*, *62*(3), 515–524. doi:10.1016/S0016-7037(97)00341-4

Lal, R. K. (1993). Internally consistent recalibrations of mineral equilibria for geothermobarometry involving garnet-orthopyroxene-plagioclase-quartz assemblages and their application to the South Indian granulites. *Journal of Meteorology and Geology*, *11*(6), 855–866. doi:10.1111/j.1525-1314.1993.tb00195.x

Lamb, W. M., Brown, P. E., & Valley, J. W. (1991). Fluid inclusions in Adirondack granulites: Implications for retrograde P-T path. *Contributions to Mineralogy and Petrology*, *107*(4), 472–483. doi:10.1007/BF00310681

Lee, H. Y., & Ganguly, J. (1988). Equilibrium compositions of coexisting garnet and orthopyroxene: Experimental determination in the system FeO-MgO-Al_2O_3-SiO_2 and applications. *Journal of Petrology*, *29*(1), 93–111. doi:10.1093/petrology/29.1.93

Morra, V., Secchi, F. A. G., Melluso, L., & Franciosi, L. (1997). High-Mg subduction-related Tertiary basalts in Sardinia, Italy. *Lithosphere*, *40*, 69–71.

Newton, R. C., & Perkins, D. III. (1982). Thermodynamics calibration of geobarometers based on the assemblage garnet-plagioclase-orthopyroxene (clinopyroxene)-quartz. *American Journal of Mineralogy*, *67*, 203–222.

Ohta, H., Maruyama, S., Takahashi, E., Watanabe, Y., & Kato, Y. (1996). Field occurrence, geochemistry, and petrogenesis of the Archean Mid-Oceanic Ridge Basalts (AMORBs) of the Cleavenville area, Pilbara Craton, Western Australia. *Lithosphere*, *37*, 199–221.

Pattison, D. R. M., & Newton, R. C. (1989). Reversed experimental calibration of the garnet clinopyroxene Fe-Mg exchange thermometer. *Contributions to Mineralogy and Petrology*, *101*(1), 87–103. doi:10.1007/BF00387203

Pearce, J. A., Harris, N. B. W., & Tindle, A. G. (1984). Trace element discrimination diagrams for the tectonic interpretation of granitic rocks. *Journal of Petrology*, *25*(4), 956–983. doi:10.1093/petrology/25.4.956

Perchuk, L. L., & Lavrent'eva, I. V. (1983). Experimental investigation of exchange equilibria in the system cordierite-garnet-biotite. In S. K. Saxena (Ed.), *Kinetics and equilibrium in Mineral reactions* (pp. 199–239). Springer-Verlag. doi:10.1007/978-1-4612-5587-1_7

Perchuk, L. L., Podlesskii, K. K., & Aranovich, L. Y. A. (1991). Thermodynamics of some framework silicates and their equilibria: application to geothermometry. In L. L. Perchuk (Ed.), *Progress in metamorphic and magmatic petrology* (pp. 131–164). Cambridge University Press. doi:10.1017/CBO9780511564444.009

Perkins, D. III, & Chiper, A. (1985). Garnet-orthopyroxene-plagioclase-quartz barometry: Refinement and application to the English river subprovince and the Minnesota river valley. *Contributions to Mineralogy and Petrology*, *89*(1), 69–80. doi:10.1007/BF01177592

Philibert, J. (1963). *X-ray optics and X-ray Microanalysis*. Academic Press.

Powell, M., & Powell, R. (1977). Plagioclase-alkali feldspar geothermometry revisited. *Mineral Magnetics*, *41*(318), 253–256. doi:10.1180/minmag.1977.041.318.13

Price, J. C. (1985). Ideal site mixing in solid solution with an implication to two-feldspar geothermometry. *American Journal of Mineralogy*, *70*, 696–701.

Provost, A. (1990). An improved diagram for isotope data. *Chemical Geology*, *80*, 85–89.

Rama Rao, P., Ramana Rao, A. V., & Poornachandra Rao, G. V. S. (1999). Geological, geochemical and geochronological and palaeomagnetic studies on the rocks from parts of East Antarctica. *Scientific Report. Department of Ocean Development Technical Publication*, *8*, 97–106.

Rameshwar, R. D., & Subba Rao, T. V. (1994). AKP: A two-feldspar geothermometry using Lotus 1-2-3 software. *Himalayan Geology*, *5*, 93–101.

Rameshwar, R. (1995). BGT - the macros driven spreadsheet program for biotite-garnet thermometry. *Computers & Geosciences*, *21*(4), 593–604. doi:10.1016/0098-3004(94)00098-F

Rameshwar, R. D., Vijian, A. R., & Prabhu, B. N. (1995). Geochemical and Rb-Sr isotope studies of the felsic rocks from the Schirmacher region, East Antarctica. *Geosciences Journal*, *20*, 41–59.

Rameshwar, R. (1996). AMPH - a program for calculating formulas and for assigning names to the amphibole group of minerals. *Computers & Geosciences*, *22*(8), 931–933. doi:10.1016/S0098-3004(96)00018-0

Rameshwar, R. (2000). Metamorphic evolution of charnockites and felsic gneisses from the Schirmacher region, East Antarctica. *Gondwana Research*, *3*(1), 79–89. doi:10.1016/S1342-937X(05)70059-9

Rameshwar, R. (2000). Origin of high magnesian metatholeiites of the Schirmacher region, East Antarctica: Constraints from trace elements and Nd-Sr isotopic systematics. *Gondwana Research*, *3*(1), 91–104. doi:10.1016/S1342-937X(05)70060-5

Rameshwar, Rao, D. (2006). The evolutionary history of the Schirmacher region, East Antarctica. *Himalayan Geology*, *27*, 81–94.

Rameshwar Rao, D., Sharma, R., & Gururajan, N. S. (1996). *Mineral Chemistry, Fluid Inclusion Studies and P-T path of Gneiss-Granulite Rocks of Schirmacher Region, East Antarctica.* Twelfth Indian Expedition to Antarctica, Scientific Report, Department of Ocean Development, Technical Publication.

Rameshwar Rao, D., Sharma, R., & Gururajan, N. S. (1997). Mafic granulites of the Schirmacher region, East Antarctica: Fluid inclusion and geothermobarometric studies focussing on the Proterozoic evolution of the crust. *Transactions of the Royal Society of Edinburgh*, *88*(1), 1–17. doi:10.1017/S0263593300002285

Rameshwar Rao, D., Sharma, R., & Gururajan, N. S. (1998). Geothermobarometry and fluid inclusion studies of leucogneiss from Schirmacher region, East Antarctica. *Journal of the Geological Society of India*, *51*, 595–607.

Ravikant, V. (2005). Metamorphism of Ultramafic and mafic enclaves within granulites, Schirmacher Oasis, East Antarctica. *Journal of the Geological Society of India*, *65*, 279–290.

Ravikant, V., & Kundu, A. (1998). Reaction textures of retrograde pressure-temperature deformation paths from granulites of Schirmacher Hills, East Antarctica. *Journal of the Geological Society of India*, *51*, 305–314.

Ringwood, A. E. (1974). Petrological evolution of island arc systems. *Journal of the Geological Society*, *130*(3), 183–204. doi:10.1144/gsjgs.130.3.0183

Rogers, J. J. W. (1996). A history of the continents in the past three billion years. *The Journal of Geology*, *104*(1), 91–107. doi:10.1086/629803

Santosh, M. (1986). Carbonic metamorphism of charnockites in the southwestern Indian Shield: A fluid inclusion study. *Lithosphere*, *19*, 1–10.

Saunders, A. D., Tarney, J., & Weaver, S. D. (1980). Transverse geochemical variations across the Antarctic Peninsula: Implications for the genesis of calc-alkaline magmas. *Earth and Planetary Science Letters*, *46*(3), 344–360. doi:10.1016/0012-821X(80)90050-3

Sen, S. K., & Bhattacharya, A. (1984). An orthopyroxene-garnet thermometer and its application to the Madras charnockites. *Contributions to Mineralogy and Petrology*, *88*(1-2), 64–71. doi:10.1007/BF00371412

Sengupta, P. K., Dasgupta, S., Bhattacharaya, P. K., & Hariya, Y. (1989). Mixing behaviour in quaternary garnet solid solution and an extended Ellis and Green garnet-clinopyroxene geothermometer. *Contributions to Mineralogy and Petrology*, *103*(2), 223–227. doi:10.1007/BF00378508

Sengupta, S. (1988). History of successive deformation in relation to metamorphic-migmatitic events in the Schirmacher Hills, Queen Maud Land, East Antarctica. *Journal of the Geological Society of India*, *32*, 295–314.

Sengupta, S. (1993). Tectono-thermal history recorded in mafic dykes and enclaves of the gneissic basement in the Schirmacher Hills, East Antarctica. *Precambrian Research*, *63*(3-4), 273–292. doi:10.1016/0301-9268(93)90037-3

Sharma, R., & Rameshwar, R. D. (2006). Fluid evolution in the granulite-amphibolite metamorphism of the Schirmacher region, East Antarctica. *Indian Journal of Geochemistry*, *21*, 103–117.

Stackebrandt, W., Kampf, H., & Wetzel, H. U. (1988). The geological setting of the Schirmacher Oasis, Queen Maud Land, East Antarctica. *Zeitschrift für Geologische Wissenschaften*, *7*, 661–665.

Sterne, R. S. M., & Bodnar, R. J. (1989). Synthetic fluid inclusion VII. Re-equilibration of fluid inclusions in the quartz during laboratory simulated burial and uplift. *Journal of Meteorology and Geology*, *7*, 465–484.

Stormer, J. C. (1975). Practical two-feldspar geothermometers. *American Journal of Mineralogy*, *60*, 667–674.

Sun, S. S., & Mc Donough, W. F. (1989). Chemical and isotopic systematics of oceanic basalts: implications for mantle composition and processes. In A. D. Saunders & M. J. Norry (Eds.), *Magmatism in the ocean basins. Journal of Geological Society of London Special Publications*. doi:10.1144/GSL.SP.1989.042.01.19

Thomas, R. J. (1989). The petrogenesis of the Mzumbe Gneiss suite is a tonalite-trondhjemite orthogneiss suite from the southern part of the Natal structural and metamorphic province. *South African Journal of Geology*, *92*, 322–338.

Thompson, A. B. (1976). The mineral reaction in pelitic rocks. Calculations of some P-T-x (Fe-Mg) phase relations. *American Journal of Mineralogy*, *276*(4), 425–454. doi:10.2475/ajs.276.4.425

Thorpe, R. S. (1982). *Andesites: orogenic andesites and related rocks*. Wiley.

Wells, P. R. A. (1979). Chemical and thermal evolution of Archean Sialic crust, Southern West Greenland. *Journal of Petrology*, *20*(2), 187–226. doi:10.1093/petrology/20.2.187

Whitney, J. A., & Stormer, J. C. Jr. (1977). The distribution of $NaAlSi_3O_8$ between coexisting microcline and plagioclase and its effects on geothermometric calculations. *American Journal of Mineralogy*, *62*, 687–691.

Chapter 3
A Comprehensive Study of the Structural Geology of the Schirmacher Hills

Sudipta Sengupta
Jadavpur University, India

ABSTRACT

The Precambrian basement of the Schirmacher Hills records multiple episodes of deformation, metamorphism, migmatization, and emplacement of successive generations of mafic and felsic bodies. The earliest tectono-thermal event (D1/M1), preserved in some mafic and ultramafic enclaves, indicates deformation at great crustal depth. The mineralogical assemblage of these enclaves indicates early high temperature (900o C) and high-pressure (10 Kbar) granulite facies conditions. The second tectono-thermal event also showed deformation under granulite facies metamorphism (D2/M2) under 800-850oC and 8 Kbar. The third group of events (D3/M3) is the most dominant in this region and involved deformation under amphibolite facies conditions with synchronous emplacement of granites and mafic dykes and culminated in regional thrusting, producing a regional inversion where the granulates were emplaced over the amphibolite facies rocks. The later events created upright folds and vertical shear zones under amphibolite facies conditions.

INTRODUCTION

Ravich and Kamenev (1975), Grew (1983), and Grew and Manton (1983) initiated the geological study in the Schirmacher Hills region of East Antarctica. Their research laid emphasis mostly on petrological aspects. Since 1983, Indian and

DOI: 10.4018/978-1-6684-4078-0.ch003

German scientists have carried out detailed work in this region. Most of the works are on petrological, geochemical, geochronological, and glaciological aspects. Paech and Stackebrandt (1995) carried out a study on preliminary structural aspects of the area. Jadavpur University took part in five expeditions to Antarctica. These are the third (1983-84), ninth (1989-90), fifteenth (1995-96), sixteenth (1996-97), and seventeenth (1997-98). This area's detailed structural, petrological, and geochemical analyses were carried out (Sengupta 1986, 1988 a and b, 1993, 1997 Sengupta and Bose 1997, Dasgupta *et al.,* 2001, Bose and Sengupta 2003).

The tectonic of Paech and Stackebrandt (1995) did not report the first deformation (Sengupta (1993).

Furthermore, the initial metamorphism of the granulite facies and a phase of deformation (D1) were contemporaneous. On the contrary, this event corresponds to the second granulite facies metamorphism and related deformation (D2) of Sengupta (1993).

A complete geological and structural map of the Schirmacher Hills area was prepared for the first time during the third expedition. During the ninth expedition, the emphasis was on the comprehensive study of shear zones and identifying an early basement as enclaves within the granulites. A detailed geological and structural map of the Schirmacher Hills was prepared during the last three expeditions, emphasising the area's tectono-thermal evolution.

Five episodes of deformation and three episodes of metamorphism were identified from the Schirmacher Hills. Identification of successive generations of ductile, brittle-ductile, and brittle shear zones formed an important aspect of the research work. A detailed microstructural study of high-temperature mylonites was carried out. This study has yielded essential indicators for understanding the deformation mechanisms of the constituent minerals at great crustal depth. Such detailed work on shear zones was carried out on Antarctic mylonites. Another important outcome of the research work was recognising an earlier basement, which occurs as enclaves within the host granulites of the Schirmacher Hills. It has a significant geological consequence on the evolution of the Precambrian basement of this region. The gradual transformation of these early basement rocks by later tectono-thermal events were studied in detail.

SCIENTIFIC IMPORTANCE

Antarctica is a crucial piece in Gondwanaland's jigsaw puzzle. It occupies the central position and shows paleogeographic and tectonic continuity across its ancient common borders with Africa, Australia India. Antarctica is the least known among all the continents. Apart from the late discovery of the continent and its inhospitable climate, the regional geologic studies in Antarctica face the problem that a major part of the rocky surface is covered by a thick ice sheet. Hence, an extensive survey of the available rock surfaces gives us vital information for extrapolation on other unexplored areas to understand the crustal evolution of that part of the continent.

The study carried out by Jadavpur University aims to reconstruct the tectono-thermal history of the area to decipher the course of the region's crustal evolution for a meaningful correlation with other parts of East Antarctica as well as with similar Precambrian terrain of the Gondwanaland supercontinent. The multiple episodes of deformations and related metamorphism were identified and characterised; successive generations of ductile and brittle shear zones were studied. Their time-correlation ship was established, microstructural variations of the high-temperature mylonites and the rocks' geothermobarometry were studied in detail. Expedition wise details of geoscientific investigations are outlined below.

1. **3rd expedition (1983-84):** A complete geological map of Schirmacher Hills was prepared for the first time during 1983-84 (Sengupta 1986, 1988 a and b). This map is the basis of all the later maps where later workers made only minor modifications. An area of 35 square Kilometres was mapped on a scale of 1: 25,000. Six major lithological units: banded gneiss, alaskite, garnet-biotite gneiss, calc gneiss, khondalites, pyroxene granulites, and associated migmatites, augen gneiss, and streaky gneiss were delineated. It was reported that the rocks suffered an early granulite facies metamorphism and an early migmatization. It led to the development of charnockitic stones. The amphibolite facies metamorphism, extensive granitisation, and concomitant deformation with isoclinal folds foliation-parallel ductile shear zones are superimposed.
    a. It is the most dominant tectono-thermal event in the area. The circumstances produced two sets of vertical folds and subvertical flexible shear zones under amphibolite facies conditions.
    b. Mafic dykes, discrete bodies of granite and brittle shear zones characterise the area's latest geological activity.
2. **9th expedition (1989-90):** An important finding of this expedition was identifying an early basement preserved as enclaves within the current basement. These enclaves are composed of interlayered mafic and ultramafic rocks. Accordingly, five major groups of events were recognised with the help of structural relationships of the enclaves and dykes with the associated host gneiss. It is a significant finding in this area that provides a clear picture of the tectonic evolution of the area under investigation (Sengupta, 1993).
3. **15th, 16th and 17th expeditions (1995-96, 1996-97, 1997-98):** During this period, detailed mapping of the Schirmacher Hills (Scale 1: 12,500) for three consecutive years under 'Comprehensive geological study of the rocks

of the Schirmacher Hills, East Antarctica,' was completed. Subsequently, a mineralogical, petrological, and geochemical study was carried out on the rocks of Schirmacher Hills (Dasgupta *et al.,* 2001). (Sengupta, 1994, Bose and Sengupta, 2003)

MAIN SCIENTIFIC FINDINGS

The Precambrian basement of the Schirmacher Hills has evolved through a complex geological history of multiple episodes of deformation, metamorphism, and migmatization with the emplacement of granitic and mafic bodies of different ages. Moreover, the fabrics of rocks have undergone several stages of modification due to multiple folding, shearing, and the development of successive generations of foliations under different metamorphic conditions.

The hill range was mapped on a scale of 1:12,500 (Figures 1 and 2). The eastern part primarily consists of banded gneiss of interlayered enderbite and pyroxene granulites. This unit has several mafic and ultramafic enclaves. The western part mainly consists of Augen gneiss and streaky gneiss. A thick unit of interbanded calc gneiss, khondalite, pyroxene granulite, and quartzofeldspathic gneiss occurs in the central part. The eastern part of the area is granulitic mainly, whereas the western part is granitic. The deformation history of the Schirmacher Hills can be divided into five groups of events. Each group consists of folding, shearing, metamorphism, and emplacement of quartzofeldspathic and mafic bodies.

The first recognisable group of events consists of an early deformation D1 under a granulite facies metamorphism. This episode's signature is found in some isolated mafic and ultramafic enclaves, which occur as xenoliths within the gneissic basement of Schirmacher Hills (Figure 2). The chromitite bearing enclaves' blocks are distributed over a large area in the eastern Schirmacher Hills, shown as banded gneiss in the geological map. The enclaves' different bands comprise high Mg-gabbro, metanorite, websterite, olivine spinel bearing enstatite, and spinel lherzolite. D1 foliation within the enclaves is marked by the parallel arrangement of granulite facies minerals, indicating that the first recognisable metamorphism M1 is syntectonic with D1. M1 took place at ca. 10 Kbar and 900°C. The mafic magma, parental to the enclaves, crystallised at 11.2 Kbar. It would imply the emplacement of the mafic magma in the lower crust at approximately 35 Km depth. Following post-peak isobaric cooling, the mafic granulites were subsequently transported to shallower crustal levels by the enderbite magma, most likely by a thrust.

Figure 1. Geological map of the Schirmacher Hills

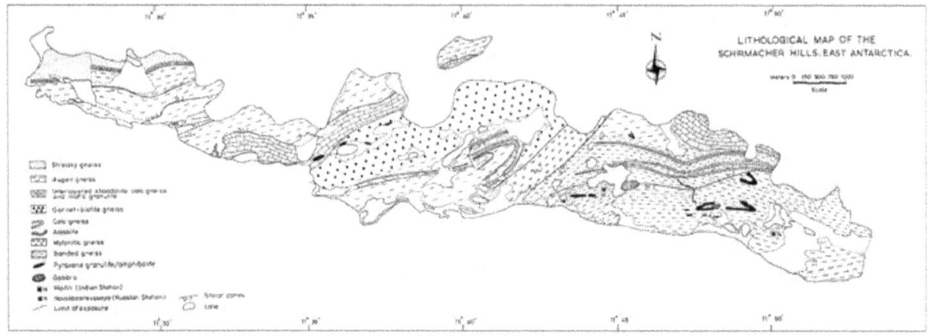

Figure 2. Structural map of the Schirmacher Hills

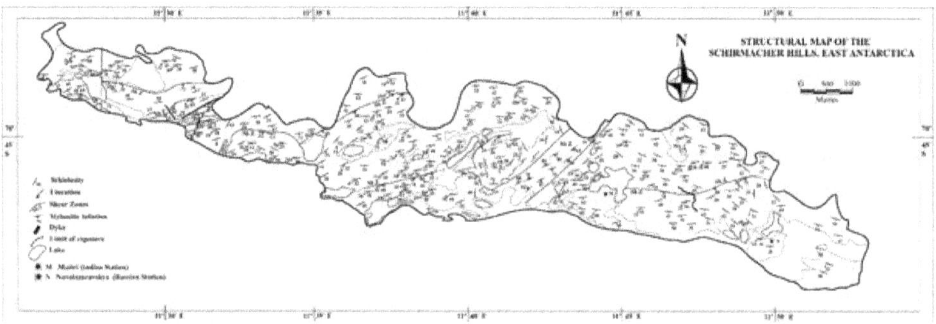

Successive later tectono-thermal events have reworked this older basement. Various steps of the dismemberment of an initial continuous layered basement and its gradual incorporation within the host rock by fragmentation, deformation and subsequent metamorphism has been traced by examining the transitional stages. A narrow mylonite zone in the host rock has developed at the contact with the enclaves. It occurred mainly because of the extreme rheological contrast between the rigidly behaving enclaves and the host rock. Incidentally, the enclaves which are the least deformed have been least affected by later metamorphism.

The second group of events includes D2 deformation. It occurred under granulite facies metamorphism (M2) and emplacement of enderbite/charnockite. The D1 foliation of the enclaves is truncated by the D2 foliation of the enveloping granulites (Figure 3). The enclaves were deformed to various stages by the D2 deformation (Figure 4). The more equant enclaves were rotated or strained and gradually brought into parallelism with the D2 foliation. The narrow-elongated enclaves along with

D1 foliation were isoclinally folded. D2 isoclinal folds are mostly reclined with an axial planar foliation, and Granulite-facies minerals marked it. It's the dominant foliation of the eastern part of the Schirmacher Hills. Ductile shearing took place in the continuation of this deformation under granulite facies conditions. The product of this shearing is not a typical mylonite but a fine to medium-grained gneiss. Syntectonic emplacement of the enderbitic or charnockitic pegmatite is common among these shear zones. Mafic dykes, broadly syntectonic with D2 deformation, were emplaced during this event. These were deformed mainly by D2 and later deformations and became nearly parallel to regional foliation (Figure-3.5). However, discordant relationships are still preserved in some places.

Figure 3. Partially migmatized banded enclaves within enderbitic gneiss. The D1 internal foliation and banding occur at an angle with the D2 external foliation. Late pegmatites transact the enclave

Figure 4. Enclaves showing the diverse relationship between inner and outer foliation. The elongate enclave with layer-parallel early D1 foliation is folded by D2 fold. The internal foliation of the upper enclave is at an angle with the foliation of the host gneiss. In the lower part, the folded enclave shows concordant foliation

Figure 5. Early mafic dyke deformed by D2 and later deformations show broad parallelism with D2 foliation of the associated granulites

M2, coeval with D2, is recorded by all the units but for the mafic granulite enclaves. Due to dehydration-melting of biotite on the prograde path of M2 in khondalite, porphyroblastic garnet (with the inclusion of spinel and biotite), sillimanite perthite, plagioclase, and quartz were stabilised. An early assemblage of scapolite, clinopyroxene, calcite, and plagioclase in the calc gneisses was stabilised. In enderbite, porphyroblastic garnet, orthopyroxene, plagioclase, and quartz were stabilised at the peak M2 condition. Application of several geothermobarometers for the core compositions yields peak P-T conditions for M2 as 8- 8.5 Kbar, 800°C.

The third group of events includes isoclinal folding (Figure 6) and a dominant axial planar foliation development. It took place under amphibolite facies conditions (M3). The D3 deformation transposed and retransposed all the earlier foliations. The D3 folds are inclined to recline with a moderate plunge of the fold axis towards the southwest. The D3 foliation has an easterly strike with a moderate dip towards the south. D3 deformation is also associated with the extensive emplacement of granitic materials. Mafic dykes, broadly synchronous with D3, are found to cross-cut the granitic gneiss with D3 foliation. These dykes are folded independently by the same deformation and show axial planar foliation parallel to the outside host gneiss (Figure 7). Synchronous ductile shear zones developed subparallel to the D3 foliation (Figure 8). Zones of intense non-coaxial deformation were often produced by shearing out from the middle limb of the asymmetric folds. Some of these shear zones may have developed by reactivation of earlier shear zones. There was a syntectonic emplacement of granitic material along these shear zones. The shear zone rocks show excellent development of mylonites. These shear zones have a thrusting sense of movement and a moderately large pitch of the stretching lineation on the mylonitic foliation. Along with a thrust zone in the central part, the granulites were stacked over the amphibolite facies rocks to produce a tectonic inversion.

Figure 6. Isoclinal D3 fold with axial planar cleavage

Figure 7. Mafic dyke broadly contemporaneous with D3, itself folded by D3 folds with axial planar foliation

Figure 8. Ductile shear zone parallel to D3 foliation within Augen gneiss

M3, synchronous with D3, is an amphibolite facies overprint on the granulite facies rocks. It resulted in the formation of cummingtonite- anthophyllite amphibole in mafic granulite enclaves. It took place at the expense of pyroxenes, biotite-quartz symplectite at the expense of garnet in khondalite, and epidote-zoisite in calc gneisses. All these reactions involve hydration. Garnet-biotite thermometry in the khondalite attests to 600-550°C at an assumed pressure of 5.5 Kbar.

Ductile shearing under granulite facies conditions produced gneisses, not mylonite. The very well foliated shear zone gneiss was made either from a massive rock or weakly foliated rock. Although there was also a reduction in the grain size of the constituent minerals, it was not as intense as to produce a mylonite. On the other hand, the ductile shearing under amphibolite facies conditions often made a mylonite. The detailed microstructure of the mylonites indicates the deformation of minerals by crystal plastic processes. The mylonitization was associated with a drastic grain size reduction. The grain size of the upper amphibolite facies mylonites was distinct. It was larger than commonly described mylonites which develop under greenschist facies conditions. Mylonitisation, under amphibolite facies conditions, involves an intense hydration reaction. Besides grain refinement, it enhanced the deformability of the sheared rock to a great extent.

Partly the variation in the microstructure of the sheared rock is due to the heterogeneous intensity of strain from domain to domain. It produced protomylonites, orthomylonites, and ultramylonites are produced.

The microstructural variation resulted from syn- to post-tectonic recrystallisation. Grain growth of constituent minerals, both quartz and feldspar, have deformed by crystal plastic processes with dominant grain boundary migration (Figure 9). The presence of weakly strained or unstrained long quartz ribbons characterises these mylonites (Figure 10). The development of quartz ribbons with the absence of significant strain suggests grain recovery and grain growth during high-temperature mylonitization.

Figure 9. K-feldspar megacryst with quartz ribbon swerving around it. The quartz ribbon does not show any strain. Length of photo 3mm

Figure 10. Strain free quartz ribbon with large aspect ratios. Length photo 3mm

Ductile shearing under amphibolite facies conditions has also taken place in the unfoliated massive rocks. In these rocks, a new foliation was produced by shearing itself. In the initial stage, a diffuse zone of non-coaxial deformation was created within the massive rocks. With progressive deformation, a narrow zone of intense strain concentration developed in the central part of the diffused zone. It is likely that with progressive ductile shearing, the width of the shear zone increases, and the shear zone foliation makes a lower angle with the shear zone border.

The fourth group of events shows upright folds with a moderately plunging fold axis towards the southwest. Two dominant sets of shear zones developed during this period (Figure 11). One set has an easterly trend and a steep dip of mylonitic foliation. The mylonitic lineation has a moderate to gentle plunge towards the southwest. The second set has a northeasterly trend with a steep dip. A wide shear zone of this type occurs in the central part of the Schirmacher Hills. The mylonitic lineation has a variable plunge from subhorizontal to gentle. It indicated g a significant component of strike-slip movement.

Figure 11. Brittle-ductile shear zones at a high angle to D3foliation

The last group of events in the Schirmacher Hills includes the development of easterly trending open upright folds. The folds developed at the waning phase of amphibolite facies metamorphism. Discrete bodies of granites and pegmatites were emplaced along some brittle-ductile shear zones that developed during this period.

The last stage of the deformation history of Schirmacher Hills is represented by late fractures and faults, which cut across all the earlier structures. A set of mafic dykes were emplaced after the cessation of all tectonic activity (Figure 12).

Figure 12. Post tectonic discordant mafic dyke cutting across early gneissic foliation

The tectono-thermal history of the Schirmacher Hills is summarised below.

1. Granulite facies metamorphism (M1) is syntectonic with D1 deformation. The signature of this event is now recognised in some isolated enclaves. The enclave rocks showed the imprint of the earliest metamorphism, M1 at ca. 10 Kbar and 900°C
2. The second phase of granulites facies metamorphism (M2) is associated with 800°C temperature and 8 to 8.5 Kbar pressure. The associated enderbite emplacement is syntectonic with transposition of an early foliation by D2 and emplacement of enderbite/charnockite and synchronous mafic dykes.
3. Emplacements of granites and pegmatites associated with amphibolite facies metamorphism (M3) are under a temperature of 600-550°C and a 5.5 Kbar pressure, broadly coeval with isoclinal folding and syntectonic development of axial planar foliation (D3). A set of broadly contemporaneous mafic dykes were emplaced during this event.
4. Later deformations further affected the rocks, producing two open upright folds (D4 and D5) under amphibolite facies of cross-cutting ductile shear zones.
5. The sheared rocks under granulite facies conditions do not produce mylonites. The amphibolites facies mylonites show charactcristics of high-tcmpcraturc mylonites.
6. Mafic dykes were emplaced with each tectono-thermal event and even after cessation of all tectonic activities, which led to brittle shear zones.

Given the foregoing, the following are the significant scientific achievements:

1. Detailed structural and lithological mapping of the Schirmacher Hills.
2. Identification of different deformation episodes, metamorphism, migmatization, and emplacements of several generations of mafic and felsic bodies.
3. Correlations of successive generations of ductile shearing with other tectono-thermal events.
4. Identification of enclaves of early basement as enclaves within the present basement

The multiple episodes of deformations and related metamorphism were identified and characterised; successive generations of ductile and brittle shear zones were studied in detail, and their time correlations were established. The structural study of the Schirmacher hills and the rocks' geothermobarometry was also studied in great detail. The tectono-thermal history of the Schirmacher Hills area was established. This study has formed the basis of all later studies. Identification of an early basement as an enclave and detailed research on the shear zones were also carried out.

REFERENCES

Bose, S., & Sengupta, S. (2003). High-temperature mylonitization of quartzofeldspathic gneiss: An example from the Schirmacher Hills, East Antarctica. *Gondwana Research*, *6*(4), 805–816. doi:10.1016/S1342-937X(05)71026-1

Dasgupta, S., Sengupta, S., Bose, S., Fukuoka, M., & Dasgupta, S. (2001). Polymetanorphism in the Schirmacher Hills granulites, East Antarctica: Implication for the tectono-thermal reworking of an isobarically cooled deep continental crust. *Gondwana Research*, *4*(3), 337–357. doi:10.1016/S1342-937X(05)70334-8

Grew, E. S. (1983). Saphirine-garnet and associated paragneisses in Antarctica. In R. L. Oliver, P. R. James, & J. B. Jago (Eds.), *Antarctic Earth Science* (pp. 40–43). Cambridge University Press.

Grew, E. S. (1983). Saphirine-garnet and associated paragneisses in Antarctica. In R. L. Oliver, P. R. James, & J. B. Jago (Eds.), *Antarctic Earth Science* (pp. 40–43). Cambridge University Press.

Grew, E. S., & Manton, W. I. (1983). Geochronologic studies in east Antarctica: Reconnaissance uranium-thorium-lead data from rocks in Schirmacher Hills and Mount Steiner. *Antarctic Journal of the United States*, *18*, 6–8.

Paech, H. J., & Stackebrandt, W. (1995). Geology. In P. Borman & D. Fritzsche (Eds.), *The Schirmacher Oasis, Queen Maud Land, East Antarctica, and its surroundings. Gotha-Perthes.*

Ravich, M.G. & Kamenev, E. N. (1975). *Crystalline basement of the Antarctic Platform.* Academic Press.

Sengupta, S. (1986). *Geology of Schirmacher Range (Dakshin Gangotri), East Antarctica.* Department of Ocean Development, Technical Publication No. 3, 187-217.

Sengupta, S. (1988a). History of successive deformations in relation to metamorphism-migmatitic events in the Schirmacher Hills, Queen Maud Land East Antarctica. *Journal of the Geological Society of India, 32*, 295–319.

Sengupta, S. (1988b). Precambrian rocks of the Schirmacher Range, East Antarctica, Z. Geol. *Wiss. Berlin, 16*, 647–660.

Sengupta, S. (1993). Tectono-thermal history recorded in mafic dykes and enclaves of the gneissic basement in the Schirmacher Hills, East Antarctica. *Precambrian Research, 63*(3-4), 273–291. doi:10.1016/0301-9268(93)90037-3

Sengupta, S. (1994). *Mylonites of the Schirmacher Hills, East Antarctica.* Department of Ocean Development, Technical Publication No. 6, 173- 187.

Sengupta, S. (1997). Contrasting fabrics in deformed dykes and host rocks: natural examples and a simplified model. In S. Sengupta (Ed.), *Evolution of Geological Structures in micro to macro-scales* (pp. 293–319). Chapman and Hall. doi:10.1007/978-94-011-5870-1_18

Sengupta, S., & Bose, S. (1997). Development of successive generations of shear zones in the Precambrian Basement of Schirmacher Hills, East Antarctica. *Gondwana Research, 1*(1), 35–45. doi:10.1016/S1342-937X(05)70004-6

Chapter 4
Pollen-Spore Transport Into Antarctica and Possible Past Climatic Oscillations

Samir Bera
Birbal Sahni Institute of Palaeobotany, India

ABSTRACT

During XIX-XX Indian Antarctic Expedition (IAE), two sediment cores were collected. The occurrence of the exotic pollen taxa like Larix, Ulmus, grasses, few herbs, local moss spores, and other cryptogams prove the activity of the palaeowind, which in turn caused transportation microbiota from a long distance. The study of pollen-spores accumulation in polar air (deduced from slide exposures from 40° S) to Antarctic mainland studied for two years (2000 and 2001) as well as surface deposits (45 moss tuft, 15 lichen samples, 10 frozen soils, 10 moraine matrix) is well corroborated with lake sediment study. Ten bulk ice samples from the Antarctic ice cap (5 litres of ice melt) from Schirmacher Oasis were studied to recover trapped palynodebris to understand the depositional pattern of various microbiota in the Antarctic ice sheet.

INTRODUCTION

The glaciation and deglaciation of Antarctica (presently covered by ice 3000-4000m thick) plays a vital role in the Earth's climatic history. The Schirmacher Oasis lies between 70°46'40" and 70°43'50" S latitude and 11°22'40" and 11°54'25" E longitude in Eastern Antarctica, covering about 35 sq. km area ranging between 0-228 m altitude with an average of about 100 m above sea level. It is an ice-free Oasis that a huge

DOI: 10.4018/978-1-6684-4078-0.ch004

Continental Ice Sheet bounds to the South and Ice Shelf to the North (Figure 1). The gentle slopes and plain areas are covered with a thin blanket of the moraine matrix. The Precambrian crystalline basement of the East Antarctic platform is exposed in the Oasis. Several distinctly younger basalt (dolerite) dykes (Ravindra, 2001; Sengupta, 1986). Surface water is restricted within small to large water bodies like pools or lakes in the Oasis, more than 100 in number. Based on the aerial survey, lakes can be grouped into three classes,' i.e. Proglacial lakes, inland lakes & epishelf lakes (Ravindra *et al.*, 2001). Antarctica is floristically barren, except for only two vascular plant species like *Deschampsia antarctica* and *Colobanthus quitensis*, which grow in the Antarctic Peninsula zone. Other lower groups of terrestrial plants like mosses, viz., *Bryum* sp, *Polytrichum alpinum, Drepanocladus uncinatus,* etc., and lichens, both crustose, viz., *Acarospora, Rhizocarpon, etc.,* along with foliose type (*Umbilicaria* spp) can grow especially on frozen pattern ground, rocky substratum & nearby lakes. Despite these, other aquatic elements within the available water bodies are alga, desmids, diatoms, and dinocysts. By understanding the ecological background (e.g. varying pollen-spore production of different species, dispersal, and transport mechanism), the information in the pollen diagram can be interpreted in terms of climatic change, changes in ecology, and the result of human interference. Determining the absolute frequency of pollen-spores in the deposit within a given time interval interprets the data more reliable and more detailed. The present study may help make a palyno-dataset to reconstruct palaeoclimate in the polar region.

The ocean, atmosphere, and cryosphere interplay in the polar regions make Antarctica a key modulator of rapid climatic change. The subtle signatures of climate change occurring over these polar regions are preserved in different natural deposits such as ice cores and lake sediments. The Schirmacher Oasis in East Antarctica contains terrestrial sediments (especially frozen peat deposits), mosses (in moss bank), microbial mats and algal flakes, along with polar ice and marine sediments. It, therefore, represents the best climatic archive of the past, which retain the centennial to millennial year's climatic record in them. This chapter aims to understand the distribution of airborne palynoflora in Antarctic terrestrial deposits.

In the first phase and subsequently understand the transported pollen–spore contents from deep lake core sediments that could be used in deciphering Holocene climatic oscillations along with vegetation successions in and around Schirmacher Oasis, Eastern Antarctica.

METHODOLOGY

Palynological samples including moss turfs (45), lichen patches (15), frozen peat soil (10), polar ice (10) and moraine matrix were procured from near lake sites, dry lake beds and nunatak (Veteheia) around Schirmacher Oasis of Eastern Antarctica (Figure-4.1). 50 grams of moss soils and 5-10 grams of lichen soil were analysed for pollen to assess accurate relative frequencies of various microbiota in the measured sediment. Two lakes, namely Zub lake (70°45'39.4"11°44'8.6" E) and Long lake (70°45'20"11°4'E), were chosen for coring lake sediments during XIX and XX Indian Scientific Expeditions for palynostratigraphical studies using HYDROBIOS gravity corer (Kiel, Germany). Special care was taken for samples used for radiometric dates collected at an interval of 10 cm. The extraction of pollen and spores was done by the standard maceration technique (Erdtman, 1943). Besides, the heavy liquid method is also followed, where very little organic matter and pollen were found. The morphological study of palynodebris was made by using the Olympus BX 50 Microscope. The total pollen count in this study ranges from 90-150, depending upon the productivity of the samples. The Burkard volumetric air sampler was employed for air sampling from 40ºS of Cape town to Antarctica mainland and from December 2000 to March 2001.

Figure 1. Location of sampling sites, Schirmacher Oasis, East Antarctica

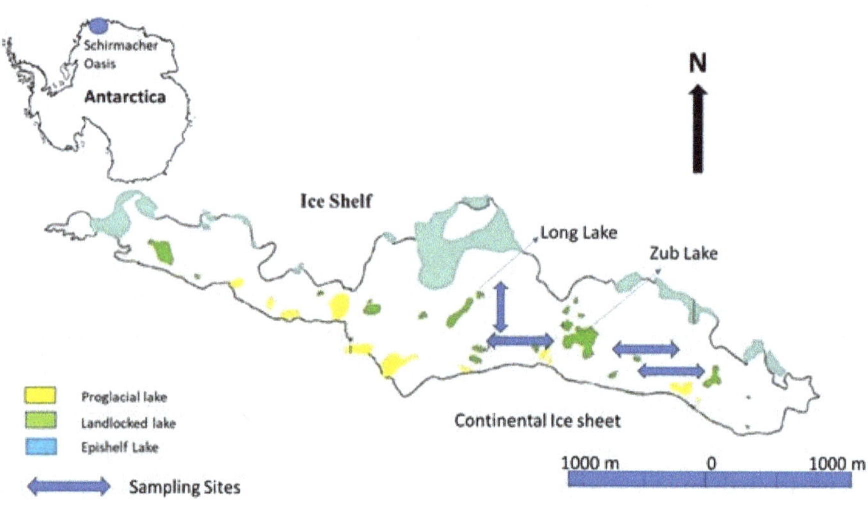

MAIN FINDINGS

The significant scientific findings of the palynological investigations of Schirmacher Oasis are detailed below

Spatial Variations and Assessment of Modern Environmental Conditions

Pollen – Spore Transport in Air Samples

The pollen grains recovered from the air catches belong to the families like Rosaceae, Oleaceae, Fabaceae, Chenopodiaceae-Amaranthaceae, Asteraceae, and Poaceae.

In low concentration Alternaria, Helminthosporium, Curvularia, Cladosporium, and Cercospora, we found the major fungal groups and small hyaline and coloured spores. The study is indicative of the regular transport of microbiota in the polar air across the Antarctic mainland. A comparative frequency of various aero-palyno debris from different air samplers and snow samples indicates the influence of local and exotic plant taxa. It further suggests that long-distance transport of small particles has occurred frequently in recent times.

Terrestrial Deposits and Pollen Analysis: Assessment of pollen-spore dispersal and deposition formulation of the pollen deposition model is a prerequisite in the first phase where the generated pollen data is designated as the proxy climate signal for the interpretation of past vegetation and environment covering forty-five moss turfs, fifteen lichen samples, ten each of frozen soil, and moraine matrix. The pollen productivity in decreasing order is moss peat, lichen, frozen soil, and moraine matrix, respectively. Non-arboreal taxa were found dominant over arboreals like Larix, Pinus, Podocarpus, Betula, Ulmus, etc., which mainly belong from subtropical to the temperate region a few are of tropical as well. As the study area is devoid of any plant taxa recorded from sediments, it may be inferred that the occurrence of the same indicates long-distance pollen transport by upthermic winds from distant islands and subcontinents. The occurrence of Antarctic grass sp. *Deschampsia Antarctica* and the herb *Colobanthus quitensis* in the sediments in low frequency are significant as they are not growing in the study area. Although detailed studies are needed to confirm this observation, the pollen morphological characters of the grass in the present sediment resemble the Antarctic grass of the peninsular region. On the other hand, *Cosmarium, Navicula, Nitzchia, Hantzchia, Pinularia, Fragilaria*, etc., among cryptogams, are local in origin.

The palynodata generated from the present study reflects that the accumulation of pollen grains is comparatively more in the moss sediments procured from valley areas than near the shelf. The reason might be the prevalence of heavy air mass blowing over towards valleys across the Southern Ocean, which do not allow deposition of most of the palynodebris immediately near the shelf. Only foliose form (Umbilicaria sp.) is more productive than the crustose form in lichen-based soil, which holds more minor underneath the soil. The study of bulk ice samples has generated fascinating pollen data. The possible source of pollen–spores trapped under the huge Antarctic glacial ice sheet- gets released after melting in summer and flows through a glacial stream, and gets deposited under various surface deposits under moss & lichen patches and finally to the bottom of the lake. The presence of fungal spores and varied dispersed organic matter is indicative of their saprophytic nature, which could have been transported along with the drifted elements and been incorporated into the sediments. The occurrence of airborne microbiota in the air catches strongly support the pollen dispersal & deposition in surface deposits of the polar region.

TEMPORAL VARIATIONS AND PALEOCLIMATOLOGY

In the present chapter, some efforts have been made to reconstruct past vegetation succession with corresponding climatic oscillations during the recent past by studying lacustrine sediments from the polar region. Out of two major lakes, the Zub lake has an average depth of 6-7 meter, and it is about 0.75 sq. km in area, and Long lake comprises the approximately 0.55 sq. km area having an average depth of 4-5 meter. These two sediment cores from Zub lake (60 cm) and Long lake (50 cm) were radiometrically dated. Out of which the Zub lake comprises 7130±140 years BP at 55 cm, 5110±110 years BP at 35 cm, and 3000 years BP at 15 cm level (calibrated) based on the sedimentation rate and Long lake shows 2000 ±90 years BP at 50 cm and 1500 ±110 years BP at 30 cm level, respectively (Table-4.1). The sedimentation rate is assessed to be 4.02 cm / 200 years. The core comprises alternate silty clay layers with organic matter and algal layers with pH range from 7-8 (Figure-4.2). The presence of Illite, Chlorite and a minor amount of mixed layered minerals in the lake sediments are due to weathering in the lake catchments (Sinha and Chatterjee, 2000). Based on the fluctuation in the frequency of different microbiota preserved in lake sediments during the last 8000 years of BP (Figure-4.2 and 4.3), three distinct palaeoclimatic oscillations are reported.

Table 1. Late Holocene climatic history and vegetation succession

Sampling site	Water depth (m)	Sample interval (cm)	Deposition type/texture	Radiocarbon dates (YBP)	Climate
Zub Lake	6.5 to 7.2	0 – 10	Algal layer		
		11-15	Silty clay	3000-Recent	Warm &
(70o45'39.4" S: 11 o 44'8.6" E)		16-30 31-40	Algal layer		more humid
				5000	Warm &
		41-55 56-60	Algal layer		humid
			Silty Clay	8000	Cold & Dry
Long Lake			Algal layer		
(70o45'20" S:	4.8-5.5	0-10	Silty clay	1500-Recent	Warm &
11 o 40' E)		11-20	Algal layer		humid
		21-25	Silty clay		
		26-35	Algal layer		Cold &
		36-45	Silty clay	2000	humid
		46-55			

Figure 2. Texture of sediment and Radiocarbon dates of polar lakes (BP), Schirmacher Oasis

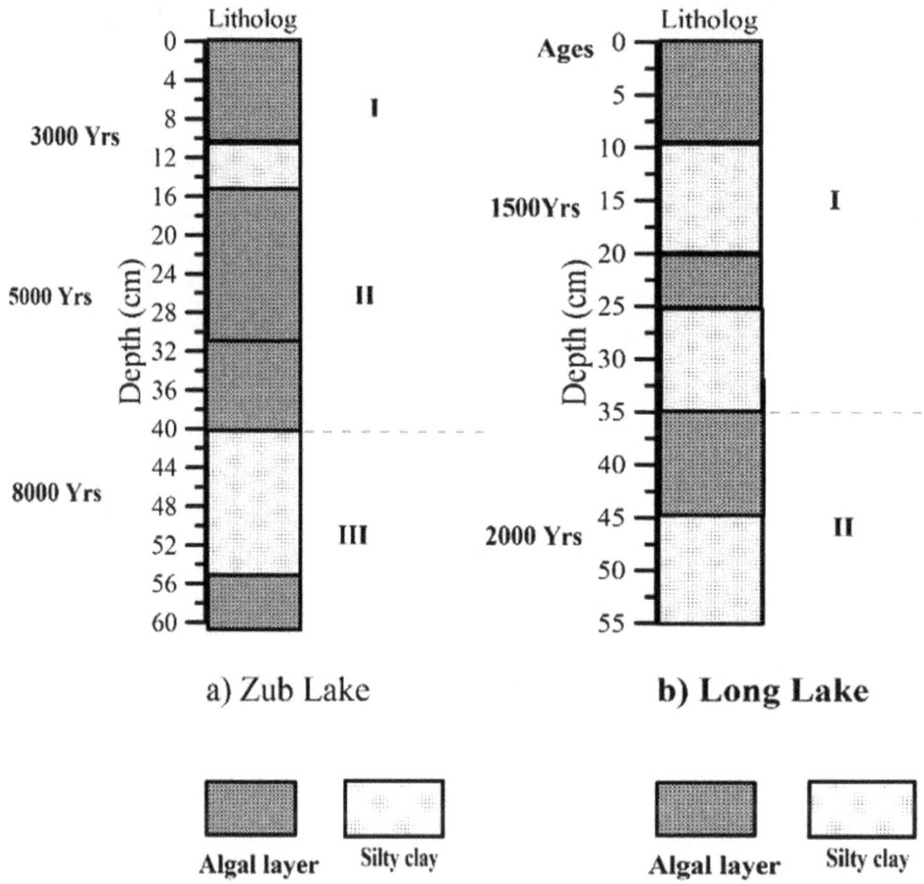

Climatic Oscillation for 8000 Years BP (Zub Lake) Pollen Zone I

This phase is characterised by the occurrence of low to a moderate value of *Cosmarium* (Av 8%), followed by a preponderance of Acritarch's (Av 40%). Tree taxa, like Larix, along with other herbs, are recorded in high value (Av 30%). The grass pollen recorded is recorded to have an average value of 7%. Moss spores are restricted to 10%. Fungal remains (Av 5%) encountered in low value are indicative of being saprophytic. The overall assemblage that has emerged from this phase indicates an arid climate. Due to which the lake became shallow and small in expansion for 8000 years BP.

Climatic Oscillation during 5000 Years BP (Zub Lake) Pollen Zone II

Compared to the preceding phase, this phase is characterised by the increased value of moss spores (Av 25%) along with other angiosperm pollen such as *Larix,* followed by *Betula, Podocarpus,* and Ttubuliflorae (Av 15%). The value of *Cosmarium* (Av 22%) has improved, whereas diatom and other algal elements appeared for the first time in low values. The decreased value of Acritarchs (Av 5%) and the high value of *Cosmarium* (Av 22%) indicate a widening of the lake. During this phase, grasses have improved their values (Av 25%) compared to the preceding phase. The occurrence of fungal fruiting body/spores (Av 8%) and moss spores supports humid conditions prevailed during this phase. The overall palynoassemblage is indicative of warm and humid climatic conditions during 5000 years BP.

Climatic Oscillation during 3000 Years BP (Zub Lake) Pollen Zone III

This phase is characterised by the invasion of more angiosperm broad-leaved taxa like Moraceae and Colobanthus, along with Betula and Larix (Av 35%), that support the amelioration of climate. The high value of local moss spores (45%) and fungal elements (Av 8%) indicates more humid conditions during the phase. *Cosmarium* (Av 10%) and Acritarchs (2%) are represented in low profile. The palynoassemblage suggests a prevailing warm climatic period and more humid conditions for 3000 years BP and lasts till the present.

Climatic Oscillation during 2000 Years BP (Long Lake) Pollen Zone I

The occurrence of a moderate value of *Cosmarium*, algal cysts, along diatom (Av 15%) and Acritarchs (Av 10%) is the main characteristic of this phase. Arboreal taxa like Betula, Moraceae are recorded to have a value of %. Whereas, among non-arboreal taxa, grasses dominated by 20%. Chenopodiaceae and Caryophyllaceae are recorded as herb elements which are very low-frequency moss spores (Av 30%) and are represented in high value. Fungal spores and fruiting bodies have also been found (10%). The palynoassemblage indicates a cool & humid climate, in which conditions the lake became shallow and small in expansion for 2000 years BP.

Climatic Oscillation during 1500 Years BP (Long Lake) Pollen Zone II

This phase is characterised by the increased value of Moss spores (Av 40%) compared to the preceding phase. The widening of the lake is reflected in the increased value of *Cosmarium* (up to 18%) and the low value of Acritarchs (up to 6%). The grasses Deschampsia & Poaceae have improved (Av 25%) compared to the preceding phase during this phase. The total assemblage indicates a cool & more humid climate. During this phase, the appearance of *Larix, Betula, Podocarpus*, Oleaceae, Moraceae, and few herbaceous taxa in low values (Av 9%), depicts amelioration of the climate at the upper zone. Fungal fruiting body and spores were encountered in the frequency of 2%. High moss spore's occurrence in the assemblage during this phase supports highly harsh climatic conditions since 1500 years BP. It lasts till the present with a little alteration of climate, i.e. the onset of warming. Though the database from only two shallow lake core sediment profiles is not adequate to understand the past vegetation history in the polar region, the preliminary findings could provide a clue to deduce the palaeoclimatic oscillations since the Last Glacial Maxima (LGM). The database which has emerged from the present study supports the data of a few earlier Scientists who worked on the transport of exotic palynomorphs into the other parts of the Antarctic region (Scott and Venzinderem Bukker, 1985; Kappen and Straka, 1988; Smith, 1991; Wynn-Williams, 1991; Van der Knaap *et al.*, 1993; Bera and Khandelwal, 2003 and Bera, 2004, 2005).

CONCLUSION

It may be concluded that little is known about the aerial transport of microbes, plants, and other biotas into and between Antarctic terrestrial and freshwater habitats. However, a recent study has yielded valuable terrestrial proxy signals to decipher the reconstruction of the polar climate. An assessment of the pollens and productivity of lake sediments was made. The long-distance transport of thermophilic bioparticles (Figure-4.4) has occurred in the polar region. Now, we seem to be in a climatic period of very strong circulation with frequent episodes of alien pollen and dust influx. Even if the Holocene climate history is not detailed, we know that the air circulation pattern has fluctuated. Hence, in an extreme case, we cannot exclude that events such as those described here have occurred periodically since the last glacial retreat, and meteorological situations making such events possible must occur several times a year. The aerobiological require international cooperation. The Scientific Committee on the Antarctic Research (SCAR), Biological Investigation of Terrestrial Antarctic

Systems (BIOTAS) research network have identified aerobiology as a significant component of their international research program.

On the other hand, the problem of obtaining radiometric dates (i.e., close in age to the actual time of deposition) from the Antarctic Continental Margin has been discussed by few workers (e.g., Stuiver *et al.*, 1981). There is no scarcity of potential tools for past climatic studies. Many lakes, significantly deeper landlocked and epishelf lakes (Figure-4.1), moss peat, sediment traps, and glacial clay verves exist in palaeovalley areas in vast Schirmacher Oasis and higher mountains for understanding local peat development, nature of palaeowind current and plant colonisation in and around the polar region. Samples of moss bank peat have been found to give the most reliable radiocarbon ages in Antarctica (Bjorck *et al.*, 1991). They are thus, optimal for the construction of polar palaeoclimate up to the past 5000 years. It is found that the overall extent of ice cover in Antarctica during LGM is not well known, and existing reconstructions are controversial. Anderson *et al.* (2002) reported that the East and West Antarctic Ice Sheet (WAIS) have not advanced and retreated in the concert.

In contrast, in the LGM, WAIS advanced for the most part to the outer shelf, and the EAIS did not expand to the continental shelf edge during the LGM. Another critical issue is the dating of terrestrial materials from the polar region, as evidenced by several sources of contamination when dating bulk sediments from Antarctic lake basins, often resulting in ages too old (Adamson and Pickard, 1986). Now what we need is the study of better chronologically resolved sequences and the refinement of existing ones. These would undoubtedly provide more clues for a complete climatic scenario for the last 20,000 years (Last Glacial Maxima). In recent years, models of present-day atmospheric circulation in the higher Southern latitudes stressing, for example, the importance of Antarctica and its apron of sea ice (Simmonds, 1981), illuminate our understanding of the climate system and warrant attention. Many Climate Models apply to the late Quaternary (Manabe and Broccoli, 1985) and can be the basis for palaeoclimatic theory (Kutzbach, 1985).

Figure 3. Larix; 2. and 7 Oleaceae; 3. Ulmus; 4. Caryophyllaceae; 5. Urticaceae-Moraceae; 6. Poaceae; 8. Unknown pollen (x 2000); 9. Spore of Mycobiont of Lichen; 10. Asteraceae (Tubuliflorae); 11. 16&20. Different stages of Alternaria; 12. Cosmarium (x500); 13. Unknown; 14. Curvularia; 15. Algal filament (Oscillatoria); 18. Unknown (x 2000); 19. Fungal spore

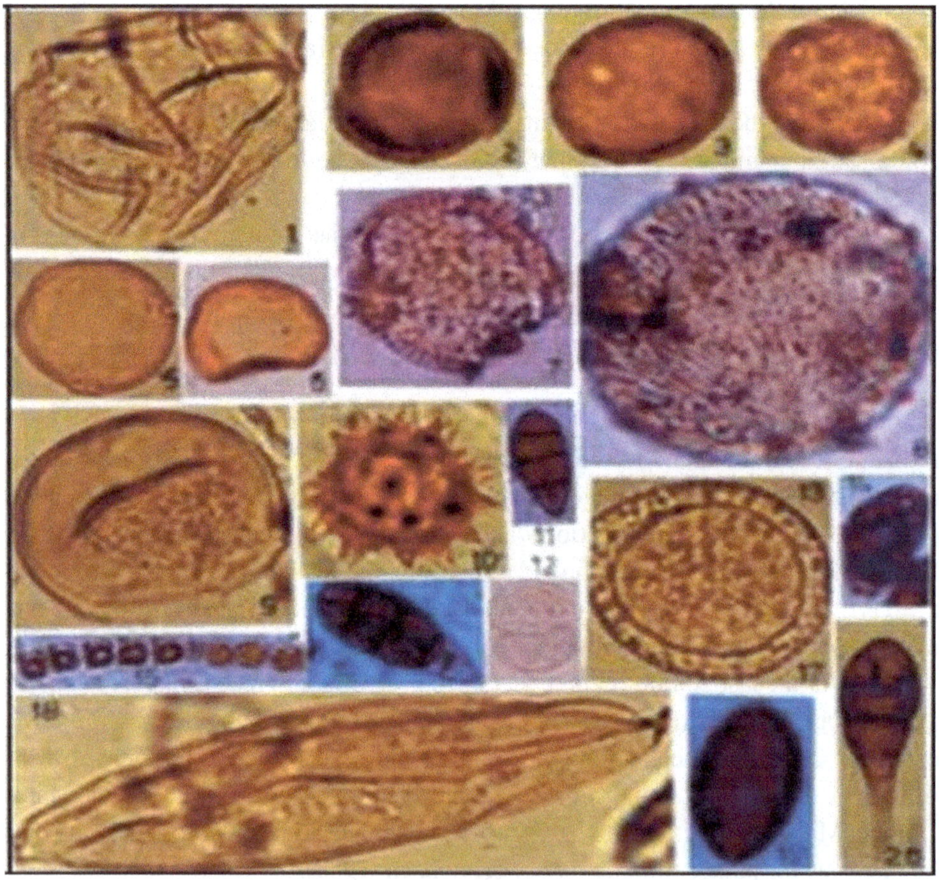

All Figures x 1000 otherwise stated

ACKNOWLEDGMENT

The author thanks the Director of BSIP for constant support to conduct the research project. The author also thanks to the Director of NCAOR, Goa and the Secretary, Ministry of Earth Sciences, New Delhi, for their kind support of this study. The author expresses his gratitude to Dr Arun Chaturvedi and Mervin J. D'Souza, leaders of expeditions and Navy personnel, for extending all possible cooperation during the coring operation of the lake sediments. The team of NZ Helicopter (AS350 SQUIRREL) is acknowledged for flying to the interiors of the mountain areas to collect samples and carry boat and coring apparatus to the lake sites. The author is also thankful to Senior Technical Officers, BSIP, namely Dr B. Sekar and Mr Indra Goel, for providing radiocarbon dates and chemical analysis of the polar lake sediments.

REFERENCES

Adamson, D., & Pickard, J. (1986). Cenozoic history of the Vestfold Hills. In J. Pickard (Ed.), *Antarctic Oasis* (pp. 63–98). Academic Press.

Anderson, J. B., Shipp, S. S., Lowe, A. L., Wellner, J. S., & Mosola, A. B. (2002). The Antarctic Ice Sheet during the Last Glacial Maximum and its subsequent retreat history: A review. *Quaternary Science Reviews*, *21*(1-3), 49–70. doi:10.1016/S0277-3791(01)00083-X

Bera, S. K., & Khandelwal, A. (2003). Modern pollen – spore rain in Schirmacher Oasis, East Antarctica. *Current Science*, *85*(2), 137–140.

Bera, S. K. (2004). Late Holocene palaeo-winds and climatic changes in Eastern Antarctica as indicated by long-distance transported pollen-spores and local microbiota in polar lake core sediments. *Current Science*, *86*(11), 1485–1488.

Bera, S. K. (2005). Recovery of airborne palyno debris from continental ice sheet qasis, East Antarctica. *Current Science*, *88*(10), 1550–1552.

Björc, K., Hjort, S., Ingólfsson, C., & Skog, G. (1991). Radiocarbon dates from the Antarctic Peninsula region - problems and potential. *Quaternary Proceedings*, *1*, 55–65.

Erdtman, G. (1943). An introduction to pollen analysis. Academic Press.

Kappen, L., & Straka, H. (1988). Pollen and spores Transport into the Antarctic. *Polar Biology*, *8*(3), 173–180. doi:10.1007/BF00443450

Kutzbach, J. E. (1985). Modelling of palaeoclimates. *Advances in Geophysics*, *28A*, 159–196. doi:10.1016/S0065-2687(08)60223-X

Manabe, S., & Broccoli, A. J. (1985). The influence of continental ice sheets on the climate of the ice age. *Journal of Geophysical Research*, *90*(D1), 2167–2190. doi:10.1029/JD090iD01p02167

Ravindra, R. (2001). Geomorphology of Schirmacher Oasis, East Antarctica. In *Proc. Symp. On snow, Ice and Glaciers*. Geological Survey of India, Spl. Publication. No. 53, 379-390.

Ravindra, R., Chaturvedi, A., & Beg, M. J. (2000). Meltwater lakes of Schirmacher Oasis – Their genetic Aspects and classification. In D. Sahoo & P. C. Pandey (Eds.), *Advances in Marine and Antarctic Sciences* (pp. 301–314). APH Publishing Corporation.

Scott, L., & Van Zinderen Bakker, E. M. Sr. (1985). Exotic pollen and long-distance wind dispersal at a sub-Antarctic Island. *Grana*, *24*(1), 45–54. doi:10.1080/00173138509427422

Sengupta, S. (1986). *Geology of Schirmacher Range (Dakshin Gangotri), East Antarctica*. Department of Ocean Development. *Technical Publication*, *3*, 187–217.

Simmonds, I. (1981). The effect of sea ice on a general circulation model of the Southern Hemisphere. *International Association of Hydrological Sciences Publication*, *131*, 193–206.

Sinha, R., & Chatterjee, A. (2000). Lacustrine sedimentology in the Schirmacher Range Area, East Antarctica. *Journal of the Geological Society of India*, *56*(3), 9–45.

Smith, R. I. L. (1991). Exotic sporomorphs as indicators of potential immigrant colonists in Antarctica. In M Hjeimroos, S. Nilson, & G.E. Ghazaly (Eds.), *Proc. 4th International Conference of Aerobiology* (pp. 313-324). Grana.

Stuiver, M., Denton, G. H., Hughes, T. J., & Fastook, J. L. (1981). History of the Marine Ice Sheet in West Antarctica during the last glaciation: A working hypothesis. In G. H. Denton & T. J. Hughes (Eds.), *The last great Ice Sheets* (pp. 319–439). Wiley & Sons.

Van der Knaap, W. O., & Jacqueline, F. N. (2019). A recent pollen diagram from Antarctica (King George Island, South Shetland Islands). *The Holocene*, *3*(2), 169–173. doi:10.1177/095968369300300209

Wynn–Williams, D. D. (1991). Aerobiology and colonisation in Antarctica-the BIOTAS program. *Grana*, *30*(2), 380–393. doi:10.1080/00173139109431994

Chapter 5
Integrated G3 Investigations in Antarctica

H. K. Gupta
National Geophysical Research Institute, India

G. S. Mittal
National Geophysical Research Institute, India

S. K. Verma
National Geophysical Research Institute, India

D. V. Divakaria Rao
National Geophysical Research Institute, India

M. S. Joshi
National Geophysical Research Institute, India

R. L. Dhar
National Geophysical Research Institute, India

S. C. Jain
National Geophysical Research Institute, India

A. V. Ramana Rao
National Geophysical Research Institute, India

P. Rama Rao
National Geophysical Research Institute, India

T. Madoom Hussain
National Geophysical Research Institute, India

D. V. Subba Rao
National Geophysical Research Institute, India

S. S. H. Jafri
National Geophysical Research Institute, India

V. N. Sarma
National Geophysical Research Institute, India

B. Vijayagopal
National Geophysical Research Institute, India

DOI: 10.4018/978-1-6684-4078-0.ch005

K. N. S. Reddy
National Geophysical Research Institute, India

M. B. Vyageshwar Rao
National Geophysical Research Institute, India

V. M. Tiwari
National Geophysical Research Institute, India

Md. Ghous Uzzaman
National Geophysical Research Institute, India

ABSTRACT

CSIR-NGRI has been carrying out integrated G3 investigations in Antarctica since the second IAE. The geophysical studies of the initial 25 years of IAE included surface and helicopter-borne magnetic, EM, seismic, gravity, and paleomagnetic surveys. A total of 60 line-km magnetic profiles over the ice-shelf revealed the magnetic characteristics of the bedrock beneath the ice cover. Based on these and in conjunction with the early seismic studies, a crustal structural model was evolved. Secondary sulfide mineralisation near a suspected fissure zone in Schirmacher Oasis (SO) was inferred. It was based on the multi-frequency EM and radiometric measurements. The helicopter-borne magnetic N-S profiles over an area of 100x100sq.km with a spacing of about 3.0 km between SO and Wohlthat Mountains (WM) yielded gross features of sub-glacial topography nunataks (exposed peaks of sub-glacial hills).

INTRODUCTION

National Geographical Research Institute (NGRI), Hyderabad, has been participating in all the Indian Antarctic Expeditions (IAE) since the second (IAE) to carry out integrated geophysical, geological and geochemical (G^3) investigations in Antarctica encompassing the entire gamut of Earth Sciences. The geophysical investigations were launched from the second (IAE) (1982-83) in the first Indian Station, "Dakshin Gangotri" (DG), over the ice-shelf. After commissioning India's second station, "MAITRI" in Schirmacher Oasis (SO), the investigations carried out magnetic and seismic surveys over the ice shelf around DG (Verma *et al.,* 1987a; 1987b; Verma and Mital, 1988). Subsequently, heliborne magnetic surveys were conducted over the glaciated region between SO and Wohlthat Mountains (WM) (Verma *et al.,* 1988, Mital and Verma, 1989, Mital and Verma, 1991, Verma *et al.,* 1999). It was followed by EM, seismic, and gravity surveys in the subsequent expeditions and paleomagnetic, radiometric and geochronological studies on the rock samples obtained from around the SO region (Verma *et al.,* 1987c, Verma *et al.,* 1987d). Geological and geochemical studies were carried out in the Schirmacher Oasis and parts of Wohlthat ranges viz., Gruber Massif, Petermann I and Humboldt regions to develop an evolutionary model and finally compare these results with the Indian counterparts (Verma *et al.,* 1988). The geochronological studies were also conducted to supplement the results obtained from geological and geochemical analyses (Verma, 1988). Further details of these investigations are summarized by Verma and Mital, (1988), Verma and Mital (1994), Verma (1993), Gupta and Verma (1995) and Verma *et al.,* (2004).

During the XXIII IAE, NGRI carried out Absolute Gravity measurements for the first time in Antarctica. Since earlier gravity surveys were mainly dependent on the measurements by gravimeters that are known to have drifted with time, it was felt that a Reference Absolute Gravity Station had to be established at Maitri. The gravity field of the Earth is sensitive to both vertical position and mass distribution. Thus, it provides a powerful tool to study crustal dynamics. High precision (~5 μGal) absolute gravity measurements are essential in the time and space domain. These are essential to understand the redistribution of mass and vertical crustal motions caused due to several geodynamic processes. Some of them are earthquakes, plate convergence, isostatic rebound, etc. A Portable and high precision (1μGal) absolute gravity meter (FG5 measuring system) has been developed during the last decades. Thus, absolute gravity measurements are widely used to address different geophysical problems. For example, the crustal deformation studies in conjunction with GPS for complimentary verification or the analysis of mass and elevation change due to glacial rebound (Wahr *et al.,* 2000). In addition to the uses mentioned above, the essential requirement of any gravity survey is the absolute value at least at one point. The antarctic landmass has very few absolute gravity values, and none of them is near the Indian station (Maitri). This chapter discusses India's significant scientific efforts and achievements in the contemporary fields of integrated geophysics, geology and geochemistry (G3) during the initial twenty-five years of the annual launching of expeditions to the icy continent.

The Scientific Importance of Antarctica

The history of the evolution of Antarctica itself makes an exciting and compelling scientific study. To quote a few interesting facts, about 200 million years ago, Antarctica was near the equator and surrounded on three sides by Australia, India, Africa, and South America.100 million years ago, the Gondwana broke up, and the continents began shifting in opposite directions. 60 million years ago, Antarctica moved close to the South Pole, and oceans developed between Antarctica, Africa and India. Australia and Antarctica began to separate, and 45 million years ago, Antarctica continued to drift towards the South Pole and became surrounded by oceans. Understanding geodynamics is essential. It includes plate tectonics, seismicity, plate boundary reorganisations, and crustal deformation in the south of the Indian peninsula and the Antarctic region. The driving mechanisms and the response of the Indian Ocean Lithosphere are the primary interests of NGRI and -is its major thrust area.

Plate-boundary deformation occurred a long time ago. Recent research shows that the Antarctic continent is still tectonically active, with recent (post-glacial) crustal movement in the Trans-Antarctic Mountains occurring along fault structures formed during the earlier Ross Orogeny. During Gondwana formation, the voluminous intrusive and extrusive magmatism that occurred during the Jurassic period of the Mesozoic era, approximately 180 million years ago, are essential to understand. The vast volumes of magma were produced in many southern hemisphere continental areas during precursor events leading to the supercontinent's fragmentation. The very formation of Antarctica - is the basis to investigate the geological development of the plate margins of Gondwana and the evolution of the unique flora and fauna of the southern lands and oceans. The outcrops in Antarctica are good sites to study mid-crustal igneous and metamorphic processes at an ancient plate boundary. The geology of the landmasses around the Southern Ocean chronicles the development and breakup of the Gondwana Supercontinent and the subsequent development of their unique flora and fauna. The island arcs and microcontinents, which created the Pacific margin of Gondwana, are well displayed in Antarctica's geology and other remains of the Gondwana split. Fossil remains in deposits from these landmasses chronicle the evolution of the southern fauna and the development of the Southern Ocean. The superb outcrop in Antarctica reveals details of geology more vividly than anywhere else on Earth.

Given the above scientific importance, the following activities have been undertaken under broad themes of Geophysical, geological and geochemical investigations; GPS and Seismic Studies; Geodynamical Processes and Absolute Gravity Measurements

- Seismic and magnetic investigations on the ice-shelf
- Magnetic, radiometric and multi-frequency measurements in and around the Schirmacher Oasis region
- A helicopter-borne magnetic survey between Schirmacher Oasis and Wohlthat Mountain
- Helicopter supported gravity surveys around the Schirmacher Oasis region.
- Establishment of gravity stations around Maitri
- Establishment of a permanent seismological station at Maitri
- Establishment of a permanent GPS station at Maitri
- Paleomagnetic studies on lamprophyre dykes of Schirmacher Oasis region
- Geochronological studies on garnet biotite gneiss and leucogneisses from Schirmacher Oasis region

- Geological and geochemical investigations in Schirmacher Oasis, parts of Wohlthat ranges viz., Gruber Massif and Petermann 1
- Major and trace elemental chemistry of the acid and basic volcanic rocks from Littlewood, Bertrab, and Moltke Nunataks, and the Weddell Sea
- Establishment of Absolute Gravity Reference Station at Maitri.

Understanding the igneous, metamorphic and structural processes involved in a major mountain building episode of the Ross Orogeny, which occurred approximately 500 million years ago, is a challenging scientific quest. Subsequent uplift resulted in the Trans-Antarctic Mountains that uplifted this orogenic belt to expose the deep-seated products of ancient plate boundary interaction. The extensive, clean, un-weathered outcrop on the Antarctic continent enables us to make very detailed observations. It helps not only to understand the evolution of Antarctica but the future of planet Earth.

Geophysical Investigations

Geophysical investigations have been an integral part of India's research program in Antarctica from its initial stages. The initial studies were confined to the ice¬shelf region since the first Indian station, 'Dakshin Gangotri', was located on the Ice-shelf. Subsequently, the emphasis shifted to the Schirmacher Oasis (SO) region, where India's second station, 'Maitri', was constructed. Helicopter-borne surveys have also been conducted over the glaciated area between the SO and the Wohlthat Mountains (WM). Geophysical studies carried out by Indian scientists in various expeditions can be classified into the following three categories:

Investigations on the ice-shelf: The first geophysical survey over the ice shelf was conducted during the second IAE (1982-83), in which eight magnetic profiles were recorded to study the magnetic characteristics of the bedrock beneath the ice cover. During the third IAE (1983-84), the first Indian station, 'Dakshin Gangotri' (DG), was constructed on the ice-shelf, and a total of 60 line-km of magnetic data was collected around the DG station.

The above work around the DG station was further extended during the fourth IAE (1984-85). A 35 km long magnetic traverse was measured, and seismic studies were carried out over the ice-shelf. Based on these studies, a structural model around the DG was presented. The seismic and magnetic surveys carried out during the fifth IAE provide further supportive evidence for this model. During the fourth expedition, the relative performance of conventional gelatine-based explosive, plastic explosive and low-temperature plastic explosive in the Antarctic field environs was also studied.

Studies in and around the SO region: Geophysical investigations in the SO region were initiated during the third IAE when three magnetic traverses totalling about 3-line km with an average station interval of 7 m and 6 short crisscrossing magnetic profiles totalling over 2 km were measured. During the fourth IAE, a short spectrometric profile cutting across a few geological contacts close to the Priyadarshini Lake was measured. A study of the (low energy) radioactivity of water samples collected from the lake also revealed that the water did not contain any hazardous, radioactive contaminations. Geochronological studies were also carried out on 9 rock samples collected from the SO region. Four of these samples belong to the crystalline basement forming part of the East Antarctic shield, while the other samples were of the dolerite dykes occurring in the region. These results contribute to the study of the evolution of the polymetamorphic crystalline basement forming part of the Dronning Maud Land (DML). The ages of Dolerite dykes support the paleo-relationship between the DML and the Mozambique region in Africa proposed by earlier workers.

During the fifth lAE, multi-frequency EM, magnetic and radiometric measurements covered an area between longitudes 11° 26 'E, 11° 50' E, and 13 traverses in the SO region. Out of these, two traverses were laid across the reported secondary sulphide mineralisation. Two other transects in the area were laid where secondary sulphide mineralisation was noticed near a suspected fissure zone. However, no appreciable EM anomaly was found along all the traverses. It could be attributed to a very low percentage of sulphide mineralisation that does not yield any significant EM response. Magnetic and gamma-ray spectrometry measurements carried out along all the 13 traverses also did not yield any significant anomaly.

Helicopter-borne Magnetic Survey: During the seventh IAE, an area measuring 100 x 100 sq km, lying between the SO and the WM, was covered by a helicopter-borne magnetic (heli-magnetic) survey. Figure 1(a) shows the photograph of the Heli-borne survey, while Figure 1(b) depicts the routine positioning of the magnetic sensor before starting the survey. Figure 2 shows the area of the survey. The N-S profile at the spacing of about 3.0 km approximately yielded a station interval of 40 m. While the total magnetic intensity was sampled at a rate of 2.0sec that Figure 3 represents the magnetic intensity contour map. This heli-magnetic survey yielded gross features of the sub-glacial topography (Figure 4). The results mostly agreed with the known presence of nunataks (exposed peaks of sub-glacial hills). However, an extensive magnitude anomaly was found to be associate with the SO region. A high anomaly occurred on the ice-shelf just north of Schirmacher Oasis, and a low anomaly started from the SO and extended to few tens of km south of the SO.

The observed total intensity of magnetic data was corrected for diurnal and IGRF variations. It was then subjected to 2-D spectral inversion to yield gross features of the topography of the base of the glacial region lying between the SO and WM. The results revealed the basement depths associated with various nunataks and the Schirmacher Oasis region, providing supportive evidence for the correctness of the results. It was also found that a ridge-like structure runs in E-W direction between 10 and 20 km south. Although only the SO shows exposed rocks in this ridge structure, the glacial thickness along the axis of this ridge is likely to be very small. The ridge is interpreted to mark the continental margin in the area, and the region immediately north of it represents the continental shelf.

The maximum glacial thickness about 35 km south of SO had the maximum basement depth in the shape of a bowl-like depression.

Figure 1. (a) Helicopter–borne Magnetic survey in Antarctica carried out by NGRI in January-February 1988 around Dakshin Gangotri, using a Chetak Helicopter of the Indian Navy (b) Routine positioning of the magnetic sensor bird under the helicopter before starting the survey

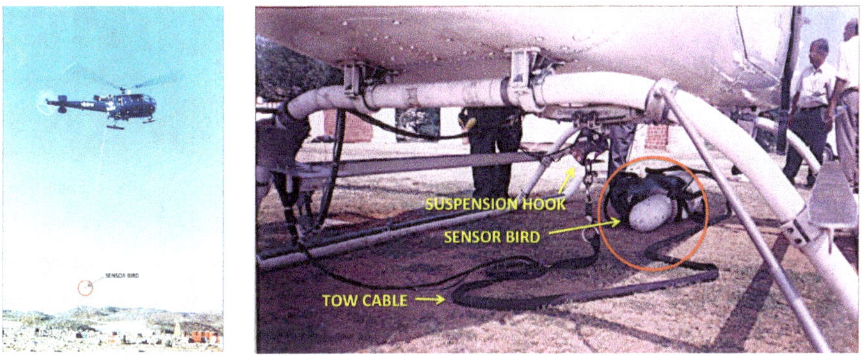

Figure 2. helicopter-borne profiles flown by NGRI during the seventh IAE

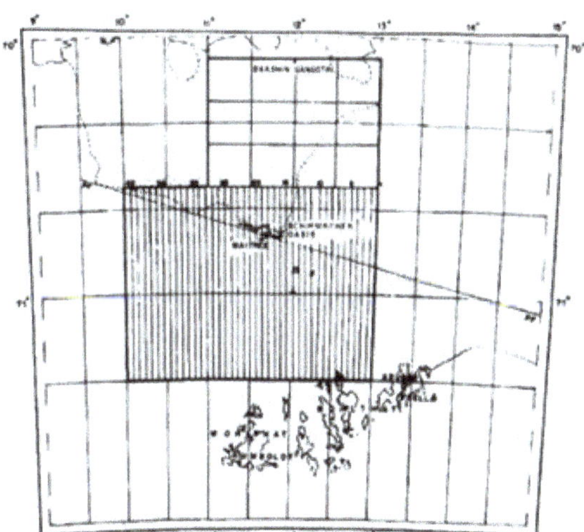

Figure 3. Magnetic intensity contour map of the region between Schirmacher Oasis and Wolthat mountains

Figure 4. Subglacial topography along the profiles A-A, B-B, and C-C as shown in Figure 3

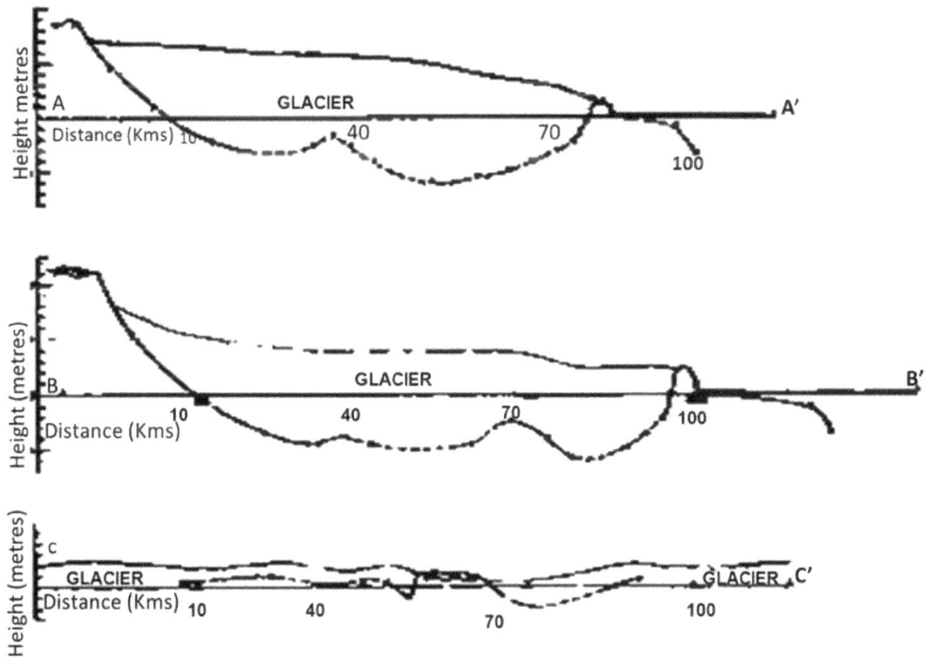

Magnetic Studies

During the thirteenth Indian expedition to Antarctica, complete geomagnetic field measurements were done along nine traverses totalling about six-line kilometres across known geological contacts. It inferred photo-geologic faults in the Schirmacher area. The study showed several prominent anomalies and provided the ground truth for various geological and tectonic features. Thus, the magnetic method seems to be an excellent tool for extensive, systematic mapping of the region for reliably delineating subsurface structures and understanding tectonics.

A portable proton precession magnetometer (Geometrics; Model G 816/826A) with an accuracy of 1nT was used for field measurements. Another proton precession magnetometer (NGRI, Model 600R), also with the same precision, was constantly run at Maitri coupled to a poly-chart recorder (TOA, Model ETR 200A) for monitoring the diurnal variation of the earth's total magnetic field.

Prominent anomalies across lithological/ geological contacts, faults, and shear zones were observed. It confirmed the faults inferred from photogeology by Bormann *et al.* (1986) and thus revealing the ground truth. Further, qualitative and quantitative interpretations are expected to provide deeper insights and information.

Helicopter-Supported Gravity Surveys Around the SO Region

Several interesting features revealed by the helimagnetic survey conducted during the seventh IAE prompted me to complete a gravity survey. While the nunataks between the Schirmacher Oasis and WM showed high magnetic anomalies, the SO was associated with a magnetic low. Also, an intense magnetic high was observed just northwest of the SO. To further study this strong magnetic anomaly of more than 600 nT associated with the SO, a gravity survey was carried out during the ninth IAE. Measurements were taken along with 4 profiles near the SO. After standard data reduction and processing, the observed gravity values were interpreted considering the region's available geological and geophysical information.

It is worth mentioning that the glacial thickness obtained in the graben region by the interpretation of aeromagnetic data is almost identical to that obtained by the gravity data. After applying the correction for the magnetic anomaly caused by the fault (or intrusive body) N-W of the SO, the reinterpreted glacial thickness from the magnetic data also yielded shallower glacial thickness in 0 to 500 m in the region just south of the SO. Based on the regional gravity anomaly, the Moho thickness varied. It is reported about 38 km below the Humboldt Mountains to about 32 km below the SO. It reflects the normal behaviour of crustal thickness with a gradual reduction from the continental margin (below the WM) towards the oceanic crust.

Establishment of gravity stations: As a sequel to the participation of NGRI in the ninth IAE for Antarctica, during the tenth IAE, it was decided to launch a program to establish gravity stations around Maitri in collaboration with the Survey of India (SoI). Accordingly, NGRI provided gravity values at selected locations while the SoI provided the positions and elevations using a Global Positioning System. NGRI's interest was establishing gravity stations in the SO region and various nunataks located south of the SO. During the survey, a permanent station with a benchmark was also established close to the Maitri station. A total number of 18 stations were set. Sarma (1990) gives the details of these stations with geological descriptions, position location, and elevation. Figure 5 shows the locations of gravity stations

Gravity readings were recorded at the expedition's beginning and end at an important station in the Goa National Institute of Oceanography. The Survey of India established this base station. Its absolute value, 976344.71mgal, was taken for computing the gravity values in Antarctica. Thus, all the gravity values in Antarctica, repeated in this work, are preferable to the base at NIO's main entrance. Depending on the weather conditions and the availability of helicopter transport, the time taken to occupy one station was 1/2 h to 1 h each. The recording was taken approximately at 5 metres intervals. A minimum of 10 such readings was recorded at each station. During all the observations, consistency and repeatability were observed to be satisfactory. Each recording of the station was followed by a base station recording immediately. It minimised the effect of instrument drift of readings and diurnal variations. A permanent station opposite Maitri, adjacent to Indira Priyadarshini lake,

Figure 5. Gravity stations established by NGRI around Schirmacher Oasis, Antarctica

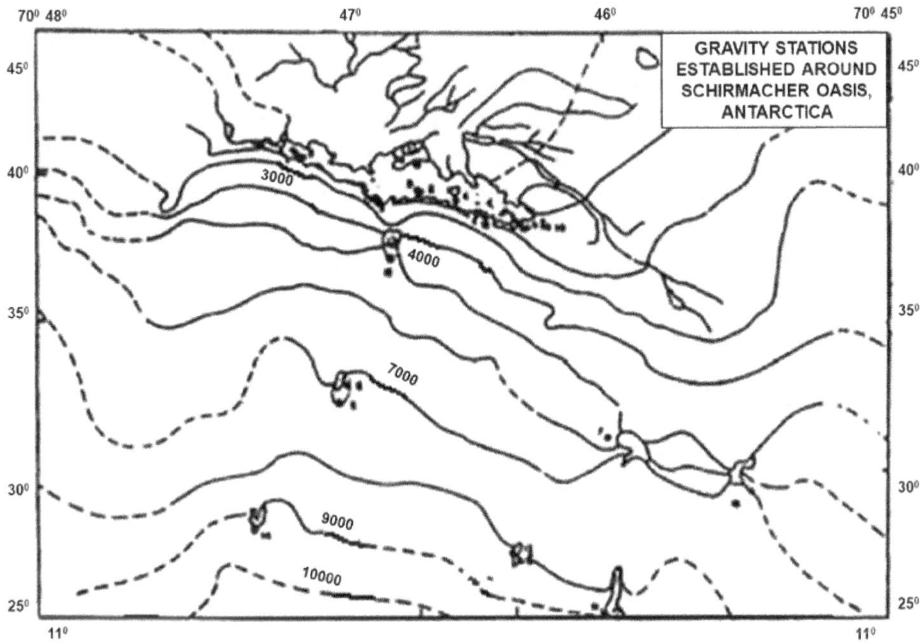

which serves as a base station in and around Maitri, was established by constructing a sound platform. For this purpose, Lacoste and Romberg make S. No. 954 Model was used. The instrument was kept at 51°C using the thermostat supplied by the manufacturer for temperature compensation throughout the stay in Antarctica. SOI provided the position location and elevation required for further computations in gravity values using the global positioning system.

Based on NGRI's requirement and the SOI experience in establishing various survey stations in India, several stations were established in Schirmacher Oasis and various Nunataks, between Wohlthat Mountains and Schirmacher. The accuracy of gravity values for stations established in Antarctica is better than 0.2 mgal. However, since the gravimeter was not kept at 51°C during the journey from Goa to Antarctica, the accuracy may not be better than 1.0 mgal. However, all the measured values are accurate to within 0.2 mgal with respect to the Maitri value. Gravity values are computed for 18 stations in Antarctica and with respect to the absolute value recorded in Goa (at the National Institute of Oceanography). Table-5.1 shows the gravity values for all the stations established in Antarctica by NGRI during Tenth Scientific Expedition.

Palaeomagnetic studies: Palaeomagnetic studies were carried out on lamprophyre dykes. It intrudes the Precambrian crystalline garnet biotite gneisses of the Schirmacher Oasis. The palaeomagnetic results of these dykes reveal similar characteristic remnant magnetic vectors. The lamprophyre dyke is having a mean remnant vector of Dm 29.6°, 1m = + 66°, (K=246.9, α_{95} = 3.9°) while the other dyke is having a mean remnant vector of Dm = 35.0°, Im = + 64.9° (K=57.8, α_{95} = 8.23°). These remnant magnetic vectors and their corresponding Virtual Geomagnetic Pole positions indicate their contemporaneity. These studies also indicated an age of late Ordovician (439 m.y.) λp = 31° S and Lp = 37°E with a palaeolatitude of 47.6°S.

Palaeomagnetic studies were also carried out on basic dykes and sills of different areas from SO. The studies indicated an age of middle Cretaceous with λp = 47.3°S

Table 1. Gravity Values Recorded in Tenth Indian Expedition to Antarctica

S. No	Station	Date	Observed gravity	Free-air Bouguer		Remarks
			absolute			
			value in			
			mgals.			
1.	Maitri	25-12-90	982576.55	-40.0	-53.3	
2.	ThrisQule	25-12-90	982552.33	-30.3	-55.9	
3.	West Lake	26-12-90	982576.20	-44.5	-55.7	
4.	East Hill Top	28-12-90	982569.32	-36.2	-53.7	All the values are
5.	Point 5	29-12-90	982568.35	-35.7	-53.5	computed on the basis of
6.	Nunatak old (9)	31-12-90	982442.63	5.6	-75.6	Absolute value recorded at N.LO's main entrance
7.	N unatak (14)	3-1-91	982491.32	-14.0	-70.6	by Survey of India
8.	Nunatak New (9)	4-1-91	982430.56	10.9.	-76.9	978344.71 IGSN
						The base for the entire work
9.	Offset (I) from	5-1-91	982576.02	-30.2.	-47.1	
	Trishule					
10.	Nunatak (15)	6-1-91	982438.14	5.1	-78.8	

Continued on following page

Table 1. Continued

S. No	Station	Date	Observed gravity	Free-air	Bouguer	Remarks
11.	Nunatak (13)	7-1-91	982371.55	0.8	-107.5	
12.	Offset (2) from	8-1-91	982578.17	-40.0	-52.8	
	Hill Top					
13.	Nunatak (12)	10-1-91	982362.55	14.9	-10 1.1	
14.	Nunatak (lO)	10-1-91	982353.27	23.4	-98.7	
15.	Nunatak (8)	19-1-91	982496.71	.1	-59.8	
16.	Nunatak (2)	20-1-91	982568.16	43.6	-58.1	
17.	Offset for (5)	23-1-91	982573.16	-37.3	-54.0	
18.	Eastern Tip	9-2-91	982567.73	-43.1	58.7	
	Point 7					

and Lp = 97.1°W. This palaeomagnetic data partially fills the existing large gap in the Polar Wander Path for Antarctica.

Geological and Geochemical Investigations

Geological and geochemical studies in Antarctica aim to reconstruct Greater Gondwanaland by finding out the nature and composition of various litho units, their field configuration, the P- T conditions, and the nature of metamorphism, deformation, and time of formation. The above studies have been carried out in the Schirmacher Oasis and parts of Wohlthat ranges viz., Gruber Massif, Petermann I and Humboldt regions to work out an evolutionary model and compare these results with the Indian counterparts. The results of the studies are given below.

Gneisses: Garnet-biotite gneisses (GBG), augen gneisses (AG) and quartzo-feldspathic gneisses (QFG) constitute the main litho-units in the granulite terrain of Schirmacher Oasis, East Antarctica. The GBG and AG show chemical and mineralogical similarities. The QFG differs from GBG and AG, both in major and traces elements. The overall composition of these different polyphase gneisses indicates an increase in silica and alkalis from GBG to QFG and a decrease in CaO, FeO and MgO. Most of the gneisses exhibit negative Eu anomalies. There is variation in the trace and REE from GBG to QFG. Overall, major trace and REE abundance in the garnet-biotite and augen gneisses suggest their origin by progressive metamorphism of a pelitic source. Partial melting of GBG locally appears to have formed quartzo-feldspathic gneisses. These gneisses resemble the metapelites of the Eastern Ghats granulite terrain and Kerala Khondalite belt of India

Granites: The dominant rock formation of Petermann I, which occupies the high hills and the bedrock, is a pinkish greasy looking porphyroblastic (3-5 cm in length) rock. It consists of feldspar megacrysts, orthoclase micro-perthite, plagioclase, myrrnekite, quartz, hornblende, and biotite. The rock has been classified as 'quartz-alkali feldspar syenite' to 'alkali-feldspar granite' with a chemical composition of more than 6% K20 and more than 3% Na20. We report the evidence of soda metasomatism after the crystallisation of perthitic K-feldspar megacrysts.

Anorthosites: Anorthosites, occurring as enclaves in granitoids in the Petermann I area, were studied. These are coarse-grained and contain megacrysts of plagioclase set. And in a granoblastic groundmass consisting of fine-grained plagioclase, clinopyroxene, amphibole, biotite and ilmenite. The average content of megacryst plagioclase is 52%, while that of groundmass is 70%. Clinopyroxene is of salite-ferrosalite composition with Mg/Mg +Fe ++ ranging from 44% to 61%. The amphibole belongs to the pargasite-ferrohastingsite series. Metamorphism of anorthosites in an alkaline environment has been suggested.

Lamprophyres: Several lamprophyre occurrences are reported from the Schirmacher Oasis. The lamprophyres occur as dykes in the garnet-biotite gneisses. That show polyphase deformation and metamorphism from higher amphibolite to granulite facies. These lamprophyres have been identified as 'Minette' based on mineralogical chemical composition and texture. The whole-rock chemistry shows very high MgO (15.00%), high K_20 (5.60%), iron (7.00%), Ba (6500 ppm) and Sr (800 ppm). These rocks exhibit fractionated REE. These lamprophyre dykes could result from the metasomatic enrichment of the source with LREE due to higher hydroxyl content. The source magma appears to have been derived by 30% partial melting of the mantle at 1375° C and 30 Kb pressure.

Basaltic dykes: Studies on two basaltic dykes (new occurrences) in the Schirmacher Oasis have also been carried out. One dyke shows several morphologies of quench olivine textures, indicating alkalinity and fast cooling. Based on its trace and REE composition, it is suggested to have been derived from a mantle plume. The other dyke shows microspinifex texture and komatiitic chemistry (a common phenomenon of early Precambrian terrain). The occurrence of this dyke in Schirmacher Oasis in the Phanerozoic era is an uncommon phenomenon.

Ist Weddell Sea (Western Antarctica) Expedition: Two scientists from NGRI participated in the 1st Indian Scientific Expedition to the Weddell Sea, Western Antarctica. In this surveillance, basalts and andesites were collected from Moltke, Littlewood and Bertrab nunataks, from the east of the Weddell Sea area, and the western margin of the peninsula. The chemical and petrological characters of these samples were estimated. Overall major and trace elemental chemistry of the acid and basic volcanic rocks from the Littlewood, Bertrab and Moltke nunataks, coupled with the systematic compositional variation from basic to the acid end, suggest a common source of origin for this bimodal volcanism. The source magma, a melt of the lower crust/upper mantle, appears to have evolved in a compressional tectonic regime. It intruded/extruded in an extensional rift tectonic setting.

P-T studies: P-T studies have been carried out on the anorthosites of Petermann I, which occur as in the alkali feldspar granite. The formation temperature has been 520°-545°C for alkali-feldspar granite at an assumed pressure of 5-7 Kb in Petermann I. It was based on Clinopyroxene-ilmenite geothermometry and plagioclase geobarometry. It coexisted K-feldspar and plagioclase in the 'alkali-feldspar granite.'

Lake-water studies: Studies on the lake water of SO have shown that the water is mildly acidic (pH 6.15-7.15) and potable. Higher values of pH, EC and IDS in Priyadarshini lake samples are due to human activity. Calc-silicate pockets in the bedrock and their interaction with water could explain the enriched dissolved ionic contents.

During the thirteenth expedition, geological investigations in the Schirmacher range revealed several new occurrences (s) of lamprophyre dykes, basaltic dykes, and a rhyolitic dyke.

The lamprophyre dykes were noticed in the eastern part (near Novo station) and the range's central features (near DG glacier). These dykes ranged from 1 m to 100 mm in length and 3 cm to 125 cm in width. Based on the Megascopic studies, it is suggested that the lamprophyre dykes of the eastern and central regions are different. It indicates that they are the product of different volcanic episodes. The new occurrence of basaltic dykes is also confined to the eastern and central portions of the range. Megascopic studies of these dykes suggest that at least one of the basaltic dykes (about 1.5 km west of Novo station) is different in composition and may be of different volcanic activity than the other basaltic dykes of the range. A rhyolitic dyke was also noticed in the eastern part of the range and was found to be about 100 metres in length and 2 metres in width, showing a clear-cut discordant relationship with the gneiss. Many rock samples of the above intrusive rocks, as well as other rock types (e.g., gneisses, amphibolites, calc-silicates and basic/ultra-basic sills and dykes) of the area, were collected for geochemical studies. Twenty oriented rock samples from three basic dykes were collected for paleomagnetic studies to substantiate the paleo-magnetic results obtained earlier by NGRI better to understand the Antarctic continent's polar wandering curve.

Detailed Petrological and Geochemical Studies on The Dominant Basic Rocks from Schirmacher Oasis

The chemical composition of the amphibolites shows an alkaline to tholeiitic trend. On the contrary, the pyroxene granulites show an alkaline trend. Gabbro and dolerite show an alkaline trend, while basalts show an alkaline to calc-alkaline nature. The amphibolites have total iron more than Ti02' A1203', CaO, and less MgO and alkalies than pyroxene granulites and other basic rocks of the area. The amphibolites have moderate LREE and low HREE with negative Eu anomalies.

The pyroxene granulites show low REE content with almost flat patterns compared to amphibolites. The gabbro, dolerite, and basalt do not offer much variation in their compositions. An exception would be that alkalies are slightly higher in the gabbro due to the metamorphism of the country-rock. The gabbro and basalt dykes show similar REE contents with higher levels of REE with no Eu anomalies. The basalt and gabbro dykes (-300 Ma) intrude the late Proterozoic gneisses of the area. The lamprophyres, which are "Minette", show an alkaline trend and have contents of MgO (15%), K20 (5.60%), iron (7.00%), Ba and Sr. The lamprophyres exhibit highly fractionated REE (both LREE and HREE) without prominent Eu anomalies. Based on chemical parameters, the source magma of lamprophyre appears to have been derived by 30% partial melting of the mantle at 1375°C and 30 Kb pressure.

Geochronological Studies

Geochronological studies on the garnet-biotite gneisses and leucogneisses from the Schirmacher Oasis by Rb-Sr method have given an age of 853±51 Ma with an initial Sr ratio of 0.7085 ± 0.0018 for the former, and 773 ±26 Ma with an initial Sr ratio of 0.7079 ±0.0026 for the leucogeneiss. These indicate a late Proterozoic age for these gneisses.

The Rb-Sr whole rock-mineral ages of two lamrophyre dykes which intrude the garnet biotite gneisses of Schirmacher Oasis are 455±12 Ma (Sfi=0.70886±0.00005) and 458±6 Ma (Sfi = 0.71388 ± 0.00098). These dykes are the manifestation of post-orogenic alkaline magmatism. Another larnrophyre dyke from the eastern part of SO has yielded an Rb-Sr and Sm-Nd isotopic age of 439±10 Ma.

Absolute Gravity Measurements

The gravity field of the Earth is sensitive to both- vertical position and mass distribution. It provides a powerful tool to study crustal dynamics. High precision (~5 μGal) absolute gravity measurements in the time and space domain are essential for understanding mass redistribution and vertical crustal motions. It is caused due to earthquakes, plate convergence, isostatic rebound etc. A Portable and high precision (1 uGal) absolute gravity meter (FG5 measuring system) has been developed during the last decades. Thus, absolute gravity measurements are widely used to address different geophysical problems. For example, crustal deformation studies with GPS for complimentary verification or mass and elevation change analysis due to glacial rebound. In addition to the abovementioned uses, the essential requirement of any gravity survey is an absolute value at least at one point. The antarctic landmass has very few absolute gravity values, and none of them is near the Indian station (Maitri). Earlier, the gravity survey was mainly dependent on the measurements by gravimeters that are known to have drifted with time. Therefore, in collaboration with the Department of Ocean Development (DoD), NGRI acquired the first Absolute Gravity Meter (AG) meter to address vertical deformation near tide gauges along the Indian coast and establish reference gravity stations in India and Antarctica.

Methodology

The functioning of the FG-5 absolute gravimeter is based on precisely measuring the time and distance of the free-falling object inside a vacuumed chamber. Figure 6 is a schematic diagram of the instrument, and Figure 7 is a photograph of the system in operation at Maitri, Antarctica. The interference fringes between the 2 light beams are counted and timed for time and distance measurements, and the distance-time pairs are collected during the 20cm drop. The acquired data is used in the equation of motion to solve the acceleration due to gravity 'g'. A computer automatically controls the whole system, and typically 100 drops are recorded for one set of gravity observations. Since the absolute gravimeter usually produces 100 values per hour, it is common to measure gravity at a site by measuring for 2-3 days. Detailed procedures of measurements and uses of absolute gravity measurements are provided below.

Measurements at Maitri, Antarctica

Absolute Gravity (AG) measurements and some repeat gravity measurements were carried out at Maitri in January 2004. Preliminary checking of instruments revealed some problems that might have arisen due to transportation. The dropping chamber of the AG meter was pumped and maintained at the required vacuum. A concrete platform for AG measurements was constructed and covered under the 'Aravali hut' with the needed sealing by 28th January 2004. Before putting the hut, GPS observation for about four hours was made on the reference point of the pillar for precise height (Figure 8).

Figure 6. Schematic diagram of the Absolute Gravity instrument

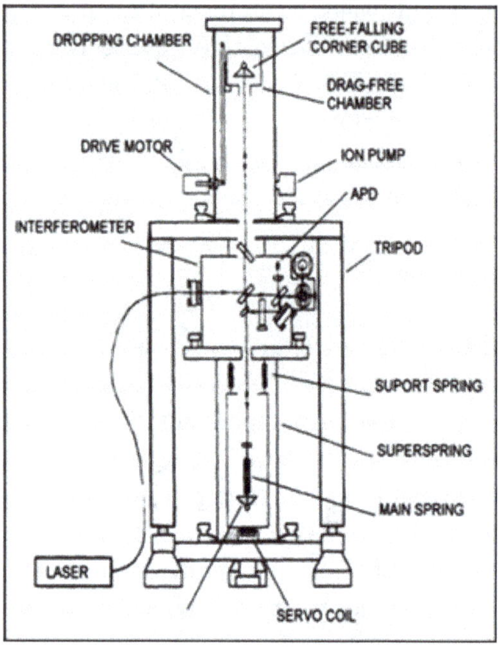

Figure 7. Photograph of the gravity system in operation at Maitri, Antarctica

Actual observations took off on 4th February, following the construction of a good platform in a temperature-controlled hut. Before measurements, the stability of all the components in general and super-spring and laser sources were tested independently; a required temperature (15-25degree Centigrade) was maintained by the auto cut off heater during measurements. Since then, until 27th February 2004, continuous measurements were made except during awful weather days. Once, the observation was interrupted due to a major blizzard but resumed after two days. Earlier gravity observations with AG reference points were also connected. The same observational procedure mentioned above was followed, and sets of observations of 100 drops were recorded. The number of adequate data recorded is 365sets*100

Figure 8. GPS Measurements to aid gravity measurements

drops*700 = 25550000, and these provide the absolute gravity at the reference pillar with an uncertainty of ~ 11uGal. Monitoring of temporal variation of the gravity field can constrain elastic rebound due to changes in ice mass. Any change in the ice will deform the Earth, with results that could be evident in observations of crustal displacements or surface gravity (Wahr *et al.*, 2000). Also, continuous magnetic observations for 20 days and a magnetic traverse cutting all the exposed litho units of the Schirmacher area were measured. The photo of taking magnetic measurements is shown in Figure 9.

Long Term GPS Measurements

Long term GPS measurements for geodetic typing of India and Antarctica were started by CSIR-NGRI from 2004 onwards. A very useful time series of the related data is now available. It has been used for 1) geodynamic and strain accumulation studies (Ravi Kumar *et al.,* 2008 a,b), 2) Crustal deformation around East-Antarctica and Southern Ocean (Akilan and Balaji, 2013), 3) Plate motion predictability using Hurst exponent (Akilan *et al.,* 2013), 4) Changes in atmospheric water content associated

Figure 9. Photograph showing the magnetic measurements being made

to an unusual high snowfall during June 2004 at Maitri station (Schirmacher Oasis, East Antarctica) and the role of south west Indian ridge geodynamics (Akilan *et al.,* 2016), 5) Study of volcanic passive continental margin beneath the Maitri station in central DML, East Antarctica using constrains from the crustal shear velocity structure from receiver function modelling (Gupta *et al.,* 2017) and 6) Perturbations in atmospheric gaseous components over coastal Antarctica detected in GPS signals and its natural origin to volcanic eruption (2019).

Most Significant Achievements

The significant scientific findings are briefly provided below;

- Geophysical studies: These were helpful to delineate the sub-glacial basement of the topography of the region around the Indian station at Dakhin Gangotri and the region lying between the exposed Schirmacher and Wohlthat mountains to the south.
- Geological and geochemical studies: These resulted in the reconstruction of Greater Gondwanaland by finding out the nature and composition of the litho units, their field configuration, the P-T conditions, the nature of metamorphism, deformation, and the time of formation.
- To understand the evolution of the polymetamorphic crystalline basement of Dronning Maud Land (DML). The ages of Dolerite dykes support the paleo-relationship between the DML and the Mozambique region in Africa proposed by earlier workers.
- High precision (~5 μGal) absolute gravity measurements in the time and space domain: These are helpful to understand the redistribution of mass and vertical crustal motions caused by several geodynamic processes such as earthquakes, plate convergence, isostatic rebound, etc.
- GPS measurements: Strain accumulation studies, Crustal deformation, Plate motion, Changes in atmospheric water content, Crustal shear velocity structure from receiver function modelling, Perturbations in atmospheric gaseous components over coastal Antarctica detected in GPS signals and their natural origin to volcanic eruption.

CONCLUSION

In conclusion, it may be summarised that the Integrated G^3 Investigations in Antarctica carried out by NGRI since the second IAE has resulted in acquiring a wealth of knowledge on the supercontinent. Mainly, the understanding of the formation of Antarctica, the geological development of the plate margins of Gondwana and the evolution have been valuable. The island arcs and the microcontinents created the plate margins of Gondwana and other remains of the Gondwana split.

REFERENCES

Akilan, A., Abdul Azeez, K. K., Schuh, H., & Kumar, S. (2019). Perturbations in atmospheric gaseous components over coastal Antarctica detected in GPS signals and its natural origin to volcanic eruption. *Polar Science*, *19*, 69–76.

Akilan, A., Abdul Azeez, K. K., Schuh, H., Padhy, S., & Subhadra, N. (2016). Changes in atmospheric water content associated to an unusual high snowfall during June 2004 at Maitri station (Schirmacher Oasis, East Antarctica) and the role of south west Indian ridge geodynamics. *Natural Hazards*, *83*, 563–574.

Akilan, A., & Balaji, S. (2013). Crustal deformation around East-Antarctica and Southern Ocean using GPS-Geodesy. *Current Development in Oceanography*, *6*, 41–49.

Akilan, A., Balaji, S., Srinivas, Y., & Yuvaraj, N. (2013). Plate motion predictability using Hurst exponent applied to Maitri-Antarctica GPS network. *Journal of the Geological Society of India*, *82*, 613–620.

Gupta, H. K., & Verma, S. K. (1995). *India's contribution to geophysical investigations in Antarctica during Precambrian* (Vol. 34). Mem. Geol. Soc.

Gupta, S., Nagaraju, K., & Akilan, A. (2017). Volcanic passive continental margin beneath the Maitri station in central DML, East Antarctica. constrains from the crustal shear velocity structure from receiver function modeling. *Polar Research*. Advance online publication. doi:10.1080/17518369.2017.1332947

Mital, G. S., & Verma, S. K. (1991). *Helicopter-borne magnetic survey around the Indian station in Antarctica*. Session GAM 5.14, XX General Assembly IUGG, Vienna.

Ravi Kumar, N., & Malaimani, E. C. (2008). 10 years of Continuous GPS measurements for Geodetic tying of Antarctica and India for geodynamical and strain accumulation studies in the south of Indian Peninsula. *Journal of Indian Geophysical Union*, *12*, 141–148.

Ravi Kumar, N., Malaimani, E. C., & Akilan, A. (2008b). Strain accumulation studies between Antarctica and India through geodetic tying of two continents from continuous and long-term GPS measurements. *Indian Journal of Geo-Marine Sciences*, *37*, 404–411.

Verma, S. K. (1988). Preliminary evidences substantiating the paleorelationship of Dronning Maud Land, Antarctica and Mozambique region, Africa. *Proc. Workshop on Antarctic Studies*.

Verma, S. K. (1993). Document highlighting the ongoing and future programs for the Antarctic research by NGRI. DOD.

Verma, S. K., Singh, B., Malaimani, E.C., Tiwari, V.M., Prem Kishore, L., & Rao, M.B.S.V. (2004). *India's geoscientific achievements and future programs in Antarctica*. International Scientific Conference on Antarctica and the Southern Ocean in the Global System (XXVII SCAR) and the Annual Meeting of Council of Managers of National Antarctic Programs (COMNAP XVI), Bremen, Germany.

Verma, S. K., & Mital, G. S. (1988). Geophysical investigations on the ice-shelf around Dakshin Gangotri, Antarctica. *Proc. Workshop on Antarctic Studies*.

Verma, S. K., & Mital, G. S. (1988). Geophysical studies in Antarctica with particular emphasis on the work in Dronning Maud Land. *Proc. Workshop on Antarctic Studies*.

Verma, S. K., & Mital, G. S. (1989). Helicopter-borne Magnetic Survey Between the Schirmacher Oasis and the Wohlthat Mountains, Dronning Maud Land, Antarctica. Detailed Report Submitted to the Dept. of Ocean Development.

Verma, S. K., & Mital, G. S. (1994). Geophysical Surveys Around the Schirmacher Oasis, Antarctica. *Twentieth Annual Convention and Seminar of AEG on Exploration Geophysics at Dehra Dun*.

Verma, S. K., Mital, G. S., & Dayal, A. M. (1987c). *K-Ar dating of some rocks from the Schirmarcher Oasis, Dronning Maud Land, East Antarctica*. Tech. Pub. no. 4 on Fourth Indian Scientific Expedition to Antarctica, Scientific Report, DOD.

Verma, S. K., Mital, G. S., Dixit, M. M., Reddy, K. N. S., Venkatarayudu, M., & Pantulu, K. P. (1987b). *Seismic Investigation on the ice-shelf near Dakshin Gangotri Antarctica*. Tech. Pub. no.4 on Fourth Indian Scientific expedition to Antarctica, Scientific Report, DOD.

Verma, S. K., Mital, G. S., Rambabu, H. V., Venkatarayudu, M., & Srinatha Reddy, K. N. (1987a). *Magnetic Survey over the ice-shelf around the Indian permanent station in Antarctica*. Tech. Pub. no. 4 on Fourth Indian Expedition to Antarctica, Scientific Report, DOD.

Verma, S. K., Mital, G. S., Rao, G. V., Rangarajan, R., Reddy, K. S. N., & Venkatarayudu, M. (1987d). *Radioactivity measurements on some rock and water samples from Dakshin Gangotri, Antarctica*. Tech. Pub. no. 4 on 4th Indian Scientific expedition to Antarctica, Scientific Report, DOD.

Verma, S. K., Ram Babu, H. V., & Mital, G. S. (1988). Structure of the continental margin between Wohlthat range and Sor Rondane mountains, Antarctica. *Proc. Workshop on Antarctic Studies*.

Verma, S. K., Rambabu, H. V., & Mital, G. S. (1999). *Sub-glacial imaging in Schirmacher Oasis-Wohlthat Mountains region in Antarctica employing heli-magnetic and surface gravity surveys*. IUGG 99 Workshop, Birmingham, UK.

Chapter 6
GPR Surveys in Antarctica to Map Ice Thickness and Rock Topography

Rajib Kumar Sinharay
Central Water and Power Research Station, India

ABSTRACT

Ground penetrating radar (GPR) surveys have been carried out at Schirmacher Oasis and Dakshin Gangotri located at Queen Maud Land, East Antarctica, during the 22nd Indian Antarctic Summer Expedition, 2002-2003. The present study confirmed the ability of the high-resolution GPR for monitoring the glaciers. It gives information about the health of the glaciers before it collapses. GPR survey over three frozen lakes near the Maitri Station provided the lakes' top ice thickness and bedrock depth. Similarly, the internal layers of the glaciers have been mapped in-situ using GPR. The results can be correlated with the results obtained by ice-core drill wells to understand the different field parameters (i.e., thickness of each layer, dielectric constant, etc.).

INTRODUCTION

Ground Penetrating Radar (GPR) is an advantageous exploration technique in Antarctica because of its portability, non-destructive nature, rapid acquisition facility, and excellent penetration depth through the ice. This technique is being used in Antarctica for the last several years to map the bedrock below the ice cover. However, the GPR systems which are deep penetrating have a low resolution to provide the details of the internal characteristics of the glacier. High-resolution

DOI: 10.4018/978-1-6684-4078-0.ch006

GPR surveys at Antarctica during the 22nd Indian Antarctic Expedition (Summer 2003) was conducted by the Central Water and Power Research Station (CWPRS). These high-resolution GPR systems that use very high electromagnetic signal frequencies and smaller transmitter-receiver spacing are being used in recent years (Richardson *et al.,* 1997; Vaughan *et al.,* 1999; Snisalo *et al.,* 2003; Siegert *et al.,* 2003) to overcome the limitation of the low resolution of the GPR.

GPR is significantly capable of differentiating between the warm and cold ice in a polythermal glacier. It is because of the high electric permittivity contrast between water and ice. A system of veins, moulins, channels, and conduits are present in the polythermal glacier—water forms between grain boundaries, cavities, and channels within the warm ice. The GPR profiles show multiple reflections over warm ice due to water and ice interfaces. On the other hand, ice does not contain any internal water molecules and is transparent to the GPR signal. Thus, warm and cold ice zones can be easily distinguished in a polythermal glacier using the GPR reflection technique (Moran et al., 2003). Crevasses are the fractures created on the surface of brittle ice due to a differential pressure of the flowing glacier. Often the fractures are buried below fresh snow or the thin ice surface. These crevasses may also be detected using GPR due to high reflection from the ice and air boundaries. The thickness of annual snow deposition and the depositional patterns is important to study a glacier. A High-resolution GPR survey very successfully maps these internal layers. The inner layers of the Antarctic ice sheets can be correlated with the isochronous deposition of snow (Vaughan *et al.*, 1999; Travassos and Simones, 2004). Mapping internal layers can also help calculate the amount of annual snow deposited from the GPR records.

The GPR investigations were conducted for acquiring the following subsurface information:

To Locate a vehicle hangar containing a few hundred ice core samples buried and lost several years ago below the Antarctic shelf ice during a powerful blizzard. The possible depth of the buried hangar was 10-15 m. In contrast, its rooftop entrance shaft of one square metre dimension had a probable burial depth of several metres below the snow surfaces at Dakshin Gangotri, East Antarctica. Also, the internal structure of shelf ice has been studied.

Investigations of bedrock below the thin ice cover areas where ice thickness varied from few meters to few tens of meters at the edge of Schirmacher Oasis, Shivaling, Vettiyya and Tallaksenvarden nunatak. Permafrost has also been studied at Schirmacher oasis near Maitri station.

Study of the nature of ice in the glaciers near Schirmacher Oasis, Shivling and Vettiyya. Warm and cold ice, crevasse, water channels, and meltwater pockets were detected within the glaciers.

- To Map firn and ice boundary and the internal layers of the glaciers indicating depositional pattern near Tallaksenvarden nunatak.
- Survey the frozen lakes at Schirmacher oasis to determine the depth of bedrock and thickness of the top ice cover.

The scientific objective of CWPRS was to deploy ground-penetrating radar for locating the buried hangar at 'Dakshin Gangotri' quickly and cost-effectively, along with fulfilling other scientific goals mentioned above.

Surveys have been carried out at Dakshin Gangotri, Schirmacher Oasis, Shivaling, Vettiyya, and Tallaksenvarden nunatak areas (Figure-6.1). The objectives and the locations of every survey have been further described in Table-6.1.

Figure 1. Schematic location of survey area at Antarctica; a) Antarctica and Queen Maud Land, b) Dakshin Gangotri and Schirmacher Oasis and c) Survey area around Schirmacher

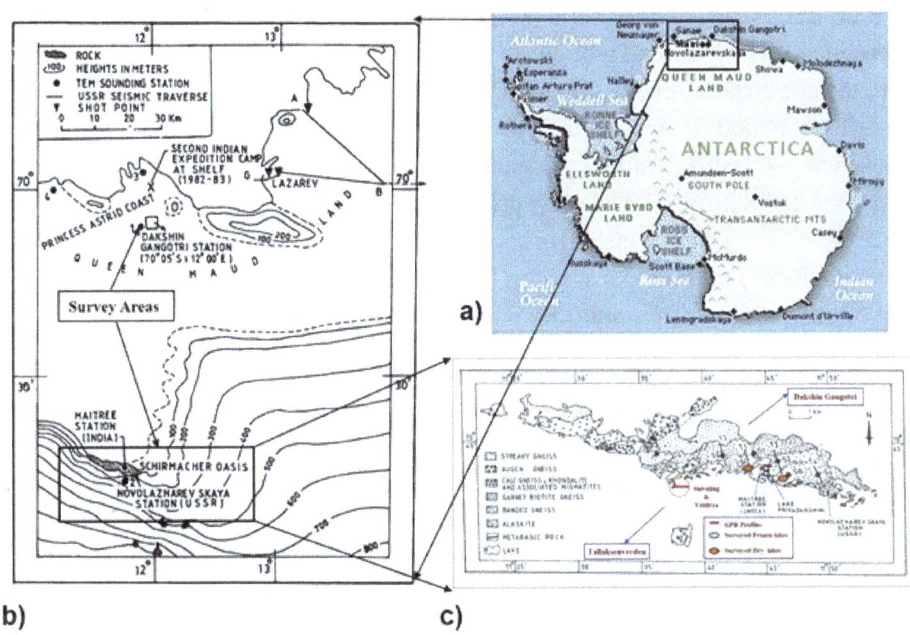

Table 1. Survey area and objectives

Sl. No.	Survey Area	Objectives
1	Dakshin Gangotri (~ 1.5 km long GPR profile)	To locate the buried hanger containing valuable ice cores. To study the yearly snow accumulation.
2	The southern margin of Schirmacher Oasis, Shivaling and Vettiyya (~ 4 km long GPR profile)	To map the ice thickness and bedrock topography. To locate water channels, meltwater pockets, warm ice and firn structures.
3	Tallaksenvarden nunatak (~ 2.5 km long GPR profile)	To map internal layers of subsurface ice structures. Also, mapping of crevasses.
4	Frozen and dry lakes and vehicle paths in and around Maitri station	To measure rigid lake parameters To undertake Permafrost studies in the dry lake bed and vehicle paths

SCIENTIFIC PRINCIPLE

GPR is an active electromagnetic (EM) geophysical technique. It uses a transmitter to produce EM waves and a receiver that records them reflected or backscattered from the subsurface objects. The GPR survey in Antarctica has been carried out in the reflection mode to achieve the present objectives with fixed transmitter-receiver separation using three mid frequencies, 250 MHz, 800 MHz, and 1 GHz (MALA Geoscience, 2002). The EM wave's propagation speed and reflection are affected by the medium's dielectric constant or relative permittivity and magnetic susceptibility. The velocity (v) of the radar wave in the medium is dependent only on the natural part of the dielectric constant (ε).

$$v = c\ \varepsilon \tag{1}$$

Where, c is the speed of light.

But, the attenuation (α) of the signal depends on both "ε" and conductivity (σ)

$$\alpha = 1635 \times \sigma\ \varepsilon \tag{2}$$

If "t" is the time interval between transmission of signal and arrival of the reflected signal, then the depth (d) of reflecting interface can be defined as:

$$d = (v.t / 2)^2 - (x / 2)^2 \tag{3}$$

Here, "x" is a separation between the transmitter and receiver. The separation varies from system to system and with frequencies for a system.

EM waves are reflected from the different internal layers and features of ice sheets. The nature of GPR reflection provides information about the pattern and depth of the subsurface objects. It efficiently identifies air, water, and ice interfaces because of a significant contrast of their dielectric constant.

METHODOLOGY

GPR data has been filtered using a band pass filter. Different frequency windows have been selected for different GPR sections depending on the frequency of the reflection feature present in the section. Time-varying gain (TVG) has been applied to the filtered section to enhance the amplitudes of reflections that arrive beyond a threshold time with a linear gain (G_L) and exploration gain (G_E) with total gain (G) given by:

$$G = (G_L \times t) + e^{(G_E \times t)} \tag{4}$$

Where, 't' is the two-way travel time in a nanosecond. The value of G_L and G_E varies from 80-120 and 15-25, respectively, in different profiles. Synthetic colour plots of GPR data were produced to present the GPR time sections.

Both the "transmitter" and "receiver" are shielded in nature and placed on a common covered platform. It is called the "GPR antenna". A wheel measures the distance of the position of the antenna for each transmission of pulses. The starting point (considered as zero distance) of the profile is attached to the antenna. The initial starting and the finishing points of the profiles have been located using a hand-held GPS, except for the area at Dakshin Gangotri. However, the precise GPS location of the excavated entrance shaft of the buried hangar has been recorded at Dakshin Gangotri. Most of the areas have been surveyed by dragging GPR antenna by hand (Figure 2), whereas some profiles have been taken using snow vehicle, "Piston Bully" (Figure 4).

MAIN FINDINGS

The pilot study unfolded several possibilities for applying high-resolution GPR in Antarctic studies, though the first application of searching for a buried hanger was purely geotechnical. The results have been described below-

Locating the Buried Hanger: GPR survey started at Dakshin Gangotri with a challenge to locate a hangar buried under the shelf ice for the last few years, which contained essential ice cores. A precisely detailed GPR survey was able to pinpoint its entrance shaft (Figure-6.2, Figure-6.3) of 1m x 1m dimension exactly at latitude 70⁰04'27.7" S and longitude 12⁰00'37.9" E. The ice cores were retrieved after excavation (Chaturvedi and Singh, 2008).

Figure 2. GPR survey with 800 MHz antenna and excavated entrance shaft at Dakshin Gangotri

Figure 3. GPR record over the hangar buried under shelf ice at Dakshin Gangotri

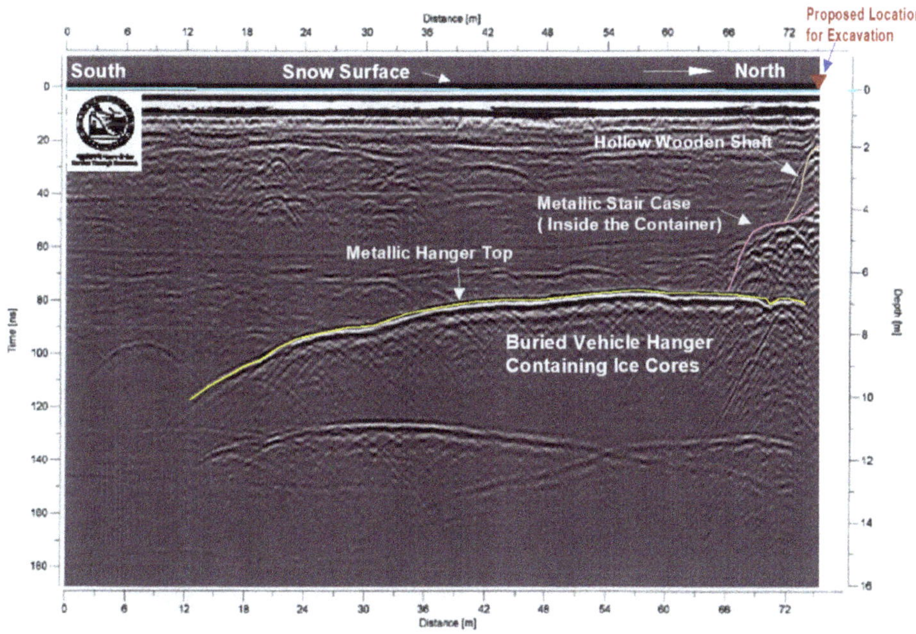

Mapping of the Bedrock

GPR surveys were carried out across several profiles to map the bedrock below the ice-covered areas near Shivaling (Figure-6.4) and Vettiyya. The bedrock has been picked up clearly in the GPR sections (Figure-6.5). It dips with an angle of 30-40 degrees in the southern part of Schirmacher Oasis, the northern part of Shivaling and the western part of Vettiyya. A buried rocky peak with similar dips has been found 102 m north of the base of Shivaling (70°46'46" S and 11°36'09" E) towards Schirmacher Oasis at a depth of 16 m. The dip of the subsurface rock topography varies from 10-30 degrees towards the east at the eastern side of Vettiyya. Bedrock is also observed north of Tallaksenvarden nunatak dips with varying angles, as steep as 47 degrees in some areas.

Figure 4. GPR survey with 250 MHz antenna with Piston Bully near Shivaling

Polythermal Glaciers

The subject GPR study revealed the possibility of assessment and monitoring of the health of any retreating polythermal glacier. The study of the continental glaciers near Vettiyya and Shivaling included the analysis of blue ice zones. A very significant anomalous hot regime was detected near latitude $70^0$46'55" S and longitude $11^0$37'05''E, where a part of the glacier is polythermal, and the density of water-filled fractures and meltwater pockets are maximum. The character of reflections obtained in GPR sections (Figure-6.6) clearly indicated water-filled fractures and meltwater pockets. A clear indication of glacier warming was detected in the region close to latitude $70^0$48'06' 'S and longitude $11^0$38'24''E (Figure-6.7).

Figure 5. Bedrock topography in the eastern part of Vettiyya

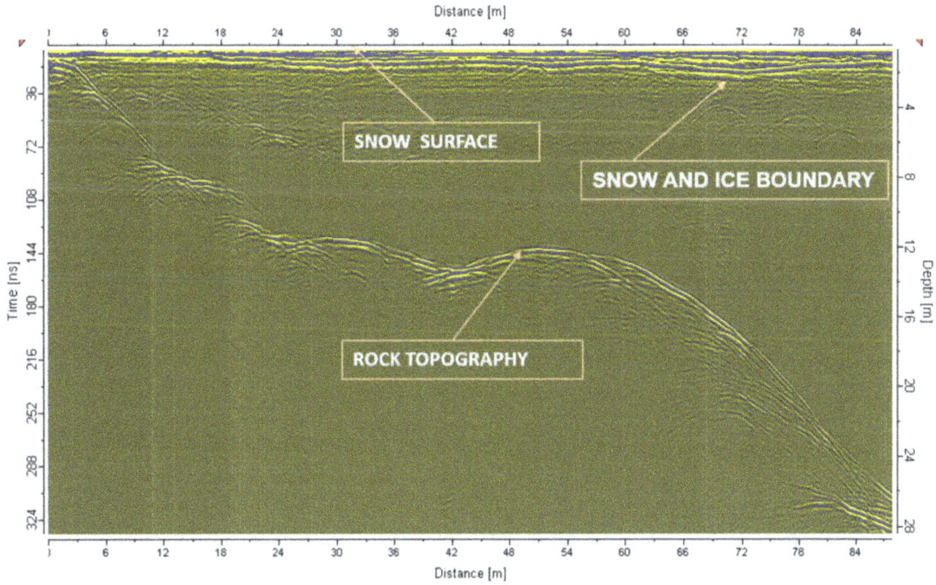

Figure 6. GPR signature of a polythermal glacier in the western part of Vettiyya

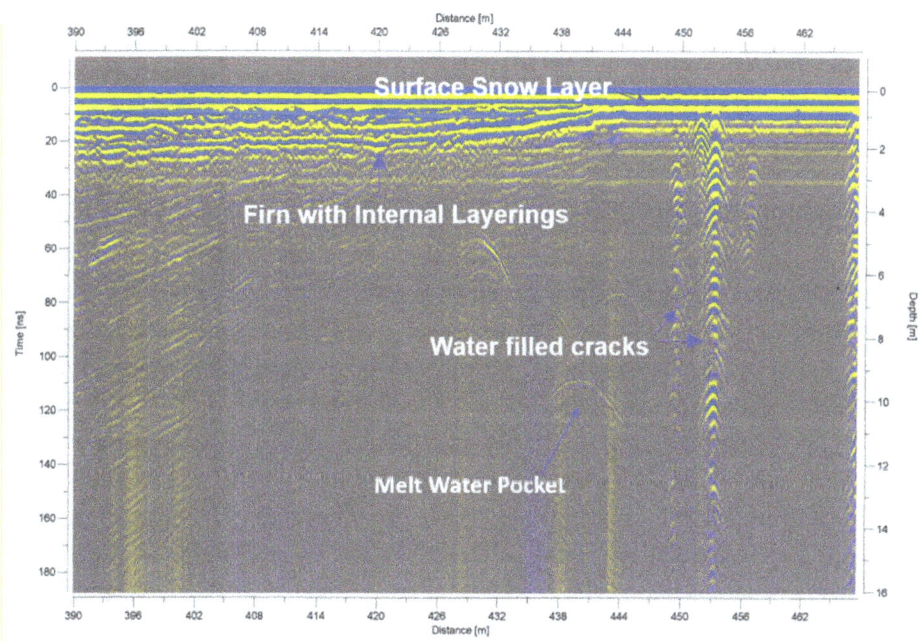

Figure 7. GPR section showing cold ice and warm ice zones of a polythermal continental glacier

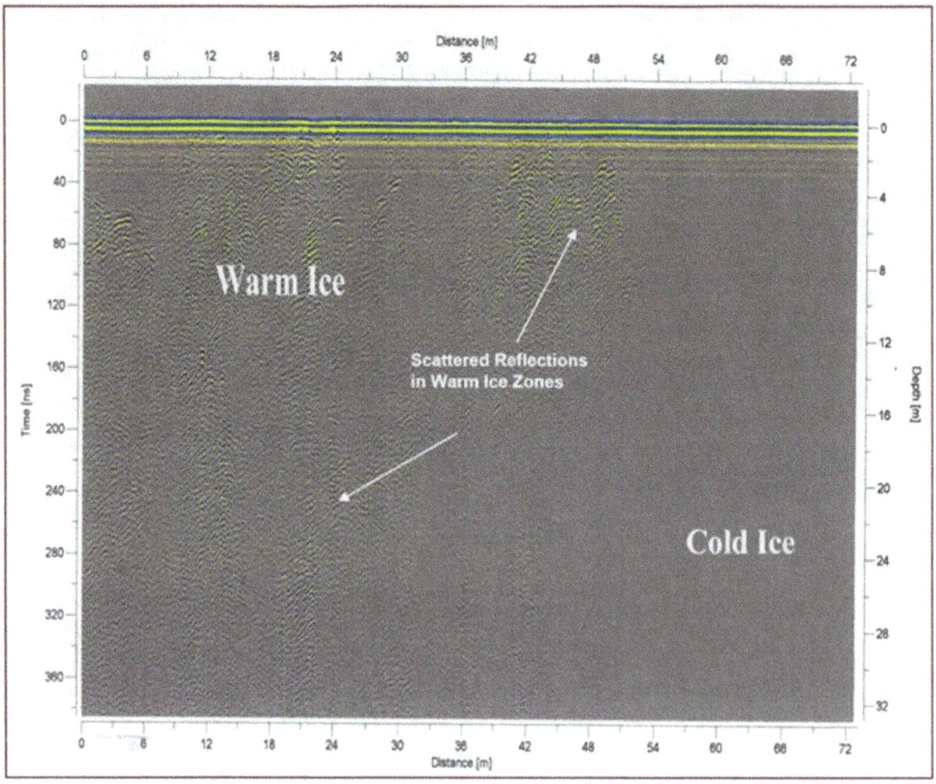

Internal Layering

The high-resolution GPR used in this survey picked up the internal layers of the ice structure on the ice shelf and continental glaciers. The GPR sections over the ice shelf near Dakshin Gangotri station show several strongly reflecting layers at various depths. These layers possibly indicate an ancient depositional environment. It requires further research to establish their relationship with ambient meteorological parameters. Ice and firn interfaces were mapped to the north of Tallaksenvarden nunatak dip by varying degrees in different directions. Firn structure has been found to the north of Tallaksenvarden. It has a thickness of more than 28m near the nunatak. It is pinching out in the northern direction with the end around 50 m south of the ice-core drill hole (latitude: 70^{0}51’38.3" S and longitude: 11^{0}32’18” E), shown in Figure-6.8.

Figure 8. Internal layering and firn structure near Tallaksenvarden nunatak

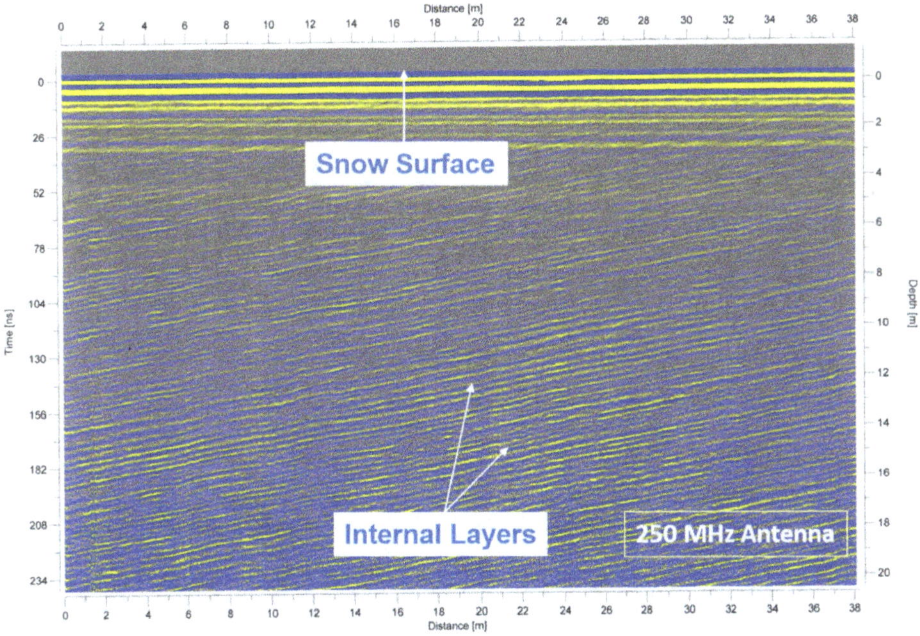

If calibrated with borehole data, the layers obtained by GPR can provide vital information quickly and continuously. A crevasse was mapped near Tallaksenvarden with the GPR (Figure-6.9).

Figure 9. Rock topography, internal layering and crevasse near Tallaksenvarden nunatak

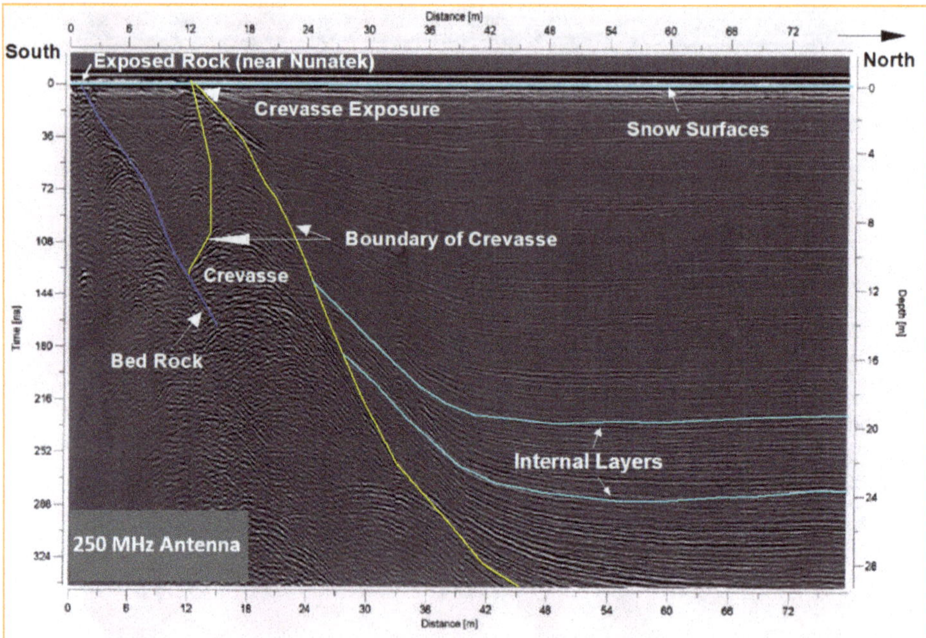

Lakes and Associated Glaciers

GPR is an efficient geophysical technique to study frozen lakes. Three frozen lakes (Figure-6.10) and their associated glaciers were studied using GPR (Figure-6.11). The results show thick gravel deposits, especially for two deeper lakes. The thickness of the ice cover and depth towards the bedrock has been tabulated below (Table-6.2). The average velocities of the EM waves for depth calculation have been assigned 160 m/µS for ice and 120 m/µS for gravel beds partially saturated with meltwater. The nomenclature of lakes (number of lakes) is referred after the lake map (Ravindra et al., 2002).

Figure 10. Location of surveyed lakes, L48, L53 and L54 at Schirmacher Oasis after Ravindra et al., 2002

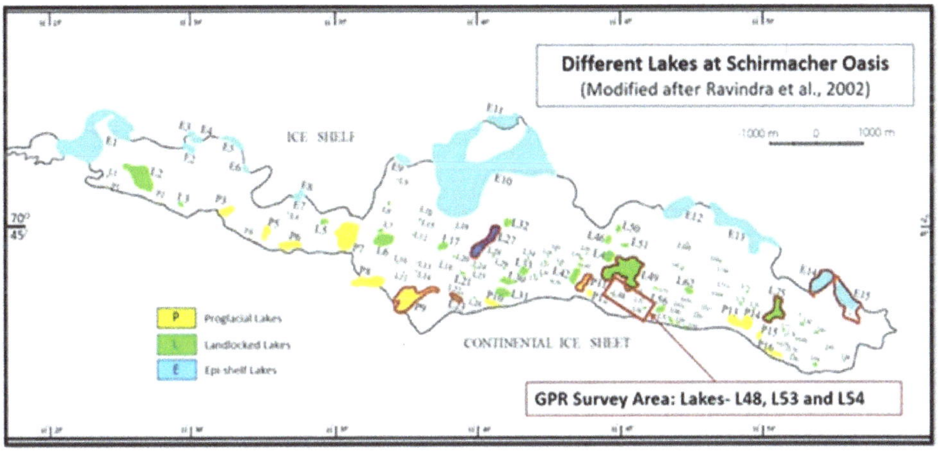

Snow

Figure 11. GPR record along with an east-west profile over lake 48 (Ravindra et al., 2002), Schirmacher Oasis

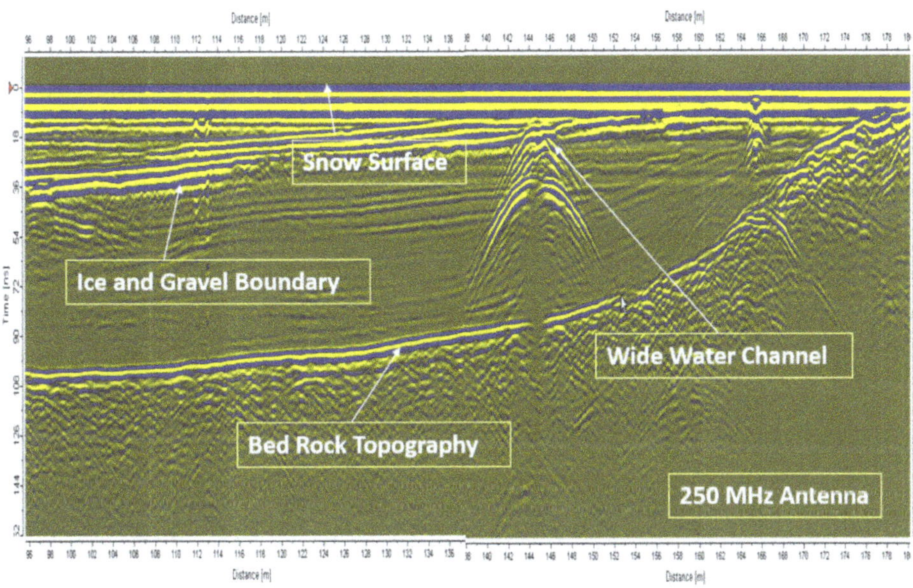

Table 2. Depth to the bedrock and the thickness of the ice cover

Lake	Bedrock	Thickness of
No. (Ravindra *et*	Depth (m)	Ice Cover
al., 2001)		(m)
L48	7.0	3.5
L53	3.2	1.5
L54	6.5	1.5

GPR survey was carried out on several dry lakes and vehicle paths around the Maitri station to find out the depth and the state of permafrost. The depth of permafrost varied from 0.5 m to 1.5 m from the surface.

India successfully conducted a high-resolution GPR survey over the Antarctic glaciers for the first time to identify the ice's nature and study the internal structures. This study has opened a possibility to assess the retreating glaciers' health in-situ and explore the ancient Antarctic environment. The depositional pattern rate of the deposition can also be studied. The recent international trend of GPR research in Antarctica is the in-situ study of the internal layers of the Antarctic ice sheets, which generally represent the layers of isochronous deposition of snow (Vaughan *et al.*, 1999). Scientists and research workers are busy calibrating and correlating the GPR records with the ice-core drilling results and finding out information about the ice flow dynamics. The high-resolution GPR study that was carried out for mapping the internal structures of the Antarctic ice sheet is perhaps just the beginning of research in this field.

In summary, the rate of increase in the warm and cold ice ratio, meltwater pockets, water channels, *etc.*, within a glacier can easily be deduced using GPR investigations. It gives information about the health of the glaciers before it collapses. It also suggests the applicability of the GPR application for Antarctic lake studies. Similarly, the calibrated layers of GPR sections can provide information about the ancient depositional pattern and the ambient environment.

IMPLICATIONS

- GPR study of glaciers will provide the necessary information for modelling and forecasting the health of the retreating glaciers.
- A map of the depositional pattern, internal layers, and bedrock depth will help select a suitable location for ice-core drilling.

- GPR study will provide necessary information about the frozen lakes for a better understanding of the Antarctic environment.
- Internal ice layers have been mapped in-situ by high-resolution GPR. After calibration by few ice-core drill holes, internal mapping layers of glaciers over a large area is possible in time and cost-effectively.
- GPR can find safer vehicle paths over the Antarctic ice sheet by delineating the crevasses and weak zones.

CONCLUSION

It may be concluded that the high-resolution GPR has been successfully used to study the Antarctic glaciers and frozen lakes. This study has opened a new area of Indian research for modelling and forecasting the health of Antarctic glaciers. Regular GPR investigation can provide information about the increase in warm and cold ice ratio, meltwater pockets, moulins and water channels within the glaciers. Unlike conventional methods, this technique gives information about the degradation of glaciers before it collapses. The work has also initiated another promising research field to study the internal layers of the Antarctic ice sheet, which will help in analysing the ancient Antarctic environment. Sincere attention should be paid to establish the relationship between the GPR and ice-core drill results and their correlation with isochronal snow deposits of earlier periods.

GPR surveys over three frozen lakes near Maitri station have provided details regarding the ice thickness and depth of the bedrock for the lakes. It has also established the possibility of a GPR application for reconnaissance surveys to select the proper ice core drilling site. A hangar containing ice cores and its entrance shaft buried under the shelf ice of Dakshin Gangotri has been located using GPR, and ice cores have been retrieved to be brought to India.

ACKNOWLEDGMENT

It's to acknowledge with thanks the Secretary, Ministry of Water Resources, New Delhi, for approving the proposal and the Secretary, Department of Ocean Development, for providing me and CWPRS the opportunity to participate in the expedition. Many thanks are also due to Mrs V. M. Bendre, Director, Central Water and Power Research Station, Pune, for encouragement and guidance. Dr P. C. Pandey, Director (NCAOR), Dr T. V. P. Bhaskara Rao, Director (Logistics) and Dr N. Khare, Director (Scientific), at the National Centre for Antarctic and Ocean Research are acknowledged. I take this opportunity in thanking Dr Arun Hanchinal, Leader of the 22nd Indian Antarctic Expedition, for providing all the logistical support and assistance to make my GPR survey a success. I also thank Dr N. Ghosh, Additional Director and Mr R. S. Wadhwa for their constant encouragement and help at all times. Special thanks go to my colleagues Mrs Maya Sinharay, Mr Subba Rao, Raja Mukhopadhyay and Aniruddha Saha for their continuous support and valuable discussions.

REFERENCES

Chaturvedi, A., & Singh, A. (2008). *First Attempt at Motorised Ice Core Drilling during the Antarctic Summer Season.* 22nd Indian Expedition to Antarctica, Scientific Report, 2008, Ministry of Earth Science, Technical Publication No.20.

Geoscience, M. (2002). *RAMAC GPR operating manual.* Mala Geoscience.

Ground Penetrating Radar. (1992). Finnish Geotechnical Society.

Moran, M. L., Greenfield, R. J., & Arcone, S. A. (2003). Modelling GPR radiation and reflection characteristics for a complex temperate glacier bed. *Geophysics, 68*(2), 559–565. doi:10.1190/1.1567225

Ravindra, R., Chaturvedi, A., & Beg, M. J. (2002). Melt water lakes of Schirmacher Oasis - their genetic aspects and classification. In D. Sahoo & P. C. Pandey (Eds.), *Advances in Marine and Antarctic Science* (pp. 301–313). APH Publishing Corporation.

Richardson, C., Aarhalt, E., Hamran, S. E., Halmlund, P., & Isaksson, E. (1997). Spatial distribution of snow in west Dronning Maud Land, East Antarctica, mapped by a ground-based snow radar. *Journal of Geophysical Research, 120*(No. B9), 20243–20359.

Siegert, M. J., Hindmarsh, R. C. A., & Hamilton, G. S. (2003). Evidence for a large surface ablation zone in Central East Antarctica during last ice age. *Quaternary Research, 59*(1), 114–121. doi:10.1016/S0033-5894(02)00014-5

Sinisilo, A., Grinsted, A., Moore, J. C., Karkas, E., & Petterson, R. M. (2003). Snow-accumulation studies in Antarctica with ground-penetrating radar using 50, 100 and 800 MHz antenna frequencies. *Annals of Glaciology, 37*, 194–198. doi:10.3189/172756403781815825

Travassos, J. M., & Simoes, J. C. (2004). High-resolution radar mapping of internal layers of a subpolar ice cap, King George Island, Antarctica, *Pesquisa Antarctica Brasileira, 4*, 57-65.

Voughan, D. G., Corr, H. F. J., & Doake, C. S. M. (1999). Distortion of isochronous layers in ice revealed by ground-penetrating radar. *Nature*, *398*(6725), 323–326. doi:10.1038/18653

Chapter 7
Studies on Seismotectonics and Geodynamical Processes Between India and Antarctica:
A Review

E. C. Malaimani
CSIR-National Geophysical Research Institute, India

G. B. Navinchander
CSIR-National Geophysical Research Institute, India

N. Ravikumar
CSIR-National Geophysical Research Institute, India

G. S. Srinivas
CSIR-National Geophysical Research Institute, India

S. V. R. R. Rao
CSIR-National Geophysical Research Institute, India

U. Gowrishankar
CSIR-National Geophysical Research Institute, India

S. Padhy
CSIR-National Geophysical Research Institute, India

I. P. Raju
CSIR-National Geophysical Research Institute, India

A. Akilan
CSIR-National Geophysical Research Institute, India

Y. V. V. B. S. N. Murthy
CSIR-National Geophysical Research Institute, India

K. Abilash
CSIR-National Geophysical Research Institute, India

M. Venkatrayudu
CSIR-National Geophysical Research Institute, India

L. Prem Kishore
CSIR-National Geophysical Research Institute, India

A. Nageshwara Rao
CSIR-National Geophysical Research Institute, India

DOI: 10.4018/978-1-6684-4078-0.ch007

ABSTRACT

Seismotectonic and the geodynamical processes between India and Antarctica are essential. During the initial 25 years of Indian expeditions, India's National Geophysical Research Institute (NGRI) launched a GPS-Geodesy programme by establishing a permanent GPS tracking station and a state-of-the-art Digital Broad Band Seismological Observatory Maitri during 1997. Both these stations have been continuously operational. It addresses the crustal deformation in the south of the Indian peninsula, the driving mechanisms, and the response of the Indian Ocean lithosphere bringing a deep understanding of the causes of the accumulation processes of strain in the Indian Ocean and the northward movement of the Indian plate.

INTRODUCTION

National Geophysical Research Institute (NGRI) is one of the primary CSIR Research Organisations that has participated in all the Indian Antarctic Expeditions since the first expedition in 1981, focusing its research on Earth Sciences. Of late, the interest in Geodynamical Processes to the south of the Indian Peninsula that are of prime concern to the Indian Plate kinematics and Indian Plate motion has increased. Since the Indian Plate is bound by the Plates of Arabia, Somalia, Antarctica, and a diffuse India-Australia boundary in the Indian Ocean Basin, it becomes significant to kinematically describe the plates and their boundaries to address the following queries:

1. How rigid is the Indian Plate?
2. Does its relatively high level of Intra Plate Seismicity indicate internal deformation more than other plates?
3. Is this related to the Indo-Eurasian collision and generation of the Himalayas?

Hence NGRI has launched the GPS-Geodesy and Seismic Studies from XVII IAE onwards, and it is continuing

A permanent GPS tracking station was established in 1997, and it is upgraded with GNSS Geodetic receiver during XXV IAE, and the station is continuously operational since its inception.

Similarly, a permanent Seismic Observatory with the state-of-the-art Broadband Digital Seismometer was established at Maitri in 1997 to monitor the seismicity in and around Antarctica. The observatory houses GURALP CMG-3ESP Broadband Seismometer of Reftek make, with a processor-based Data Acquisition System. During XXV Indian Antarctic Expedition (IAE) (January 2006), the upgraded observatory had a new Broadband Seismometer, Geotech KS-2000M with Geotech-

Smart 24R Digitizer. It was further heightened by installing Reftek130 DAS and GURALP 3-T seismometer in 2007. Data acquisition is at a sampling rate of 50 samples per second. The seismic data is analysed using SEISAN software. It is also used in studies regarding space and time distribution of earthquake occurrences to obtain hypocental parameters, magnitudes of earthquakes, the release of energy, strain accumulation and stress drop, velocity inversion for underground structure and physical properties, earthquake source mechanism, receiver function analysis, and attenuation of seismic waves and anisotropy. The collocation of GPS and Seismic stations would mutually aid the studies on seismotectonics, plate kinematics, plate boundary reorganisations, and crustal deformation between India and Antarctica. This chapter discusses India's significant scientific efforts and achievements in the contemporary fields of Seismotectonics and Geodynamical Processes during the initial twenty-five years of the annual launching of expeditions to the icy continent.

CONTINENTAL DYNAMICS AND CRUSTAL DEFORMATION BY GPS-GEODESY

Seismicity in and around Antarctica and Seismotectonics: The history of the evolution of Antarctica itself makes an exciting and compelling scientific study. About 200 million years ago, Antarctica was near the equator and was surrounded on three sides by Australia, India, Africa, and South America. Later, 100 million years ago, Gondwana broke up. The continents began drifting in opposite directions, and 60 million years ago, Antarctica moved closer to the South Pole and the ocean developed between Antarctica, Africa and India. Australia and Antarctica began to separate, and 45 million years ago, Antarctica continued to shift towards the South Pole (Arus and Dorden, 1991; Bendick and Bilham., 1999; Billham and Gaur, 2000; DeMets et al., 1990; Gorden et al., 1998; Gorden and Stein., 1992; Larson et al., 1999; Malaimani et al., 2000; Subrahmanya, 1996; Ravi kumar et al., 2007). Research on the geodynamical processes is essential. It includes plate tectonics, seismicity in and around Antarctica, plate boundary reorganisations, crustal deformation in the south of the Indian peninsula, the driving mechanisms and the response of the Indian Ocean Lithosphere are the prime interests of NGRI and form a major thrust area. The facility targeted the following objectives:

1. To link IGS GPS station at Hyderabad and Maitri
2. To estimate very long baseline lengths, their changes and velocity vectors with the inclusion of other IGS stations in the Indian Ocean
3. To precisely estimate the crustal deformation between India and Antarctica

4. To holistically understand the seismicity in and around Antarctica, tectonic activity, plate boundary reorganisations, crustal deformation in the south of the Indian peninsula, the driving mechanisms and the response of the Indian Ocean Lithosphere

Continuous GPS Monitoring between India and Antarctica: NGRI has initiated a GPS-Geodesy programme by establishing a permanent GPS station at Maitri in 1997 (Ravi kumar et al., 2004; 2006; 2008; Malaimani et al., 2006). It was upgraded with GNSS geodetic receiver. It helps to track both GPS and GLONASS signals to understand the tectonic activity and crustal deformation in the south of the Indian peninsula, the driving mechanisms and the response of the Indian Ocean Lithosphere and holistically compound the Indian Plate kinematics (Armijo et al., 1986; Bouin and Vigny, 200; Holt et al., 2000; Shen et al., 2000). As shown in Figure 1 and the permanent station, the upgraded permanent GNSS antenna is continuously operational since its installation. By these, Maitri remains one of the SCAR stations and remains in the global scenario of GPS- Geodesy. Figure 2 shows the time series of Maitri till 2009. Maitri also continues to contribute to IERS in recognising ITRF Global Reference Frames.

Figure 1. GPS choke–ring antenna mounted on a pillar and seismic vault at the rear

Figure 2. The time series of Maitri till 2009

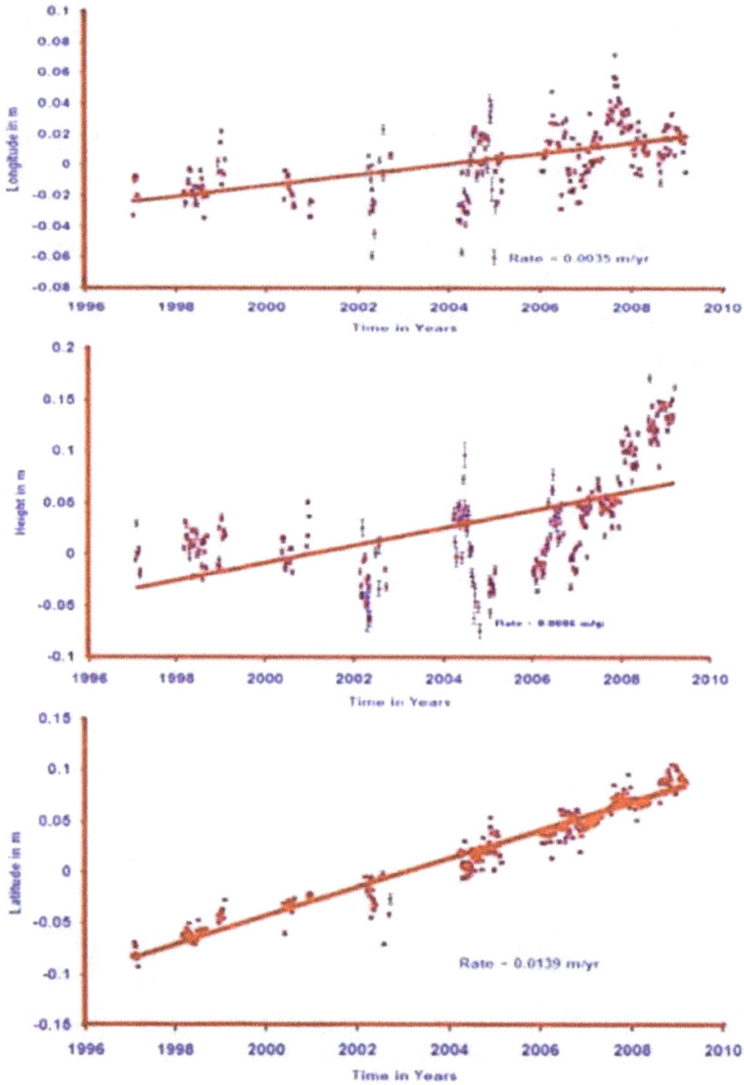

With these studies, NGRI could participate and contribute to the Scientific Committee on Antarctica Research (SCAR) GPS campaign and Maitri data, which are included in the Global Data Archive. Since Maitri GPS Station has become a permanent geodetic Marker in Antarctica, the data can be accessed at the following websites:

1.http://tpg.geo.tudresden.de/ipg/forshung/scargps/stations//MAIT.
html#MAIT_07http://tpg.geo.tu-

2. dresden.de/ipg/forshung/scargps/stations/ /MAIT.html#MAIT_08
3. http://tpg.geo.tu-dresden.de/ipg/forshung/scargps/stations//MAIT.
html#MAIT_09

The online station description of Maitri, SCAR 2001 to 2009 Epoch GPS campaigns and the other details can be accessed at the following websites:

1. http://www.tu-dresden.de/ipg/FGHGIPG/Aktuell-Dienste/scargps/MAIT.
html-MAIT_01
2. http://www.tu-dresden.de/ipg/FGHGIPG/Aktuell-Dienste/scargps/MAIT.
html-MAIT_09

Maitri continues to be in the Global Scenario, being one of the SCAR GPS stations contributing to the International Database of the SCAR Epoch GPS campaigns.

Maitri also continues to contribute to IERS and is included in the realisation of the ITRF 2005 reference frame since 2008. The Primary ITRF 2005 sites and collocated techniques wherein Maitri's estimated station positions and velocity at Epoch 2000.0 are included in the list of SCAR stations can be accessed at the following website;

https://itrf.ensg.ign.fr/GIS/index.php

NGRI plays a globally significant role in maintaining and improving a Global Reference Frame by contributing and processing high-quality geodetic data from Maitri.

Based on the Very Long Baseline lengths, velocity vectors, and their changes between India and Antarctica, signatures of strain accumulation and crustal deformation in the south of the Indian Peninsula have been obtained.

Due to its stability, Kerguelen as a reference site, the baseline lengths from Kerguelen to the Stations Davis, Casey, Maitri, in Antarctica Plate, Yaragadee, Tidbinbilla in Australian Plate, Hartebeesthoek in African plate, Seychelles and Coco near Indian Plate have been estimated by using the updated data processing software Bernese Version 5.0. (Seeber, 1993; Hofmann-Wellenhof et al., 1992; Hugentobler et al., 2001)

The time series of all these stations and the baseline lengths evolution over 1997 to 2009 have been estimated. Table 1 shows the baseline lengths and their WRMS. Table 2 shows the baseline lengths and their changes per year, and Table 3 shows the North and East velocities of all the sites and the velocity errors with 95% confidence.

Table 1. Estimated baseline lengths and their WRMS

STATION	BASELINE in m.	WRMS
DAV 1	2,172,481.3483	± 0.007612 m
CAS 1	2,933,421.2856	± 0.010758 m
MAIT	3,742,928.0214	± 0.015925 m
YAR 1	4,323,209.0566	± 0.063621 m
HRAO	4,391,931.1326	± 0.014482 m
COCO	4,677,608.2765	± 0.071555 m
SEY 1	5,005,623.2619	± 0.027006 m
TID 2	6,103,757.3093	± 0.047181 m

Table 2. Estimated baseline lengths and their changes per year

STATION	BASELINE in m.	RATE
DAV 1	2,172,481.3483	- 2.4 mm/yr
CAS 1	2,933,421.2856	- 4.8 mm/yr
MAIT	3,742,928.0214	-5.9 mm/yr
YAR 1	4,323,209.0566	+ 5.5 cm/yr
HRAO	4,391,931.1326	- 0.3 mm/yr
COCO	4,677,608.2765	+ 5.5 cm/yr
SEY 1	5,005,623.2619	+ 5.6 mm/yr
TID 2	6,103,757.3093	+ 2.8 cm/yr

Table 3. Estimated site velocities

SITE	N-Vel(mm)	E-Vel(mm)	N-Error(mm)	E-Error (mm)
Mait	11.30	2.30	1.009	1.768
Cas1	12.00	7.60	2.222	4.019
Dav1	2.90	6.40	3.495	2.542
Yar1	-50.80	-27.90	5.59	2.49
Tid2	-43.50	2.20	9.399	3.472
Coco	40.30	4.80	5.696	3.694
Sey1	3.90	17.80	22.794	11.718
Hrao	-15.80	-23.60	6.087	2.703

The objective is to establish the time evolution of baseline length between Hyderabad and Maitri.

The baseline length is more than 10,000 km between HYDE, India, and MAIT, Antarctica. These two different networks are essential for the improvement of the accuracy of the GPS's baseline measurements. Two global networks, namely IND and ANT, have been chosen to connect the two continents geodetically.

This approach is to overcome the limitation in the estimation of baseline length by GPS, the maximum of 6,900 km, due to the less availability of double difference (Hugentobler et al., 2001) observables for the GPS data analysis. Both the networks have common the IGS Station at Diego Garcia (DGAR). Ten years of data from 1997 to 2007 were used. The first network ANT includes Maitri (MAIT), Davis (DAV1), Casey (CAS1) in Antarctica plate, and several stations in the adjacent tectonic plates surrounding Antarctica, including Seychelles (SEY1), COCO, Hartebeesthoek (HARO), Yaragadee (YAR1), and Tidbinbilla (TID2). The other network IND includes Hyderabad (HYDE) in Indian Plate, COCO, DGAR, HRAO, IISC, IRKT, KIT3, LHAS, POL2, MALD, SEY1, WTZR and YAR1.

The station's HYDE, SEY1, DGAR and COCO are shown in Figure 3 whereas. Figure 4 shows the GPS derived velocity vectors.

Figure 3. Two global networks IND and ANT to geodetically tie the two large continents of India and Antarctica

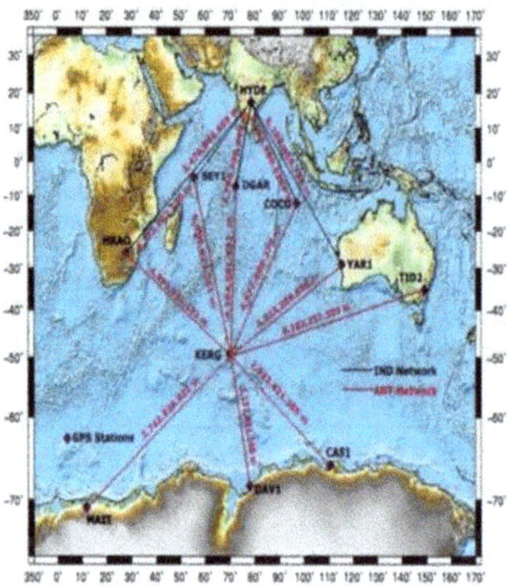

Figure 4. GPS derived velocity vector of individual sites including Maitri in ITRF 2005 reference frame

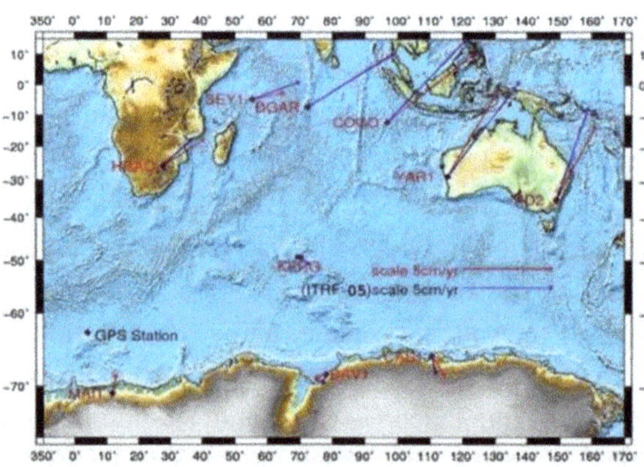

Seismological Observatory at Maitri (Malimai et al., 2008), Antarctica: Indian research base 'Maitri' continues to contribute Seismic data to ISC, UK, to remain in the Global Scenario of Global Seismographic Network at Antarctica. Also, the contribution of the permanent Seismological Observatory at Maitri, Antarctica, to

AnSWeR (Antarctic Seismic Web Resource) continues. Figure-7.5 shows the picture of the entire Broadband stations operating in Antarctica.

Figure 5. Antarctic broadband seismic

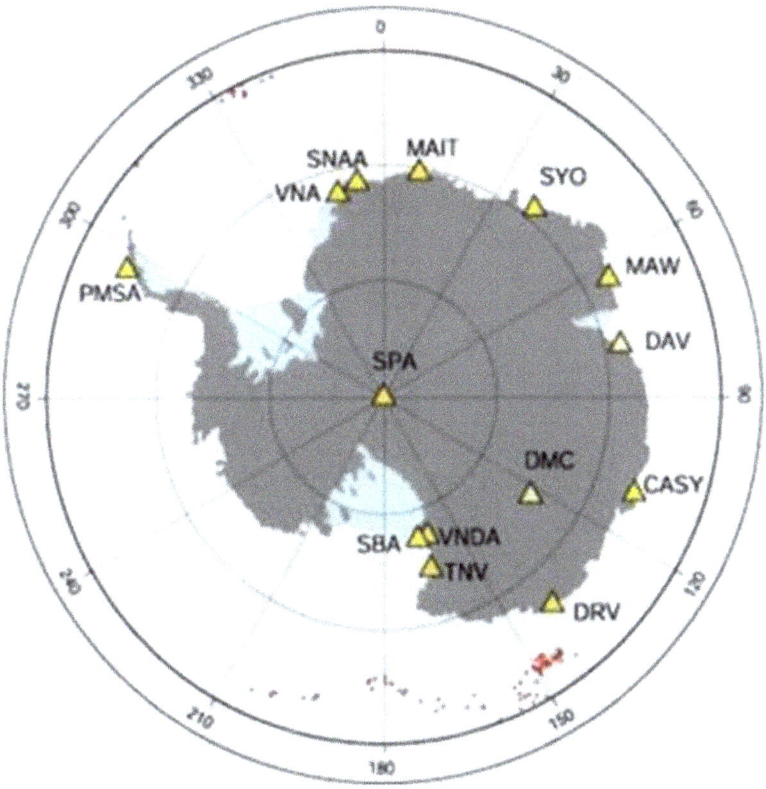

Observatories

Filled triangles=existing stations, open triangles=proposed stations.

Seismic activity in and around Antarctica, which is of interest to the study of Indian plate kinematics, continues to be recorded. (given in Figure 6 and Figure 7)

Figure 6. Geotech and Reftek digital broadband seismic sensors housed inside the vault

Figure 7. Data downloading and processing facility for both GPS and seismic stations shows the Maitri station's seismic sensors and data processing facility

The observatory has recorded 647 earthquakes during 2007. Out of these, 74 earthquakes are of a magnitude that ranges from 4.0 to 4.9. Four hundred eighty-three are of a magnitude ranging from 5.0 to 5.9. Eighty of magnitudes between 6.0 to 6.9. Eight magnitudes ranging from 7.0 to 7.9 and 2 of a magnitude above 8.0. Along with the teleseismic events and several regional earthquakes recorded within the Antarctic plate viz., 1 earthquake of Mb=5.7 from 16.4 deg. from Maitri, that is within Antarctica; 65 earthquakes from the South Sandwich Islands, the distance of 15.5 deg. to 23.2 deg. (Table 4), 4 earthquakes from Scotia sea, and 12 earthquakes from Macquarie Islands

For all these earthquakes, Maitri (MAIT) is the only data contributing agency to ISC, UK, which will help them in their final processing and publishing of ISC's

yearly seismic bulletins. This data is available on the website: http://www.isc.ac.uk as well as in the NGRI database.
Details of the large earthquakes recorded in 2007 are provided in Table 4.

Table 4.

S. No	Location	Date	Mag.	Dist (Deg,)	Lat. (Deg.)	Long. (Deg.)	Arrival time
1	Near The Coast Of Central Peru	15/08/2007	Ms-7.9	76.7	3.386S	76.603W	23 52 43.9
2	Santa Cruz Islands	2/09/2007	Ms-7.3	92.4	11.610S	165.762E	01 18 41.8
3	Suthern Sumatra, Indonesia	12/09/2007	Ms- 8.5	86.2	4.438S	101.367E	11 23 11.4
4	KepulauanMentawai Region, Indonesia	13/09/2007	Ms- 8.1	87.7	2.625S	100.841E	00 01 48.2
5	KepulauanMentawai Region, Indonesia	13/09/2007	Ms- 7.2	87.2	2.130S	99.627E	03 48 16.3
6.	South Of Mariana Islands	13/09/2007	Ms-7.1	113.6	10.454N	145.718E	02 27 11.5
7.	Auckland Islands, New Zealand	30/09/2007	Ms-7.4	58.6	49.271S	164.115E	05 33 34.4
8.	Antofagasta, Chile	14/11/2007	Ms-7.4	65.6,	22.247S	69.890W	15 51 32.8.
9.	South of Fiji Islands	09/12/2007	Mb-7.0	83.0	25.996S	177.514W	07 40 32.110
10.	Alaska	19/12/2007	Ms-7.1	159.7	51.360N	179.509W	09 50 17.1 11
11.	Antarctica	04/11/2007	Mb 5.7	-	67.034S	11.199E	-

Details of the large earthquakes recorded from the South Sandwich Islands in 2007 are provided in Table 5.

Table 5. List of earthquakes recorded from the South Sandwich Islands in 2007

S.No.	Date	Orig. Time	Arr. Time	Lat.	Long.	Depth	Dist	Mag
		hr. mn. sec	hr. mn. sec	South	West	in km.	= °	Mb / Ms
1	06-01-2007	08 32 20.3	08 36 33.4	59.723	26.682	78	18.7	5.0
2	12-01-2007	12 38 44.1	12 42 56.7	59.440	26.685	96	19.3	4.8
3	14-01-2007	11 14 18.5	11 18 36.2	59.171	24.872	10	19.2	4.7
4	20-01-2007	06 21 04.5	06 26 09.1	55.419	29.533	10	22.0	6.2
5	21-01-2007	15 03 16.3	15 08 03.8	55.764	27.016	48	22.4	5.6
6	23-01-2007	22 46 25.6	22 51 26.1	55.305	29.382	35	23.2	5.5
7	05-02-2007	20 58 06.3	21 02 19.4	59.921	28.234	112	19.5	5.2
8	16-02-2007	01 15 28.1	01 20 28.2	55.027	26.461	35	22.3	5.1
9	21-02-2007	08 40 35.1	08 44 47.1	58.923	26.351	113	20.7	4.2
10	28-02-2007	23 13 15.6	23 18 30.3	55.245	29.142	10	22.8	6.0
11	04-03-2007	07 08 01.4	07 12 45.2	56.010	27.490	151	21.9	5.0
12	08-03-2007	11 14 31.5	11 18 00.5	58.217	7.615	10	15.5	5.6
13	01-04-2007	15 24 17.4	15 28 58.5	56.582	26.497	86	22.4	4.8
14	16-04-2007	18 30 30.0	18 34 39.3	60.632	26.686	10	18.5	4.6
15	20-04-2007	11 23 11.5	11 27 28.7	58.293	25.930	134	20.0	4.7
16	28-04-2007	14 02 37.9	14 06 20.8	60.789	20.208	10	16.5	5.6
17	04-05-2007	17 37 47.8	17 42 28.4	56.460	27.322	149	22.2	5.1
18	11-05-2007	20 33 34.0	20 38 24.0	55.936	27.782	92	22.0	5.0
19	19-05-2007	10 38 33.7	10 43 14.0	56.903	24.964	10	21.8	5.5
20	30-05-2007	19 35 21.4	19 39 32.4	59.972	26.457	35	19.1	5.2

Table 6.

21	03-06-2007	01 50 17.0	01 54 48.2	58.121	25.098	10	19.2	4.7
22	03-06-2007	11 17 37.4	11 22 46.7	55.359	29.531	10	23.0	4.9
23	04-06-2007	03 06 59.0	03 11 41.8	55.986	27.696	171	22.1	5.2
24	05-06-2007	15 27 02.1	15 31 19.1	60.182	26.728	10	17.9	5.5
25	06-06-2007	02 41 01.9	02 45 50.3	55.779	26.124	35	22.0	4.5
26	06-06-2007	02 45 55.9	02 50 58.3	55.760	25.980	56	22.0	4.4
27	06-06-2007	15 03 25.8	15 07 44.2	60.150	26.280	10	17.9	4.6
28	07-06-2007	19 35 51.3	19 40 47.0	55.704	26.186	16	22.0	4.9

Continued on following page

Table 6. Continued

29	09-06-2007	10 08 36.1	10 13 27.3	55.779	26.056	37	22.0	4.2
30	12-06-2007	09 49 17.0	09 53 59.8	56.208	27.151	121	22.1	4.8
31	14-06-2007	06 58 14.2	07 02 54.7	56.346	27.283	149	22.1	5.3
32	15-06-2007	03 48 34.8	03 52 54.9	58.596	26.244	147	19.9	5.4
33	01-07-2007	23 45 28.2	23 50 31.1	55.314	28.066	21	22.6	5.2
34	05-07-2007	05 47 49.4	05 52 07.2	59.573	26.303	56	18.4	5.3
35	08-07-2007	07 28 39.8	07 32 56.4	59.131	26.228	102	17.2	5.3
36	11-07-2007	09 42 25.5	09 47 10.9	56.245	26.938	77	22.2	5.0
37	15-07-2007	12 09 48.5	12 14 04.6	59.987	26.104	11	17.2	4.9
38	30-07-2007	22 27 19.9	22 31 28.1	60.217	26.483	54	18.2	4.9
39	31-07-2007	02 42 48.0	02 47 34.7	56.058	27.733	104	21.9	5.9
40	02-08-2007	14 14 03.9	14 18 17.9	60.373	26.630	22	18.3	5.1
41	19-08-2007	00 41 28.1	00 45 39.7	60.345	26.673	10	18.4	5.3
42	19-08-2007	01 04 51.5	01 09 04.5	60.317	26.569	8	18.5	5.2
43	19-08-2007	01 32 39.9	01 36 53.0	60.276	26.473	8	18.4	5.0
44	01-09-2007	23 27 53.0	23 32 39.4	56.187	27.104	98	22.3	5.3
45	18-09-2007	19 52 48.4	19 56 50.5	60.812	24.314	10	18.4	4.9
46	02-10-2007	23 12 16.4	23 16 53.9	56.495	24.626	66	22.2	4.3
47	03-10-2007	04 56 13.3	05 01 00.7	55.931	27.732	146	22.4	4.7
48	06-10-2007	19 44 19.8	19 48 54.5	57.362	25.991	69	20.2	4.7
49	10-10-2007	10 08 43.7	10 13 05.1	58.944	25.561	46	19.2	4.7
50	15-10-2007	09 44 33.1	09 49 20.7	56.031	27.411	96	22.3	4.9
51	03-11-2007	13 52 11.4	13 57 13.6	55.284	29.439	63	22.7	4.8
52	08-11-2007	07 44 31.0	07 49 21.6	55.978	27.931	90	22.0	5.4
53	09-11-2007	21 26 48.4	21 31 28.7	56.266	27.298	155	22.1	5.0
54	11-11-2007	20 31 12.4	20 36 01.6	56.352	26.929	35	22.1	4.5
55	13-11-2007	20 20 19.9	20 24 00.4	58.301	11.416	10	15.6	4.3
56	15-11-2007	12 42 36.1	12 47 20.1	56.313	26.635	71	21.7	5.0
57	17-11-2007	02 44 55.3	02 49 07.2	59.840	25.509	35	18.7	4.4
58	20-11-2007	21 54 45.1	21 59 01.4	59.732	26.441	35	18.8	4.3
59	21-11-2007	03 24 34.3	03 28 51.5	58.983	25.663	62	19.4	4.5
60	29-11-2007	08 46 39.0	08 51 00.1	56.364	28.321	200	22.4	4.8
61	04-12-2007	17 53 46.5	17 58 42.0	56.004	27.427	121	22.1	4.6
62	05-12-2007	08 16 14.2	08 20 44.8	56.556	28.196	272	21.7	4.7
63	14-12-2007	08 28 48.4	08 34 00.3	56.298	27.200	109	22.1	4.5
64	20-12-2007	22 37 56.5	22 42 40.5	55.995	27.763	150	22.1	4.7

Continued on following page

Table 6. Continued

65	24-12-2007	21 30 03.4	21 34 28.7	56.261	27.485	35	22.1	4.5

The Epimap of the Global seismic events recorded at Maitri from 1998 to 2007 has been depicted in Figure 8.

Figure 8. Complete epimap of all the global seismic events recorded at Maitri (MAIT) from the inception in 1998 until 2007

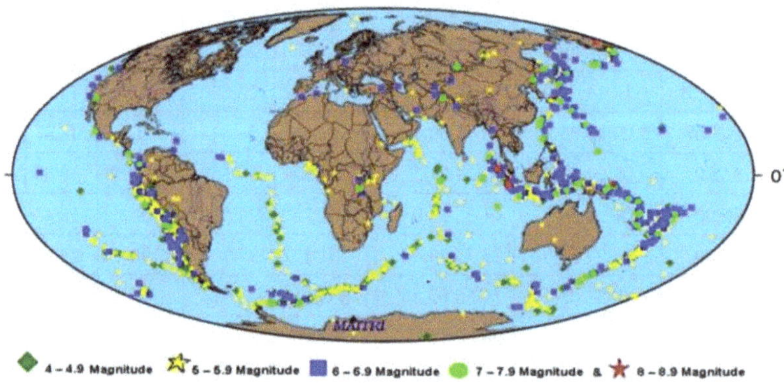

MOHO depth beneath Maitri has also been estimated to be about 40 km using Receiver Function Analysis. For this analysis, 67 receiver functions and 184 events were used. It is shown in Figure-7.9.

Along with this, the near-event earthquakes were recorded at Maitri. It also corroborates the geodynamical processes that take place continuously in the Indian Ocean. It results in interplate movements. All these dynamic processes may constitute the driving mechanisms between India and Antarctica. These forces also contribute to the non-rigid behaviour of the Indian Plate and the Indian Ocean Basin (Akilan et al., 2013).

NGRI is the only organisation in the entire country to have a Permanent GPS station as a permanent geodetic marker that is continuously operational since 1997 in Antarctica.

The presence of an IGS Permanent GPS Station at NGRI, Hyderabad, India and a Permanent GPS station at Maitri, Antarctica, positively impacts the studies on the Indian Plate Kinematics and continental dynamics between India and Antarctica. Geodetically linking the two stations and estimating the geodynamical processes continually evolving in the southern Indian peninsula. NGRI has a second permanent GPS station at Mahendragiri, the southernmost tip of India. Data from these permanent stations would add up to form the information for all the studies.

SIGNIFICANT ACHIEVEMENTS

High-precision space–geodetic data from 1997 to 2007 have been analysed to investigate the tectonic activity, plate boundary organisations, crustal deformation in the southern Indian peninsula, the driving mechanisms and the response of the Indian Ocean lithosphere.

Among the two global networks (IND and ANT) that have been chosen and analysed, the station's HYDE and MAIT are geodetically linked through DGAR.

Our analysis and results show an increase of baseline lengths between Kerguelen in the Antarctic plate and other stations. A shortening of baseline lengths between HYDE in the Indian plate and other common stations are depicted in Table-7.6

Table 7. Estimated result of geodetic tie-up between Indian and Antarctica plate

Station	Baseline length in m	Baseline change from Kerguelen	Baseline length in m	Baseline change from Hyderabad
Seychelles	5,005,623.260381	0.0049m/yr	3,476, 844.689153	-0.0117m/yr
Diego Garcia	4, 564, 601. 532	0.0340m/yr	2,790, 790.349066	-0.005m/yr
Coco	4,677,608.665675	0.0553m/yr	3,783, 980.525856	-0.0014m/yr

With this geodetic tie-up, having got the geodetic signatures of the geodynamical processes between India and Antarctica, continuous monitoring had enhanced the understanding of the crustal deformation processes between these two continents. GPS based data from 3 sites MAIT, CAS1 and DAV1, within the Antarctica plate have accumulated strain of the order of 10^{-9} yr^{-1}. GPS data at COCO suggests a high rate of movement. It could result from excessive strain accumulation due to the Indo-Australian diffuse plate boundary forms acting upon this region. The GPS analysis confirms the emergence of a diffuse plate boundary between India and Australia and relates it to the late Miocene Himalayan uplift. To the West of the Indian Peninsula, the calculated stress field roughly has a N-S directed tensional and E-W oriented compressional character. The velocity vectors of all other sites highlight the major causes of the accumulation processes of strain in the Indian Ocean and the northward movement of the Indian plate. The earthquakes recorded at Maitri and their fault plane solution analysis reveal most faults in the Indian Ocean, specifically the strike-slip flaws along the ridges. A few are thrust faults near the Indo-Australian diffused plate boundary.

The seismic data recorded since 1997 and its analysis clearly shows an increasing seismicity trend in and around Antarctica. Significant increasing seismicity along the ridges in the Indian Ocean and their spreading rates concerns the seismic tectonics study and the Indian plate kinematics. Our analysis also corroborates the evidence of the deforming zone between India and Antarctica. The stations within the Antarctic Plate, such as Maitri, Casey, and Davis, show a consistent movement. Within the Antarctic Plate itself, these stations move together as a plate.

The relevance of the achievements in Antarctic Research from the National perspective and future growth of the country:

The GPS based velocity fields of Maitri and other IGS stations (Casey, Davis, Seychelles, COCO, Hartebeesthoek, Yaragadee, and Tidbinbilla) in different plates in relation to Kerguelen show a deviation from rigid plate behaviour in a manner consistent with the mapped intraplate stress field, observations of deformations and seismicity in the region. The data show a lengthening of baselines for sites Yaragadee and Tidbinbilla in the Australian plate; and Seychelles and COCO in the Indian plate at the rates of 5.5cm/yr 2.8cm/yr, 5.6mm/yr, and 5.5 cm/yr, respectively. The rate of change of COCO is an interesting observation as this high rate of movement of COCO Island in Comparison to Seychelles could be the result of excessive strain accumulation due to the Indo-Australia diffuse plate boundary forces acting upon this region. The station Hartebeesthoek (HRAO) in the South African plate shows a trend of minimal shortening at the rate of 1.4mm/yr. The velocities and the corresponding errors are defined by a weighted least squares line that fits the daily position data for 1997 to 2008, assuming a simple white (uncorrelated) noise model.

The velocities within uncertainties (± 1) positively agree between geodetic and geologic models. Our estimated velocity vectors do conform to those of SCAR GPS campaigns consistent with rigid plate rotation. The velocity vectors of the stations Yaragadee and Tidbinbilla in the Australian plate indicate that both stations show movements away from the Indian plate. It conforms to the recent plate tectonic theory that India and Australia lie in two plates with diffuse boundaries separating them. Precisions are approximately 3-4mm (North), 5-6 mm (East), and 10-12mm (vertical

The GPS velocity fields of the global network of GPS stations in and around the Indian plate, including Maitri at the Indian Antarctic Station, indicate the computed elastic strain accumulation in the southern Indian peninsula. It shows a significant departure from rigid plate behaviour.

The results of intraplate strain accumulation within the Antarctica Plate cover three sites MAIT, CAS1 and DAV1. These are $1.8x10^{-9}yr^{-1}$,$1.6x10^{-9}yr^{-1}$ and $1.1x10^{-9}yr^{-1}$, respectively. The interplate strain accumulation between Antarctica and other plates, such as Somalia (SEY1), Africa (HARO), Australia (YAR1), and the diffuse plate boundary between India and Australia (COCO) are respectively $1.1x10^{-9}yr^{-1}$, $1.0x10^{-10}yr^{-1}$,$1.27x10^{-8}yr^{-1}$ and $1.18x10^{-8}yr^{-1}$.

These estimates are in tandem with the earlier studies on the estimation of global strain rate. GPS and seismic analysis together confirm the emergence of a diffuse plate boundary between India and Australia. It relates to the late Miocene Himalayan uplift (Molnar, 1989; 1990; Ni and Barazangi, 1984). The calculated stress field to the West of the Indian Peninsula has a roughly N-S directed tensional and E-W oriented compressional character. Also, the velocity vectors of all other sites bring significant insight into the causal linkages between the accumulation processes of stress in the Indian Ocean and the northward movement of the Indian plate.

These boundary forces may have led to frequent plate boundary reorganisations in the past, generation of small plates such as the Capricorn Plate or a diffused boundary zone between India and Australia, complicating Kinematic interpretations. These boundary forces may also contribute to the non-rigid behaviour of the Indian Plate. The net result of these complexities is that the plate kinematics of the region remains ambiguous compared to other regions, and the issue of plate kinematics is inextricably linked to the question of plate rigidity.

The GPS-Geodesy requires observations of a more extended period, measurements, and analysis for definite and tangible results concerning the tasks at hand. There is an immediate requirement to establish a GPS array consisting of a network of a minimum of 5 GPS stations in and around Maitri. The new third station will come up at Larsemann Hills.

An array of GPS stations would help to refine the estimation of deformation processes within Antarctica. The existing GPS IGS stations, for example, Casey and Davis, are far from our permanent Station, which impairs the deformation processes' critical evaluation. Hence, an array of GPS stations would further help in estimating the internal deformation processes within Antarctica.

Similarly, a seismic array with a minimum of five broadband seismic stations can help precisely locate the epicentres and hypocentres of the earthquakes, even minor magnitude ones that occur in and around Antarctica, which are the main objectives of the long-term programmes. This array would also help to assess the velocity and crustal structure beneath Maitri and the surrounding area

It may be summarised that the time series capable of providing reliable information on crustal deformation processes and the seismic activities taking place in and around Antarctica has been estimated between 1997 and 2008. These were possible only because of the GPS and Seismic Stations at Maitri, Antarctica. This study has helped to understand the driving mechanisms.

From the GPS velocity vector map, it is evident that the islands Seychelles and Coco are converging towards the Indian Plate. The station Yaragadee in the Australian Plate moves away from the Indian Plate, confirming Indo-Australian Plate's drifting, especially in Western Australia. The GPS and the Seismic results show the strain accumulation and deformation processes towards the Indian Plate. It also confirms the emergence of a diffused Plate boundary between India and Antarctica. The stations within the Antarctic Plate, such as Maitri, Casey, and Davis, show a consistent movement within the Antarctic Plate. Partly, the Indian Ocean Basin plate kinematics may be related to the unique set of forces on the boundaries. These boundary forces may have led to frequent plate boundary reorganisations in the past. As a result, the generation of either small plates such as the Capricorn Plates or a diffused boundary zone between India and Australia complicating Kinematic interpretations.

These boundary forces may also contribute to the non-rigid behaviour of the Indian Plate. The net result of these complexities is that the plate kinematics of the region remains less well described compared to other regions, and the issue of plate kinematics is inextricably linked to the question of plate rigidity. The GPS-Geodesy requires a longer period of observation, measurements, and analysis for definite and tangible results regarding the tasks at hand. Therefore, these studies, which are of national and international importance, would continue for a longer time to precisely estimate the seismicity and tectonic activity in and around Antarctica and in the Indian Ocean.

REFERENCES

Akilan, A., Balaji, S., Srinivas, Y., & Yuvaraj, N. (2013). Plate motion predictability using Hurst exponent applied to Maitri-Antarctica GPS network. *Journal of the Geological Society of India*, *82*(6), 613–620. doi:10.100712594-013-0199-z

Argus, D. F., & Dorden, R. G. (1991). No Net-rotation model of current plate velocities incorporating plate motion model NUVEL-1. *GRL*, *18*(11), 2039–2042. doi:10.1029/91GL01532

Armijo, R., Tapponnier, P., Mercier, J. L., & Tanglin, H. (1986). Quaternary extension in southern Tibet: Field observations and tectonic implications. *Journal of Geophysical Research*, *91*(B14), 803–13,872. doi:10.1029/JB091iB14p13803

Bendick, R., & Bilham, R. (1999). Search for buckling of the southwest Indian coast related to Himalayan collision. *Geological Society of America. Special Paper*, *328*, 313–321.

Billham, R., & Gaur, V. K. (2000). Geodetic contributions to the study of seismotectonics in India. *Current Science*, *79*, 1259–1269.

Bouin, M.F., & Vigny, C. (2000). New constraints on Antarctic plate motion and deformation from GPS data. *J.Geophys.Res., 105*, 279-293.

DeMets, C., Gordon, R. G., Argus, D. F., & Stein, S. (1990). Current Plate Motions. *Geophysical Journal International*, *101*(2), 425–478. doi:10.1111/j.1365-246X.1990.tb06579.x

DeMets, Gorden, Argus, & Stein. (1994). Effect of recent revisions to geo magnetic time scale on estimates of current plate motions. *Geophysical Research Letters, 21*, 2191-2194.

Gorden, R. G., DeMets, C., & Royer, J.-Y. (1998). Evidence of for long-term diffuse deformation of the lithosphere of equatorial Indian Ocean. *Nature*, *395*(6700), 370–374. doi:10.1038/26463

Gorden, R. G., & Stein, S. (1992). Global tectonics and space geodesy. *Science*, *256*(5055), 333–342. doi:10.1126cience.256.5055.333 PMID:17743109

Holt, W. E., Chamot-Rooke, N., Pichon, X., & Le. (2000). The velocity field in Asia inferred from Quaternary fault slip rates and Global Positioning System observations. *Journal of Geophysical Research*, *105*(B8), 19185–19209. doi:10.1029/2000JB900045

Kreemer, C., Holt, W. E., & Haines, A. J. (2003). An integrated global model of present-day plate motions and plate boundary deformation. *Geophysical Journal International*, *154*(1), 8–34. doi:10.1046/j.1365-246X.2003.01917.x

Larson, K. M., Burgmann, R., Bilham, R., & Freymueller, J. T. (1999). Kinematics of the India-Eurasia collision zone from GPS measurements. *Journal of Geophysical Research*, *104*(B1), 1077–1093. doi:10.1029/1998JB900043

Malaimani, E.C., & Ravi Kumar, N. (2006). *Recent results on Indian plate kinematics with the longer time span of GPS measurements in the global network solution*. EOS Special Volume, 53 B-0893.

Malaimani, E. C., Campbell, J., Görres, B., Kotthoff, H., & Smaritschnik, S. (2000). Indian plate kinematics studies by GPS-geodesy. *Earth, Planets, and Space*, *52*(10), 741–745. doi:10.1186/BF03352275

Molnar, A. (1990). review of the seismicity and the rates of active undertrusting and deformation at the Himalaya. *J. of Himalayan Geology*, *1990*, 131–154.

Molar, P., & Lyon-Caen, H. (1989). Fault plane solutions of earthquakes and active tectonics of the Tibetan plateau and its margins. *Geophysical Journal International*, *99*(1), 123–153. doi:10.1111/j.1365-246X.1989.tb02020.x

Ni, J., & Barazangi, M. (1984). Seismotectonics of the Himalayan collision zone: Geometry of underthrusting Indian plate benath the Himalaya. *Journal of Geophysical Research*, *1989*(B2), 1147–1163. doi:10.1029/JB089iB02p01147

Shen, Z.-K., Zhao, C., Yin, A., Li, Y., Jackson, D. D., Fang, P., & Dong, D. (2000). Contemporary crustal deformation in east Asia constrained by Global Positioning measurements. *Journal of Geophysical Research*, *105*(B3), 5721–5734. doi:10.1029/1999JB900391

Subrahmanya, K. R. (1996). Active intraplate deformation in south India. *Tectonophysics*, *262*(1-4), 231–241. doi:10.1016/0040-1951(96)00005-4

Seeber, G. (1993). *Text book on Sattlite Geodesy*. Walter de Gruyter & Co.

Hofmann-Wellenhof, B., Lichtenegger, H., & Collins, J. (1992). *Textbook on Global Positioning System Theory and Practice*. Academic Press.

Hugentobler, U., Cchaer, S., & Fridez, P. (2001). *Bernese GPS software version 4.2 manual*. Astronomical Institute, University of Berne.

Ravi Kumar, N., Malaimani, E. C., & Akilan, A. (2006). Geodynamical Processes between Antarctica and India as revealed by very long base lines between the continents estimated from continuous and long-term GPS measurements. *EOS Special Volume G*, *53*, B-0906.

Ravi Kumar, N., Malaimani, E.C., Prem Kishore, L., Chaitanya, T., Akilan, A., Srinivas, G.S., Navin Chander, G.B., & Rao, S.V.R.R. (2004). *Significance and importance of collocating permanent GPS and seismic observatory at Maitri, Antarctica*. Terra Nostra, Schriften der Alfred- Wegner- Stiftung 2004/4.

Malaimani, E.C., & Ravi Kumar, N., Padhy, S., Rao, S.V.R.R., Navin Chander, G.B., Srinivas, G.S., & Akilan, A. (2008). Continuous monitoring of seismicity by Indian permanent seismological observatory at Maitri. *Indian Journal of Geo-Marine Sciences*, *37*, 396–403.

Ravi Kumar, N., Malaimani, E. C., & Akilan, A. (2008). Strain accumulation studies between Antarctica and India through geodetic tying of two continents from continuous and long-term GPS measurements. *Indian Journal of Geo-Marine Sciences*, *37*, 404–411.

Malaimani, E. C., Ravikumar, N., Akilan, A., & Abilash, K. (2007). GPS-Geodesy with GNSS Receivers for revalidation of Indian Plate Kinematics with recent plate velocities. *Proceedings of Conference on Global Navigation Satellite System and Its Applications*.

Ravi Kumar, N., Malaimani, E.C., Rao, S.V.R.R., Akilan, A., & Abilash, K. (2007). Geodetic tying of Antarctica and India by continuous GPS measurements for strain accumulation studies. *Proceedings of Conference on Global Navigation Satellite System and Its Applications*.

Chapter 8
Hydrographic Surveys in Antarctica

Prashant Srivastava
Ministry of Earth Sciences, India

ABSTRACT

Hydrographic and physical oceanographic studies could enhance the national data bank. They could also greatly assist the various national agencies in correlating their work with the Antarctic waters' bathymetric and physical properties, leading to an improved understanding of the characteristics of the Antarctica waters. The bathymetric data gathered near the permanent station 'Maitri' could be utilised to exchange data with other nations like Russia and IHO. As per the charting scheme promulgated by the International Hydrographic Organization (IHO), INT Chart Nos. 9050 and 9051 are required to be co-produced by Russia and India. In the INT Chart 9050, vast blank areas exist due to the non-availability of bathymetric data. Thus, concerted efforts are required to fill in these gaps by undertaking systematic hydrographic surveys. These ocean areas are close to the permanent station in Antarctica. It is in the best interest of India to have a well surveyed navigational chart near the permanent station.

INTRODUCTION

The Antarctic is a vast but little known area of the globe which is assuming an ever increasing importance as man's frontiers of knowledge (Dixon, 1965). The Antarctic represents unique challenges for surveying and charting due to its remoteness, severe climatic conditions (Nairn, 2010). Indubitably, the need for accurate and reliable nautical charts is essential specially with the increasing number of vessels of all sizes

DOI: 10.4018/978-1-6684-4078-0.ch008

navigating the area for scientific research, in support of national Antarctic programs (Nairn, 2010). The survey and charting of bodies of water, such as seas, lakes, rivers, etc., are known as Hydrography (http://www.iho.int/srv1/index.php?option=com_content&view=article&id=299&Itemid=289). The parameters of the seas and the oceans augments the hydrographic mapping. It primarily focuses on the shape of the bottom of such water bodies. Initially, it has been associated with the safe and optimum operation of shipping. With rapid expansion in little-known territory, it has not always been easy for the ships reaching to the Antarctic waters to move for much of their time in little surveyed, and often completely uncharted, waters (Wynne-Edwards, 1960). Still, its relevance in protecting the marine environment and coastal studies has become more evident in the twentieth century. It is also essential for disaster warning and response, as learned from the Indian Ocean tsunami of December 2004. The environment in Antarctica is changing perhaps more rapidly than anywhere else in the world (Jones-Couture, 2020). Antarctic waters are one of the most challenging and fragile marine regions on the face of the globe. Accurate charts are essential to help and guide human activity in the area. Navigation data is poorly available in Antarctica. If navigational data is available it often dates back to the early 1900s with many lead line surveys (Wilkins et. al., 2013). It is estimated currently only 10% of Antarctic waters are surveyed to IHO specifications, with some charts dating as far back as the 1900s (Jones-Couture, 2020). Hydrographic training is expensive, and assets are scarce. Hence, coordination of international efforts is of the utmost importance (Report on Antarctic Treaty Consultative Meeting, 2007).

Scope of Hydrographic Survey

To publish the Navigational Chart INT 9051 and 9050. The following scope of the work involved.

1. Collection of nearly 20000 nautical miles of bathymetric data to publish the chart
2. Collection of 250 Nautical Miles (approx.) of coastline data of the Ice shelf area
3. To collect meteorological and oceanographic data in the area
4. Tide observation of the area
5. Tidal stream observation in the area
6. Collection of seabed samples if feasible
7. Sound velocity assessment.
8. Reconfirmation and Validation of the data collected during the previous expeditions.

Status of Hydrographic Survey in Antarctica

Valuable data has been collected during the last seven expeditions by the Naval Hydrographic Survey Teams. About 1000 Nautical Miles of Sounding (survey) has been completed. The teams have completed the ice shelf delineation. They have also collected valuable Meteorological and Oceanographic data collaborating with other members of the scientific groups. Tidal Stream and current observations have been undertaken at various possible locations by different teams. Tide observation and seabed sampling could not be undertaken due to the area's specifications, extreme climatic conditions, and depth. The varied data collected during The Expeditions has immensely enriched the data bank of the National Hydrography Office (NHO) and has been effectively utilised for the compilation and improvement of Navigational Chart INT 9051.

SIGNIFICANT ACHIEVEMENTS

National Hydrographic Office has been participating in the Indian Antarctic expedition since 1995. The first team from National Hydrographic Office participated in the XIV Indian Antarctic Expedition. To collect bathymetric data, a Survey Motor Boat (SMB) was mounted on The Expedition vessel. This Expedition helped study the suitability of the marine equipment to the extreme weather, the feasibility of The ship and boat operation, and human endurance in icy conditions. The team was able to collect limited bathymetric data for the preparation of the chart.

The survey team of the XVI expedition was able to map the coastline within approaches to the India Bay (Chart 9051) with the help of a modified Survey Motor Boat (SMB) onboard *MV Polar Bird.* During the XVI Indian Antarctic Expedition, the Princess Astrid coast was delineated. However, attempts made by the survey team to collect bathymetric data in India Bay did not produce the desired results due to frequent breakdown of SMB even though it was modified for polar conditions.

Desired results were not achieved from the XIV and XVI expedition. Hence, a survey launch of *MV Polar Bird* was hired during the 18th Expedition to collect bathymetric data. The team observed that the survey boat's sounding could be undertaken only in ice-free zones as a single boat could not manage the operations alone. The survey team could collect only limited bathymetric data in the area. However, the survey team collected valuable data for the Sailing Directions (a navigational publication) and attempted to delineate the coastline of the ice shelf with a helicopter. The general depth in the area ranges between 100 to 200 meters, and that an ice-class ship should be used for undertaking hydrographic surveys in the region. During the XX Expedition, The Expedition ship itself was utilised for the collection of bathymetric data. The helicopter hired for transportation of scientists to Maitri from The Expedition vessel was modified for delineating the ice shelf. As far as the bathymetric data collection was concerned, the field output was low due to unprecedented field conditions. The area was covered by fast ice for a significant part of The Expedition. However, 120-line miles of bathymetric data was collected during the time available. Delineation of nearly 130 Km of the ice shelf was also carried out using The helicopter of The Expedition vessel.

During XXI Indian Antarctic Expedition, 300 NM of the seabed was surveyed by Hydrographic Survey Team utilising The Expedition vessel for surveying. The field output in the bathymetric data collection was not up to the expected levels due to unfavourable field conditions, as 90% of the survey area was covered by ice. The survey team also carried out physical Meteorological observations.

The methodology developed in the previous expeditions was used during the XXII Expedition. The survey team was able to collect 200 miles of bathymetric data and was also able to delineate 120 Kms of ice-shelf coastline. Also, 300 miles of passage bathymetry data were compiled by the team. The non-availability of The Expedition vessel to survey due to the limitation of fuel was a significant handicap for further survey progress. Only limited flying hours of The Expedition helicopter were made available to the team. Despite these odds, the survey team collected valuable bathymetric and oceanographic data.

During the XXIII Indian Antarctica expedition (from December 2003 to March 2004), the team accomplished 310 NM of sounding and 150 NM of coastline. Indian Naval Hydrographic Teams have participated in seven validating the quality of data collected during earlier expeditions. It was achieved by running cross lines in the previously surveyed areas,

IMPORTANT OUTPUTS

XIV Antarctic Expedition (Period: December 1994 to March 1995)

The primary role was exploratory, keeping in mind the requirement to study the encompassing field conditions for deployment of Personnel/ Survey Equipment in sub-zero temperatures. A Survey Motor Boat was also mounted onboard MV Polar Bird on a trial basis to undertake sounding work. The ship left Goa on 17 December 1994 and arrived in Antarctica on 10 January 1995. Upon arriving at Antarctica, an initial survey was carried out to set up shore stations and establish Ground Control. The various survey activities commenced on 17 January 1994 (Singh, 1998).

SURVEY PREPARATIONS

The proposal for carrying out the Hydrographic Survey in Antarctica was approved by mid-November 1994. The objectives were to conduct a detailed study to complete a Hydrographic survey and operation of Survey Equipment in extreme climatic conditions. In the absence of technical support, all technical matters such as running the boat, charging of batteries, maintenance of equipment and emergency repairs were carried out by the Hydrographic team.

The time spent by the survey team residing in Goa before departure for Antarctica was effectively utilised to familiarise the crew members on various aspects involved in the operation of the Survey Motor Boat. The different types of survey equipment were calibrated and tested for their operational status before the team's departure for Antarctica.

Geodetic Control was established on 17 January using GPS 200. The only permanent Station available for use was at 'Maitri', which the Survey of India found in 1991. Survey equipment such as Trisponder, Smart acoustic current meter, Distomat, etc., was deployed in the survey grounds to monitor performance under extreme field conditions. Because of logistical difficulties in servicing ground stations, deploying GPS 200 in the kinematic mode was decided to fix the position. On a trial basis, the sounding of the India Bay was also planned.

Terrain

The landmass was found to be fully covered with snow and ice. The existing mountain ranges near Maitri station were not found to be covered with ice. Only two distinct types of ground surface existed- blue ice, which was very hard and slippery, and the snow cover, which was very soft and difficult to walk.

Mobility

The snow vehicles, which had tank tracks, followed familiar routes and rarely ventured out to the unknown territory.

Coastline

The average height of the ice shelf was about 10 meters from the sea surface and was generally composed of blue ice. The coast was like that of Konkan Coast, comprising of icy cliffs. The frozen shelf reflected about 80% of the sun's radiation. Therefore, it did not result in melting and was suitable for the stable movements of vehicles and personnel. However, as the blue ice was very slippery and posed a serious threat to personnel's safety, it was not feasible to delineate the coastline accurately by walking over the edges.

Ground Control

The inability to mark and build a suitable site on the icy continent would necessitate new ground control at the beginning of every survey. The newly inducted GPS 200 and SKI post-processing system resulted in effective and prompt Ground Control in a short time. Survey of India (SOI) had established a permanent station at 'Maitri' next to Priyadarshini Lake, and the same was used for extension of control.

Bathymetry

The SMB ran the sounding lines in the North-South axis, which was approximately 125 meters apart. Sound velocity of 1460 Meters/Sec was used for sounding. The survey was carried out on a scale of 1:25,000 using GPS 200 in Kinematic mode. The average depth in the area was observed to be 220 Meters.

Temperature

At the peak of summer, in December, the average temperature was + 5 Degree Centigrade. After that, there was a gradual drop, and by the end of February, temperatures up to –10 Degree were encountered.

Weather

The prevailing weather condition in Antarctica was the most crucial factor during the conduct of the survey. The meteorologists have been collecting data in Antarctica, but they have not analysed and predicted any fixed behavioural pattern. The boat operation was possible only from January to mid-March. An experience of the previous 14 years of Indian expeditions revealed that during Jan and Feb, a maximum of 20 clear days was generally available for any meaningful fieldwork.

Due to the prevailing weather conditions in Antarctica, only a limited Hydrographic task could be accomplished with the help of a survey motorboat.

The SMB was considered unsuitable for polar conditions. The team recommended various modifications to make it suitable for operations in the Antarctica environment. On a clear day, the chartered vessel was committed to operating near India Bay for providing logistics and helicopter support for the scientific Expedition which resulted in considerable boat time being spent in transit to and from the grounds of the survey, which severely hampered the overall survey output during The Expedition.

Predominant consideration would have to be given to the weather while preparing a long term perspective plan for charting Antarctica. As an estimate, not more than 20 clear days are likely to be available during the entire survey season from January to mid-March.

A hydrographic team from the Indian Navy for the first time participated in The Expedition provided invaluable experience to the team members in operating the survey equipment in the harsh climatic conditions of the Antarctic. The knowledge gained by the team paved the way for subsequent expeditions.

XVI ANTARCTIC EXPEDITION (PERIOD DECEMBER 1996 TO APRIL 1997)

The Hydrographic Survey was carried out as per the guidelines issued by the National Hydrographic Office. The survey team was deputed to participate in the XVI Antarctica Expedition. A Survey Motor Boat suitably modified to operate in the Antarctic region was also embarked onboard *MV Polar Bird* during the XVI Indian Antarctica Expedition. MV' Polar Bird' sailed from Mormugao on 12 December 1996 and arrived in Antarctica on 04 January 1997. As the Hydrographic Team was an integral part of The Expedition, the team was actively involved in logistic support activities per The Expedition leader's guidance, from 04 January 1997 to 31 January 1997 Chandrashekharan & Sreedharan, 2000). The ship finally reached the shelf at the Indian discharge point in the Indian Bay at Latitude 69° 56'. 77" S and Longitude 11° 54'. 87" E, on 31 January 1997 (Figure 1).

Figure 1. Map of Antarctica showing 'Maitri'

Conduct of Hydrographic Survey of Antarctic Waters

During the period of 04 January to 31 January 1997, while The ship photo of MV Polar Bird (Figure-2) was stuck in fast ice, no meaningful survey activity could be progressed. However, the period was effectively utilised to prepare and calibrate Survey Motor Boat and other equipment for deployment during the survey. On arrival at the shelf on 01 February 1997, the team concentrated primarily on undertaking and completing the objectives of the Hydrographic survey tasks at hand. The Antarctic terrain and the hostile climate was an alien experience for the team.

Table 1. The limits of the survey of the sheet 2-approaches to the India Bay was

(a)	Latitude	69° 45'.00 S	to 71°05'.00 S
(b)	Longitude 11° 30'.00 E		to 12° 00'.0 E
(c)	Scale	1: 50000	

Figure 2. Ship fitted GPS system

Table 2. The limits of the survey of chart 9051 were

(a)	Latitude	69° 45'.00 E	to 71°05'.00 S
(b)	Longitude 11° 00'.00 E		to 14° 00'.0 E
(c)	Scale	1: 200000	

Geodetic Control

The surveys were carried out on World Geodetic Systems 1984 (WGS 84) Datum, Universal Transverse Mercator Projection Grid, Zone 32 with the Central meridian 09° E. The Geodetic Control was set up on the Ice Shelf. The changes in the geomorphology and the dynamics of the Ice Shelf requires repetitive geodetic control in subsequent expeditions. As no original data existed for the area, the base measurement was the only confidence building method to judge the reliability and accuracy of GPS 200 coordinates in these latitudes. The comparisons were in good correlation and were duly documented along with the coordinates of the five stations established using GPS 200. The STN004-D atop the 'Sankalp Containerized Refuge Hut' was initially set up as the reference site for the Trisponder DGPS on 15 February 1997. During the trials, the Station was observed to be situated far inside. As the boat approached the ice shelf, the electronic link would be disconnected due to the height of the Ice shelf edge (10 to 15 meters). On 20 February 1997, the Sankalp Containerized refuge hut had to be removed to convey to Maitri. Therefore, the DGPS reference unit was relocated to STN005-E on the Fuel Tanks nearer to the Ice shelf's edge, which was subsequently used for all trials and coast lining by The helicopter using DGPS.

Digital Surveying System

Leica GPS 200, Trisponder 1009 + GPS, Tellurometer MRA-7 and Echo Sounder Atlas Deso 25 were the only digital systems successfully deployed in the field on this Expedition apart from the Wipro 486 DX II Laptop, which was selectively used in the boat for trials. The Smart Acoustic Current meter was also deployed from The ship but was unsuccessful.

Navaids

Trisponder Del Norte 1009 + GPS was used in the Differential GPS mode to conduct an excellent and practical survey. However, it was strongly recommended in light of the harsh Antarctic climate that only new sets be embarked for Antarctica. Systems that have lived in India were to be avoided due to the lack of equipment reliability resulting from the component ageing factor.

Bathymetry

Echo Sounder Atlas Deso 25 functioned successfully. However, only the 210 kHz frequency could be used as the connector of the 33 kHz transducer was rendered non-operational. No sounding was undertaken due to the problems concerning the Survey Motor Boat.

Sonar/Survey Boat

The primary objective of bathymetric sounding itself posed numerous problems. These were primarily due to setbacks of the Survey Motor Boat.

Tides and Sounding Datum

The average height of the Ice shelf edge was approximately 8-10 metres above MSL with irregular and uneven edges. The Ice shelf's edge was caving into the sea up to 3-5 metres from the border. It was impractical to go close to the Ice shelf's advantage to less than 3-5 metres. Given the circumstances and the conditions, it was not possible to erect a tide pole.

Tidal Streams

Distinctive flow patterns along the Ice shelf in the 'India Bay' were observed with a semi-diurnal variation. The bergy bits and small Icebergs would flow in and out of the Bay once every 6-8 hours. The SACM was found to be ineffective in the Antarctic environment. The queer phenomenon of an LCD screen freezing was observed.

Coastline, Topography, Conspicuous Objects and Marks

The coastline of the entire area falling within limits was calculated by helicopter using the DGPS. An attempt was also made to completely map the edges of the Ice shelf, forming the coastal features within the limits of chart 9051. However, due to the scarcity of time available for the task and DGPS logging error, the entire area within chart 9051 could not be mapped. The topography was relatively low level and slightly undulating with the increase in height. One headed further South towards the mountains and towards the Schirmarcher Oasis, where the Indian station 'Maitri' was located. The Wholthat mountain range and the Grubber Mountains were visible from the ship. The shelf was found to have areas riddled with major crevasses.

A study was undertaken of the extent of cracks in the Ice shelf to assess the continued use of the present discharge point. Although the areas of all visible cracks existing were mapped, no comments could be made on the gravity of the situation in the absence of any previous data.

XVIII ANTARCTIC EXPEDITION (PERIOD DECEMBER 1998 TO APRIL 1999)

The Antarctica Hydrographic Survey Team embarked upon MV Polar Bird and other XVII Indian Antarctica expeditions on 14 December 1998, setting sail from Goa. Regular passage meteorological observations commenced on 16 December 1998. On 17 December 1998, the team carried out the first CTD profiling using Seacat SBE 19-03 CTD Recorder. After that, CTD profiling was carried out at every 5° of latitude. After an uneventful passage through calm to moderate seas, the vessel was docked at Port Louis, Mauritius, on 24 December 1998. Rough seas and frequent engine breakdowns marked the passage from Mauritius. The ship was often set adrift in rough seas, for several hours at a stretch, as the engine was being repaired. The sea worsened as the vessel crossed the 40° south latitude to the 'roaring forties. Due to her rounded bottom (meant for riding ice), this vessel had no roll stabilising system whatsoever, hence rolled heavily in rough seas.

Icebergs/Passage to Antarctica

As the vessel headed south, the outside temperature dropped steadily. The first iceberg was sighted on 07 January 1999, after Crossing the 60° south latitude. Initially, the icebergs were far, and few, but their numbers increased steadily as the ship progressed south. Soon there were icebergs in every direction, and The ship had to navigate with extreme caution. Some of the icebergs were over a kilometre wide.

Pack Ice

In the early hours of 09 January 1999, the vessel encountered the 'pack ice'. By this time, the sea had calmed considerably. The temperature outside had dropped to near zero. The ship made its way through the gaps in the pack ice, slowly and laboriously. Often maximum engine power was required to push the large chunks of ice that came in the way. Good speed under these circumstances seldom meant more than 4 to 5 knots. The white snow-covered ice in every direction with several large icebergs stranded within was a breath-taking scene.

Fast Ice and Polynya

Around the midnight of 09 January 1999, while the sun was still shone brightly above the horizon, the vessel came to a halt on the 'fast ice', at latitude 69°41' South. The vessel's rounded and somewhat elevated bow rested on a seemingly endless sheet of ice

Several icebergs were stranded within the fast ice, yet the vessel's stern was still in reasonably ice-free water of 'polynya'. Not very far away, one could see the belt of pack ice. The vessel's engine was regularly vitalised with enough power to push out large ice blocks gathering at the stern.

All Day and No Night

During this part of the year, from September to March, the sun never sets over Antarctica. It remains above the horizon nearly 24 hours a day for most of the summertime in Antarctica, which considerably upsets the human biological clock. In a fortnight, the sun would start setting briefly, around the local midnight. As the days go by, the period for which the sun remains below the horizon will slowly increase. By April, the sun will stay below the horizon for most of 24 hours, marking the beginning of a period of darkness in Antarctica. The local midnight, which is the time when the sun is at its lowest elevation, was nearly an hour before UTC midnight. On 02 February 1999, the vessel manoeuvred its way to the ice shelf at the Indian Bay. The vessel was secured alongside the ice shelf on the large steel pegs driven into the ice. Some pegs used during the previous year were still in position and were reused. The new pegs were forced into the hard-blue ice after drilling holes that were one to two meters deep, with the help of a giant electrically operated corkscrew drill. Once the pegs were in place, there was little chance of them coming out because the ice around freezes hard. The pegs were usually left behind for use for the following year. Three large pneumatic fenders were secured between the vessel and the ice shelf. However, the gap between the ice shelf and The ship was too large to lay a gangway. Moreover, the ice at the edge of the shelf was quite unstable and would break off even with slight pressure. Hence men and material were transferred between the vessel and shelf in a nylon cage using the vessel's cranes.

Attempts to Observe Tide and Tidal Stream

By the morning of 03 February 1999, most of the ice around the vessel had drifted away, and the weather was very clear. The current meter was again lowered from the vessel's stern. The meter readings for the current were quite erratic. Hence, the observation was abandoned. Since it was pretty impossible to erect a tide pole on the shelf, it was planned to observe the tidal variations with the help of the vessel's Atlas Deso 10 echo sounder while The ship remained secured to the frame. Unfortunately, this echo sounder was found to be defective and could not be repaired by the team.

Sounding at Antarctica

On 03 February 1999, just after The helicopters had taken off, the boat was lowered for sounding. After a satisfactory bar-check, The ship proceeded to the first sounding line commencing at the fast ice edge within the India Bay area. The weather was near perfect with a glassy calm sea. The sky was cloudy but did not completely obscure the sun, and the temperature hovered around a comfortable - 1° C. The sounding lines could not be run to the northern limits of the survey area because of the intervening dense pack ice. This pack ice belt was about 15 kilometres from the shelf, and there was still another 8 to 10 kilometres of sounding to reach the limits of the survey area. Within this pack, ice belts were innumerable stranded icebergs. The sight of these numerous gigantic icebergs huddled together in the mute conference was indeed spectacular. Short of this pack ice belt was a zone where the icebergs drifted eastward at almost half a knot speed. The day sounding was terminated at 1730 hours. All preparations were then made to get on with sounding by 0700 the next day, subject to the weather.

Drift Ice Prevents Sounding

The morning of 04 February 1999 was unusually windless, but sounding could not take place. Drift ice covered almost the entire area that was available for sounding the previous day. The boat was lowered by 0600 hours to start communicating as early as possible but hastily hoisted shortly after to avoid being crushed by drifting ice sheets. Getting this 10-ton boat down to the water was a simple matter with the vessel's 40-ton crane but hoisting it and getting it to rest on the chocks was a complex operation, wrought with risks, mainly when it was windy and the vessel rolling. The utterly overcast sky also caused a near whiteout condition, but this did not affect flight operations on the day.

Coast Lining of Ice Shelf

On 04 February 1999, due to day's sounding, the team undertook the coast lining of the ice shelf near the vessel's vicinity. Before venturing out, all team members were issued snowshoes. About 7 kilometres of the ice shelf was walked over using GPS 200 in 'Stop and go' mode. The walk was indeed very arduous and often wrought with danger. The blue ice of the shelf was incredibly slippery, while areas covered by soft snow were quite treacherous, and one could quickly sink into them. Also, it was too dangerous to venture close to the edge of the shelf. The team maintained a safe distance of 20-50 metres from the edge of the shelf during coast lining. The long walk on the shelf ice causes one to feel disoriented due to the near-continuous and featureless whiteness all around. Deteriorating visibility and exhaustion forced the team to return to the vessel in less than four hours. At Antarctica, one loses much body heat through the respiratory process, which causes tiredness. Due to some procedural errors, the data from the GPS 200 could not be processed, resulting in a washout of the day's efforts.

The morning of 11 February 1999 was indeed delightful with bright and relatively warm sunshine. There was a slight breeze blowing from the shelf and no accompanying snowdrift. The India Bay was pretty clear of ice, but there was a belt of drift-ice to the north of the Bay about 2 kilometres away. The vessel's Master reluctantly allowed the lowering of the boat for sounding, as he feared that drift ice would sooner or later cut off all access into the India Bay, and the boat may be unable to return to the vessel. The team decided to sound the India Bay area and then venture out for further sounding based on the ice situation beyond the Bay. The boat was set off to sound the first line close to the shelf. No crew member from the vessel was taken along in the cabin. One had to remain in the cabin all the time, as the outside temperature was less than zero and the wind chill even worst.

Survey of Drift Ice Belt

Although the outside temperature was -8° C, good weather prevailed on 12 February 1999. The drift ice belt extended in an east-west direction along with the mouth of the Bay. The boat was already in the water. The team decided to closely survey this belt to check if there were enough gaps to venture into sounding in at least the bays to the east of India Bay. The entire length of the belt at the mouth of the India Bay was tracked. Much of the sea within and around this belt was frozen.

Ice Shelf Coast Lining by Helicopter

A helicopter was made available to the team to carry out a detailed survey and the coast lining of the ice shelf. Position fixing during coast lining was achieved using the Garmin GPS fitted in The helicopter. Sometimes when the GPS signals were getting locked onto the satellites, an aerial survey was carried out. The coast lining commenced from the eastern limit of the survey area (11^0 30' E) and was terminated at the western boundary (12^0 30' E). After a sortie of an hour and a half, The helicopter landed back on deck. The weather was near perfect during the coast lining.

Status of Survey

The planned limits of the survey area were as follows: -

Table 3.

Latitude	69° 45'S	to	71°05' S
Longitude	11° 30'E	to	12° 30'E

Due to the adverse ice conditions, the entire ice shelf line within the survey area was mapped using the helicopter. Only 2% area could be sounded.

Geodetic Control

WGS 84

Complete control was referred to as World Geodetic System 1984 Datum and Spheroid. The projection used was Universal Transverse Mercator.

Existing Control

The Geodetic Control was not established on the ice shelf adjoining the survey area by the XVI Indian Antarctica Expedition Hydrographic Survey Team. The ice shelf is very unstable geomorphologically, and control based on it would be pretty unreliable after just a single 'unfreezing and re-freezing cycle' at Antarctica.

The coordinates of the main station 'Maitri S', obtained from the Survey of India team participating in the current Expedition to Antarctica, are 70^0 45' 51.73" S 11^0 44' 02.57" E 117.068 Meters.:

Digital Surveying System

Due to technical reasons, no digital survey could be undertaken, and data were handled manually, adopting conventional recording and processing procedures.

Navaids

Afloat Position Fixing

Del Norte 586/1009 DDMU-GPS receiver was used during the survey for afloat position control in stand-alone GPS mode. Trimble Series 4000 GPS receiver was also connected as a standby and for comparison of position fixes.

Non-Deployment of Trisponder

The Trisponder Master and remotes deployment was not considered because of the inability to find the means to secure the remote stations on the ice shelf.

DGPS Reference Station and HF Data Link

A DGPS system with an associated HF data link would require a more precise afloat positioning while conducting large-scale surveys. This system was slated to carry the Del Norte DGPS HF data link held by the National Hydrographic School. However, it was left behind because the system failed to function during trials before the team's departure.

Bathymetry

Echosounder: Atlas Deso 20 echo sounder was used for the sounding survey. Both frequencies, 210 and 33 kHz were utilised during sounding. No boat squat trials were conducted before deployment of the boat. Also, no heave compensation was employed.

Sound Velocity: A sound velocity of 1500 m/s was set on the echo sounder. All sounding 200 meters and above were corrected using table 35 of NP 139. Every sounding below 200 meters were also updated using extrapolated corrections from the same tables, as no tidal corrections were being applied to these soundings.

Sounding: The sounding lines were run at 0.5 cm on the scale, i.e. 250 metres apart, up to about 5 km from the ice shelf. Only a tiny portion of the survey area could be sounded on the two days when sounding was possible. Except for the two sounding lines, which were run-up to the pack ice belt on the first day, all sounding remained restricted to the area close to the ice due to drift ice belts.

Seabed Topography

No meaningful analysis of the seabed topography could be made based on the few lines of sounding carried out. Two hundred meters depth was recorded in the ice shelf up to about 10 km from the shelf, and After that, the depths reduced quite sharply to nearly 100 metres as the pack ice belt neared. The grounded icebergs within the pack ice belt indicated that the depths within the belt could be shallower.

Tides and Sounding Datum Tidal Observation

No tidal station was established during the survey. At India Bay, the height of the ice shelves above the water level was about 10 metres. Indeed, the edge of the ice shelf was very fragile, due to which it was impossible to erect a tide pole or a tide gauge. Instead, it would be entirely foolhardy to attempt the erection of a tide pole or a tide gauge at the ice shelf. Moreover, no benchmark could be established on the ice shelf as it was precarious and dynamic. The nearest exposed land to the survey area where a permanent benchmark may be set was some 100 kilometres away at the Schirmacher Oasis.

Topography

The topography of the ice shelf was quite plain, with some undulations. The ice shelf sloped in general from the South to the north. In addition to the undulations, one can expect to see some cracks and crevasses. These cracks and crevasses could be hazardous for casual walkers on the ice shelf. The ice shelf was composed mainly of blue ice, which was very hard and slippery to walk on. A thin film of drift snow did cover most of the blue ice. There were more extensive snow deposits in depressions and on the leeward side of obstructions (abandoned shelters, empty fuel drums, and containers).

Meteorological Observations

Meteorological parameters like temperature, pressure, wind speed and direction, visibility and cloud cover were recorded during the survey operations.

Oceanographical Observations

The sea surface temperature was recorded daily at regular intervals during the passage and in Antarctica. A record of sea-state was also maintained during the course. CTD profiles were obtained using Seacat SBE 19-03 CTD Recorder during the passage at every 5degree change of latitude. The profiles taken at Antarctica were not satisfactory due to internal battery malfunctioning.

XX INDIAN ANTARCTIC EXPEDITION (PERIOD DECEMBER 2000 TO MARCH 2001)

A Hydrographic Survey of Antarctica was undertaken following the detailed instructions issued by the National Hydrographic Office. The primary aim was to collect the latest bathymetric data in the Antarctic waters to update charts 9050 and 9051.

To participate in the XX Indian Scientific Antarctic Expedition, the Hydrographic Survey team embarked on The Expedition ship *MV Magdalena Oldendorff* on 30 December 2000. The Special Purpose Ship *Magdalena Oldendorff* sailed from Cape Town on 30 December 2000 and arrived at Antarctica on 10 January 2001.

The ship sailed through the forties and fifties without much rolling and pitching to everybody's surprise. It encountered the foggy fifties at the Antarctic convergence zone, where visibility reduced to near zero as the temperature started falling. 53° S onwards, The ship encountered icebergs. Ice floes were the next to be encountered. The density of ice floes increased as The ship approached further South. Next to be faced was pack ice, followed by fast ice. The ship finally got stuck in the fast ice and could not progress any further on 10 January 2001 at 69° 20' S, 10° 34' E. after that, The ship struggled every day to break the fast ice but could not succeed.

The sounding operations could not be undertaken initially since the survey area was covered with ice. The delineation of the ice shelf was also not feasible because the shelf was not clearly defined at various places due to the merging of the shelf and fast ice by the drifting snow. The ship attempted to break the fast ice to reach the ice shelf about 10 Nautical Miles away, every day, by ramming into it. The fast ice was not broken till the end of February 2001, unlike previous years, when the fast ice used to die by the end of January. It was attributed to a blockade to the incoming swell by the grounded icebergs and thick pack ice in between them.

The survey team assisted scientists from Wadia Institute of Himalayan Geology (WIHG), Dehradun, fixing the position (using Leica GPS 200) while profiling ice thickness at Vataiyya Hills about 20 kilometres west of Maitri on 29 January 2001. They also carried out simultaneous GPS observations at Maitri and Dakshin

Gangotri from 2000hrs to 2359 hrs on 29 January 2001 to establish the reference station at Dakshin Gangotri for ice shelf delineation. The fast ice broke off only on 21 February 2001, enabling The ship to get berthed alongside the shelf at India Bay on 22 February 2001. Due to severe weather, The ship had to move away on 23 February and rebirth on 27 February 2001. The survey team conducted current meter observations from 20th to 22 February and 23rd to 26 February 2001 to observe the area's existing profile and sound velocity. The team delineated the ice shelf using Leica GPS 200 in Kinematic on the Fly mode on 27 February 2001 and 03 March 2001. The meteorological observations were carried out from 01 January 2001 to 17 March 2001 continuously. The coordinates of the land route were obtained from the German scientists accompanying The Expedition.

Specifications of Hydrographic Survey

The survey was carried out on the scale of 1:200,000 within the area bounded by the following limits: -

Latitude 69° 24'.00" S to 70°47'.00" S Longitude 11° 00'.00".0E to 14°00'.00".0 E

After disembarked fuel and containers at the ice shelf, a planned ship sounding was carried out on 02 March and 04 March 2001. The sounding was also carried out in a phased manner. While on passage to the ice shelf and leaving the Antarctic area between Latitude 69° 40'.00" S to 69°59'00".0 S Longitude 12°00'.00".0E to 12°20'.00".0 E due to limited availability of the ship, Horizontal control for the entire survey was established in WGS 84 Datum and plotted on Mercator Projection with scale true at 66°S. For ease of plotting, the area was divided into two sheets, Northern sheet from 69° 24' S to 70° 20' S and Southern Sheet from 70° S to 70° 47' S. GPS coordinates directly obtained from The ship fitted GPS "Trimble Navtrac XL", and was used for sounding. No control was established for this purpose. However, the GPS set was compared with the Leica GPS 200, which was available with the survey team and agreed. The following existing stations were recovered and used for extension of Geodetic Control for Ice Shelf delineation. The Geodetic coordinates in WGS 84 datum are as follows: -

Table 4.

Station Name	Latitude(N)	Longitude (E)	Height (in M)
'Maitri S'	70° 45' 51" .730 S	11° 44' 02" .570 E	117.068

For establishing the Station at "Dakshin Gangotri", GPS 200 observations were carried out on 29 January 2001 over the existing geodetic control station at "Maitri", and on top of 10 KL fuel tanks at "Dakshin Gangotri" (Figure 3) simultaneously for 4 hrs in Static Mode. The coordinates of the Maitri station were accepted and given as user input for computing coordinates of "Dakshin Gangotri" by baseline computation using SKI Version 2.3 software. Since the containers are shifted every year, the Station could not be used again in subsequent years. The Geodetic coordinates in WGS 84 datum are as follows: -

Table 5.

Station Name	Latitude(N)	Longitude (E)	Height (in M)
'A' Dakshin Gangotri	70° 04' 38" .297 S	11° 59' 58" .429 E	38.6946

Digital Surveying System

Leica GPS 200, Trimble Navtrac XL, and Smart Acoustic Current meter were the digital systems successfully deployed in the field. COMPAQ Presario 1255 Laptop computer was used for processing digital data

Figure 3. GPS observations at Dakshin Gangotri

Navaids

The ship with Trimble GPS Navtrac XL was used for position fixing while sounding for the entire survey.

Bathymetry

Two Russian made echo sounders were available onboard the charted ship, designed to measure up to a depth of 46 meters and model NEL M3B designed to measure up to a depth of 500 meters. With an average depth of 100 to 200 meters at Antarctica near India Bay, the only option was to use the echo sounder model NEL M3B. The echo sounder was crude, mechanical with minimal options, and no digital output. The details of the echo sounder are as follows: -

NEL M3B

The velocity of the sound used: 1500 m/s fixed

Frequency of transmission 169 kHz +/- 3.4

Scales of operation: 0 to 50, 40 to 90, and 0 to 500

Phasing of scales: Not present

Recording Paper Speed: Fixed speed as per scale (details could not be found out)

The survey team recommended replacing the echo sounder with a scientific echo sounder like Atlas Deso 20 or 25 with a digital output if the same ship was chartered for the next Expedition. Another option is to ensure the availability of a scientific echo sounder with the team having a capability to measure at least up to 1000 meters to make proper utilisation of time and ship for sounding while The ship is idling, and all other expedition members are away at Maitri. The sounding lines were run approximately 1 cm apart on the survey scale, perpendicular to the depth contours. Sounding density was not changed for the delineation of the shoals, which was considered adequate to meet the desired standards of accuracy and thoroughness on the survey scale. When The ship was not following any course, spot soundings were taken. The observed soundings were corrected for index error of the echo sounder, ship's draught, and sound velocity.

Tides and Sounding datum

No tidal observations were carried out during the survey. The ice shelf was floating on the water and was subjected to rise and fall along with the tide. The depth was recorded when the ship was secured at the fast ice and the ice shelf using an echosounder. The following inferences were drawn: -

- Neap and Spring tides are half diurnal.
- In between neaps and springs, it is mixed diurnal.
- The range of tide is between 0.9 m and 2.8 m.

Erecting a tide pole at any place near India Bay could not have solved the purpose since the shelf was not fixed. Some areas along the shelf were grounded and were not subjected to rise and fall with the tide, e.g., Leningrad Bay.

Tidal Streams

The tidal stream observations were carried out using Smart Acoustic Current Meter at three locations, as detailed here

Table 6.

Sl.No.	Period	Position		Duration
	From To	Latitude	Longitude	
A	20-21Feb 2001	69° 56'.7 S	11°54'.1E	17 hrs
B	23Feb 2001	69°56'.2 S	11° 55'.6 E	3 hrs
C	26Feb 2001	69°56'.2 S	11° 55'.4 E	12hrs

The Smart Acoustic Current meter deployment in temperatures up to –16° C without the wind chill was complicated. The LCD screen froze and was unable to show any reading. However, observations were made with extreme caution without exposing the recorder to cold atmospheric temperatures. The water flow was South Westerly with a 0.2 to 0.5 knot about 2 miles North of the Shelf. While at the shelf, it was South Easterly with speed up to 0.4 knots. The Smart Acoustic Current meter was not advisable for deployment as it was considered unsuitable in Antarctic conditions. It was recommended that only the 2 Dimension Current meter (2DACM) or any other current meter designed to work up to a temperature as low as -30°C be deployed.

Wrecks and Obstructions

No wreck was found to be in existence in the survey area. However, no side-scanning was carried out in the survey to confirm the non-existence of any wrecks.

Coastline, Topography, Conspicuous Objects and Marks

The extent of the ice shelf of 100 kilometres was delineated by the detached survey party using GPS 200 onboard helicopter used in Kinematic in the Fly (KOF) mode. The reference was set up at Dakshin Gangotri station while the rover was on The helicopter. Ice shelf from 11^0 E longitude to 12^0 56' E longitude was delineated through Russian Bay, India Bay and Leningrad Bay. The East of 'India Bay' was delineated by KOF mode, and post-processed data was accepted. The Western part could be delineated in GPS mode only due to tracking the satellites in one of the GPS 200 sets.

Recommendations

The following recommendations were made for future expeditions: -

- A ship fitted with a scientific echo sounder capable of measuring at least up to a depth of 1000 metres to be made available for the survey team. The non-availability of a suitable echo sounder was a significant drawback.
- Power requirements for shipborne equipment could be met using suitable adapters with stabilisers that satisfy the required specifications. The necessity of carrying batteries and chargers can be avoided in this manner. Continuous 220 V, 50 Hz power supply is available onboard. Since the sockets were of European style, the team had to be geared for the same.
- A survey team consisting of a survey officer and a survey recorder is essential for ship sounding and associated jobs assigned in this survey.

Meteorological Data

Meteorological data from 01 January 2001 to 17 March 2001 were collected continuously at an interval of 6 hours.

In summary, during the XX Indian Expedition, 200 Nautical Miles of sounding was completed. A total of 100 Km of coast lining was undertaken, a total of 600 Nautical Miles of the passage sounding was launched, and Tidal Stream observations were carried out at three locations for a total of 32 hours.

XXI ANTARCTIC EXPEDITION (DECEMBER 2001 TO MARCH 2002)

The Survey Team undertook the Hydrographic Survey of Antarctica during the XXI Indian Antarctic Expedition following the National Hydrographic Office's directives.

The team embarked on The Expedition ship *"MV Magdalena Oldendorff"* on 07 January 2002. The Special Purpose Ship *MV Magdalena Oldendorff* (Figure 4) sailed from Cape Town at 1930 hrs on 08 January 2002 and arrived at Antarctica on 16 January 2002.

Figure 4. MV Magdalena Oldendorff with the integral helicopter in Antarctica

During the passage, the ship sailed through the forties and fifties without much rolling and pitching. The ship encountered the foggy fifties at the Antarctic convergence zone, where visibility was reduced to near zero, and the temperature fell drastically. The first iceberg was encountered on 10 January in position 47° 38'. 5 S, 16° 23'. 7E. Ice floes were the next to be encountered. The density of ice floes increased as the ship approached further South. Next in line was pack ice, followed by fast ice. The ship finally got stuck at fast ice and could not proceed further on 16 January 2002.

The sounding operations could only be carried out in the initial stage viz on 16th, 18th, 25th, 26th, 27th, and 30th January when the ship attempted to break the pack ice.

Due to continuous bad weather, the ship could be berthed alongside the Shelf at India Bay only on 04 March but had to be cast-off on the same day. The ship berthed again for the second and last time on 17 March 2002, for decanting fuel, offloading the containers and taking onboard the containers for offloading to India.

During this period, the current meter was lowered for the Current meter/ Sound Velocity observations. The delineation of the ice shelf could not be carried due to the non-availability of the helicopter. Nevertheless, the survey was carried out on a scale of 1:200,000 within the area bounded by Latitudes 69° 45'00". 0 S to 70°47'00".0 S and Longitudes 11° 00' 00". 0 E to 14° 0'.00"E. Shipborne Sounding was conducted on 16th, 18th, 25th, 26th, 27th so that the logistics operations could be planned. Continuous Meteorological and Oceanographic observations were carried out during the entire period. The ship left Antarctica on 17 March 2002 and arrived at Cape Town on 25 March 2002. The various specifications of the survey undertaken by the team are enumerated in the succeeding paragraphs.

Geodetic Control

Horizontal control for the entire survey was established in the WGS 84 Datum and plotted on Mercator projection with scale true at 66° S. For ease of plotting. The area was divided into two sheets, Northern sheet from 69° 24' S to 70° 20' S and Southern Sheet from 70° S to 70°47' S. GPS coordinates directly obtained from The ship fitted GPS "Trimble Navtrac XL" were used for sounding. No control was established for this purpose.

Digital Surveying System

The ship fitted Trimble GPS NAVTRAC XL was used for sounding.

Navaid

The ship fitted Trimble GPS Navtrac XL for the entire survey was used for position fixing while sounding. The system performed satisfactory and reliable throughout the survey. The instrument had a digital output port, but the same was already connected to the radars.

Bathymetry

Two Russian made echo sounders were available onboard, model M4, designed for measuring depth up to 50 metres and model Nel M3B (Figure 5), designed to measure depth up to 500 metres. With an average depth of 100 to 200 meters at Antarctica near India bay, the only option was to use echo sounder model NEL M3B.

Figure 5. Echosounder fitted 'MV Magdalena Oldendorff' Model NEL M3B

The velocity of the sound used: 1500 m/s fixed
Frequency of transmission: 169 kHz +/- 2.4
Scales of operation: 0 to 50, 40 to 90, and 0 to 500
Phasing of scales: Not present

Recording Paper Speed: Fixed speed as per scale (details could not be found out).

The sounding lines were run approximately 1 cm apart on the scale of the survey perpendicular to the depth contours. Sounding density was not changed for the delineation of the shoals, as it was considered adequate to meet the desired standards of accuracy and thoroughness on the survey scale. When The ship was not following any course, spot soundings were also taken. The observed depths were corrected for index error of the echo sounder, ship's draught, and sound velocity. The accuracy of sound was within the stipulated accuracy standards. The team had strongly recommended that the echo sounder be replaced with a scientific echo sounder like Atlas Deso 20 or 25 with a digital output, in case. The ship was to be chartered for the next Expedition.

In summary, during the XX1 Indian Expedition, a total of 200 Nautical Miles of sounding was completed, 100 Nautical Miles of the ice shelf was delineated, approximately 500 Nautical Miles of Passage Sounding was undertaken

XXII ANTARCTIC EXPEDITION (PERIOD DECEMBER 2002 TO MARCH 2003)

The Hydrographic survey of India Bay, Antarctica, was undertaken from 01 January 2003 to 31 March 2003. The calibration of The ship fitted echo sounder was carried out alongside using a hand lead line. On completion of loading and defect rectification, the ship sailed for machinery trials on 14 January 2003. Upon successful attempts, the ship proceeded for India Bay, Antarctica, on 14 January 2003. The ship crossed into the Antarctic polar circle on 22 January 2003. With the drop-in temperature, icebergs were sighted frequently. The ship entered pack ice on 22 January 2003. Without any ice cutting mechanism, the ship used her engines to make way through the pack ice. Two helicopters carried out test flights on 23 January 2003. Upon completion of the test flights, the ice pilot carried out aerial reconnaissance. On ascertaining the availability of ice-free waters (polynia), The ship proceeded ahead. The survey team commenced passage sounding at 69° 24' S as the depth was below 500 meters. The ship was manoeuvring through thick pack ice, causing it to alter course and speed continuously, and spot soundings were also recorded. Although ice-free waters were expected much earlier, the ship finally cleared the pack ice at about 1800 hrs at 69° 40' S latitude. With the clearing of pack ice, the first snowfall was encountered with drastic fall invisibility.

The ship unsuccessfully attempted to free itself on 24 January 2003. With the weather not being conducive for any flying operations, no team members could be shifted to 'Maitri' Station. The weather worsened on 25 January 2003 with increasing winds, accompanied by a light snowfall. The ship managed to free itself and proceeded further into fast ice on 26 January 2003. The logistics sorties to the permanent station 'Maitri' commenced on 27 January 2003, as the ship could not manoeuvre herself out of the fast ice. During this period, The ship was held to exert pressure on the ice using engines at a fixed pitch. As and when The ship's propellers were stopped, the survey team could deploy a smart acoustic current meter daily. Bathymetric data collection also progressed on board. Most of the personnel and stores could be transferred only by 29 January 2003.

On 29 January 2003, the ship was drifting due to the breaking of fast ice. The ship was made available to the survey team for sounding operations. Sounding progressed as The ship manoeuvred near the survey area for confirmation of XXI IAE data. The ship returned to fast ice on 30 January 2003. about 08 Nm from the ice shelf. Aerial recce of the survey area and Dakshin Gangotri station was also carried out on 30 January 2003.

The ship drifted again with the breaking up of fast ice on 23 February 2003. She proceeded further into the broken ice pack and attempted to make fast/anchor on the ice shelf edge. The ship was made available for sounding tasks again on 23 February 2003. The sounding was progressed in Area' C' and 'B'of the Chart 9051

The ship returned to fast ice on 24 February 2003 to undertake flying operations. On completion of flying operations, the ship proceeded to progress sounding in Area' C'. The ship returned to fast ice at 70° 02'43" S 12° 40' E, on 25 February 2003, waiting for the load disembarkation area to be ice-free.

Despite the spring tides on 02 March 2003, the ship could not proceed into 'Leningrad Bay' to disembark fuel and stores. Although the original disembarkation area at India Bay was prone to breaking up and unstable, The ship proceeded to go alongside in position Lat. 69° 59' 43" S Long. 11° 55' 35''E on 04 March 2003. The disembarkation of fuel could only be progressed on 05 March 2003. The ship had to proceed back to fast ice before disembarkation due to winds gusting up to 30 Knots.

The survey team carried out simultaneous GPS observations at Maitri and Dakshin Gangotri on 08 March 2003. Ice shelf delineation continued to progress on 11 March 2003 using the Helicopter available onboard The Expedition vessel. Data was logged using Leica GPS 200 in KOF mode.

The specific Survey tasks that progressed during The Expedition are enumerated in the succeeding paragraphs.

Geodetic Control

Horizontal control for the entire survey was established in WGS 84 Datum and plotted on Mercator Projection with a scale true at 66° S. GPS coordinates were directly obtained from The ship fitted GPS "Trimble Navtrac XL" that was used for sounding. No control was established for this purpose. However, the GPS set was compared with the Leica GPS 200, which was available with the survey team and was found to be acceptable. The following existing stations were recovered and used for extension of Geodetic Control to carry out Ice Shelf delineation:

Table 7.

Stn Station Name	Latitude(N)/Longitude (E)	Height (in M)
'A' Maitri S	70° 45' 51" .730 S 11° 44' 02" .570 E	117.068

Digital Surveying System

Data was recorded manually as The ships echo sounder did not have any digital output. After extraction and application of corrections to the data, it was manually ported into the DSPS. The data was rendered in the form of a fair sheet, and a survey report was prepared.

Navaids

Leica GPS 200, Trimble Navtrac XL and EG & G Smart Acoustic Current meter were digital systems successfully deployed in the field. KDS Valiant Laptop computer and a DSPS comprising CARISAH menu-driven software from M/s QTC ported onto a Sun Sparc Work Station under Sun OS 4.1.3 UNIX operating system was used to process the digital and manual data. The GPS positions obtained at The ship fitted Navtrac were manually recorded. Although there exists a provision of digital output, the system is interfaced with The ship's radar.

Bathymetry

Two Russian made echo sounders were available onboard, model M4, which was designed to measure up to a depth of 50 meters and model NEL M3B, which was designed to measure up to a depth of 500 meters. With an average depth of 100 to 200 meters near India Bay at Antarctica, the only option was to use the echo sounder model NEL M3B. The echo sounder was of a crude mechanical type with minimal options and no digital output. The echo rolls do not have any graduations, and the scale is fixed on the echo sounder. Data were manually extracted and corrected for draught and calibration. The scale was manually traced on paper and used for data extraction.

Model: NEL M3B
The velocity of the sound used: 1500 m/s fixed
Frequency of transmission 169 kHz +/- 3.4
Scales of operation: 0 to 50, 40 to 90, and 0 to 500
Phasing of scales: Not present

Recording Paper Speed: Fixed speed as per scale (details could not be found out).

The sound velocity was fixed 1500m/s on the echo sounder, which was observed using the EG&G smart acoustic current meter. The sounding lines were run approximately 1 cm apart on the survey scale, perpendicular to the depth contours. The sounding density was not changed for the delineation of the shoals. It was considered adequate to meet the desired standards of accuracy and thoroughness on the survey scale. Spot bathymetry soundings were taken when the ship was not following any course and negotiating through pack ice. The observed soundings were corrected for index error of the echo sounder, ship's draught, and sound velocity. The accuracy of sound was within the stipulated 2σ level. Also, no cross lines were run owing to bad ice conditions.

Sea Bed Topography and Textures

The seabed in the general area was found to have a too gradual slope. The result of the survey indicates no significant undulations. Whilst The ship was alongside the ice shelf, the depths observed vary primarily even with minor position changes. Thus, it was inferred that the ice shelf extends into the sea beyond the visually prominent boundary.

Tides and Sounding Datum

It was not feasible to erect a tide pole at India bay as it is a mere ice shelf that rises and falls with the tide. Depth observations were carried out onboard whilst The ship was stuck on fast ice alongside India Bay.

Tidal Streams

Tidal stream observations were carried out on 08 March 2003 whilst The ship was alongside at the disembarkation point at India Bay.

Coastline, Topography, Conspicuous Objects and Marks

Delineation of the ice shelf progressed on 11 March 2003. Data was logged using Leica GPS 200 in Kinematic on the Fly mode. The coast was found to be barren and devoid of any marks or objects. Barrels and containers were placed by various expeditions in the area, mainly as route markers.

In summary, during XXII Indian Expedition, a total of 200Nautical Miles of sounding was completed, 120 Nautical Miles of the ice shelf were delineated, about 600 Nautical Miles of Passage sounding was also undertaken.

XXIII ANTARCTIC EXPEDITION (PERIOD DECEMBER 2003 TO MARCH 2004)

The Hydrographic Survey of India Bay, Antarctica, was undertaken from 05 December to 30 March 2004 as detailed instructions issued by the Hydrographic Office. The vessel proceeded for India Bay, Antarctica, at 0230 hrs on 19 December 2003.

The first iceberg was sighted in position 44^0 47'00 S, 17^0 57'00 E on 21 December 2003 at 1530 hrs. The ship passed through 50^0 S towards 55^0 S; 3/10 of the sea surface covered numerous icebergs. The Hydrographic team continued to record metrological data and undertake passage sounding. On 24 December 2003, The ship progressed rapidly through the floating pack ice around 60^0 S. Clear weather was experienced, and with sea state being slight, both helicopters were taken out from the hold and ranged on the deck. The density and concentration of packed ice grew thicker (about 2m) as the ship crossed into the Antarctic polar circle (30^0 S) on 25 December 2003. At 0600 hrs on 26 December 2003, the ship finally halted in position 67^0 53' S, 13^0 12' E as it could no longer break through the packed ice (2.5-3.0 mt thick). The survey team continued with met observations and attempted to lower SACM for observation.

The survey team shifted to the Indian permanent station 'Maitri' in Antarctica on 01 January 2004. The SOI station Maitri' S' was recovered on 02 January 2004, and Geodetic observations were carried out.

The survey team undertook sounding operations in Area' A' of Chart 9051 on 08 January and 10 January 2004. The area was observed to be in thick ice floes and pack ice. Lines were run in the North-South direction at an interval of 02' (minutes) on Longitude. Line keeping was difficult as the ship experienced tremendous ice pressure and had to manoeuvre to avoid getting stuck in ice continuously. A total of 33 NM of sounding was completed, and The ship returned to fast ice at 69^0 46' 1" S, 10^0 55'. 1" E to undertake helicopter operations. Sounding commenced on 16 January and continued until 18 January 2004. With one engine operating, an average speed of 04 knots was achieved. However, 90 miles of sounding could be completed in Area' A' by 23 January 2004. On 24 January 2004, the survey commenced at 0400 hrs as it could be approached easily, while area 'B' was still covered under fast ice. A total of 100 NM of Sounding, up to 500 metres contour, was completed in area

'C' of Navigational Chart 9051 by 27 January 2004. Every day during sounding (survey) operation, sound velocity was observed using Smart Acoustic Current Meter and CTD cast, taken at an interval of 10' (minutes) of Longitude in the survey area.

Geodetic Control

Horizontal control for the entire survey was established in WGS 84 Datum and plotted on Mercator Projection with the scale true at 66^0 S. GPS coordinates directly obtained from the ship fitted GPS 'Trimble Navtrac XL' were used for survey operations. No control was established for this purpose. However, the output of the ship held GPS set was compared with the Leica GPS 200 available with the survey team, and positions were acceptable. The following existing Station was recovered and used for extension of Geodetic Control. The coordinates in WGS 84 datum is as follows:
Station Name – Latitude - (S) Longitude - € Height (in M)
Maitri S 70° 45' 51".730 S 11° 44' 02".570 E 117.068

Digital Surveying System

Data was recorded manually as the ship's echo sounder did not have any digital output. This data, after the extraction and application of corrections, was manually ported to the DSPS.

Navaids

Leica GPS 200, Trimble Navtrac XL and EG&G Smart Acoustic Current meter were the digital systems successfully deployed in the field. KDS Valiant Laptop computer and the DSPS comprising of CARISAH menu-driven software from M/s QTC. The GPS positions obtained at the ship fitted Navtrac were manually recorded. The data was reliable and steady compared with the Leica GPS 200 and was found to be acceptable. Although there exists a provision of digital output, the system is interfaced with the ship's radar.

Bathymetry

The Echosounder with the following specifications was used for the collection of bathymetric data:
Model: NEL M3B
The velocity of the sound used: 1500 m/s fixed
Frequency of transmission: 169 kHz +/- 3.4
Scales of operation: 0 to 50, 40 to 90

Phasing of scales: 500 Nit present
Recording Paper Speed Fixed speed as per scale

The sound Velocity is fixed at 1500 m/s on the echo sounder. Sound Velocity was observed using the EG&G smart acoustic current meter. The sounding lines were run Approximately 1 cm apart on the scale of the survey perpendicular to the depth contours. Sounding density was not changed for delineation of the shoals, which was considered adequate to meet the desired standards of accuracy and thoroughness on the survey scale. The observed soundings were corrected for index error of the Echosounder, Ship's drought and sound velocity using Nautical Publication 139. One cross line was run on Latitude 69^0 38'.0 S in the area 'A and C' while another cross line was run in area 'C' on latitude 69^0 32'.0 S. The soundings (depths) on the main and cross lines were found to be within the desired standard of accuracy.

Tidal Streams

Tidal stream observations were attempted on 09 ^{March}, '04 whilst the ship was alongside at the disembarkation point at Leningrad Bay using 2 DACM. The instrument was found to be operational when being configured. Still, no data could be logged, probably due to the freezing ambient temperature (-14^0 C) discharging the internal battery of the instrument. An attempt was again made to observe tidal streams using SACM on 11 March 2004. The device was operational only for 15 minutes. Hence, no carrier signal was detected. It is expected that some water seepage into the instrument has taken place.

Coastline and Topography

The team delineated the ice shelf from 14^0 E to 11^0 E on 15 March 2004. Data was logged using Leica GPS 200 in Kinematic on the Fly mode and processed in RINEX format as no reference station was set up for the coast lining operation. The entire coast was barren and devoid of any marks or objects. Empty barrels and containers are placed by various expeditions, mainly as route markers for convoy teams. These are shifted as and when required and cannot be considered as permanent marks.

Thus, during XXIII Expedition, a total of 310 Nautical Miles of sounding was completed in areas A, B and C, a total of 600 Nautical Miles of Passage sounding was undertaken, and 150 Nautical Miles of coast lining was completed in The Expedition.

CONCLUSION

During each of The Expeditions, valuable experience was gained on using scientific equipment and available resources (survey boat, expedition vessel) for bathymetry measurements, observation of currents & tidal streams and utilisation of expedition helicopter for gathering the coastal features information. A significant amount of passage and on-site sounding data collected has been beneficial in updating the bathymetry database to publish a new nautical chart for the area around India Bay. Further, the additional data/information gathered from CTD casts, sound velocity measurements, and meteorological observations undertaken by the survey teams have contributed significantly towards a better understanding of the ocean parameters of the Antarctic waters. Keeping in mind, the challenges faced in coordinating efforts to undertake surveys and produce charts in Antarctic waters (largely undertaken by the International Hydrographic Organization's Hydrographic Sub-Committee (Nairn, 2010), this chapter provided an insight into the Indian initiatives to address the challenges of conducting hydrographic survey in extreme environmental conditions of the icy continent "Antarctica".

REFERENCES

Chandrashekharan, S., & Sreedharan, A. (2000). *Report on hydrographic survey of Antarctica by XVI Indian scientific expedition. Sixteenth Indian Expendilion to Antarctica, Scientific Report, 2000.* Department of Ocean Development, Technical Publication No. 14. http://14.139.119.23:8080/dspace/bitstream/123456789/622/3/ARTICLE+17-2.pdf

International Hydrographic Organization (IHO). (n.d.). https://web.archive.org/web/20140724100423/http://www.iho.int/srv1/index.php?option=com_content&view=article&id=299&Itemid=289

Jones-Couture, S. (2020). *Increasing our knowledge of Antarctica with hydrography Cooperation and expeditions.* IHO Hydrographic Commission on Antarctica. https://storymaps.arcgis.com/stories/e51256ee28504092accf031864ea65c2

Lieutenant-Commander, J. B., & Dixon, R. N. (1965). *Hydrographic Surveying in the Antarctic.* https://journals.lib.unb.ca › ihr › article › download

Nairn, R. (2010). *The Challenge of Hydrographic Surveying and Charting the Antarctic.* FIG Congress 2010 Facing the Challenges – Building the Capacity, Sydney, Australia. https://www.icsm.gov.au/sites/default/files/CongressPaper4531.pdf

Singh, R. (1998). *Hydrographic Survey of Approaches to Indian Bay Antarctic Region. Fourteenth Indian Expedition to Antarctica, Scientific Report, 1998.* Department of Ocean Development, Technical Publication No. 12. http://14.139.119.23:8080/dspace/bitstream/123456789/420/3/ARTICLE+13.pdf

Wilkins, S., Boulaire, Y., & Landet, F. (2013). *Different Approach to Surveying Antarctica -Operation Rock Bottom.* https://www.hydro-international.com/content/article/operation-rock-bottom

Wynne-Edwards, C. J. C. (1960). The Geographical Journal. *The Royal Geographical Society, 126*(3), 310-315. . doi:10.2307/1793634

Chapter 9
An Assessment of Geo-Scientific Investigation in the Antarctic Region

Rajesh Asthana
Geological Survey of India, India

Amit Dharwadkar
Geological Survey of India, India

Prakash Kumar Shrivastava
Geological Survey of India, India

ABSTRACT

During the initial period of about 25 years (1981-2006), systematic geological mapping covering an area of about 19,000 sq. km, continuous monitoring of Dakshin Gangotri (DG) snout, ice core drilling, and limnological investigations are some of the significant achievements of GSI in the field of Antarctic geoscience. For the first time, ground penetrating radar (GPR) survey for lake bathymetry has been introduced in Schirmacher Oasis. Collaborative work has resulted in quantifying the annual vector movement of the ice sheet in the Schirmacher region.

INTRODUCTION

Antarctica occupies a central stage in modern-day studies on continental drift, breaking and making supercontinents and understanding crustal evolution. In India, efforts towards Earth science studies were initiated by the Geological Survey of India (GSI) and other national organisations. One of them is the National Geophysical

DOI: 10.4018/978-1-6684-4078-0.ch009

Research Institute (NGRI), which marks India's foray into the frozen continent. Subsequently, several other institutions and national agencies pursuing various aspects of Earth Science studies joined in the pursuit. Still, G.S.I. remains a pioneer that has consistently associated with and is committed to India's quest for knowledge in the distant land. During this process, scientists from GSI have contributed profoundly in meeting the scientific and logistic ends of this national endeavour, which has projected India into the group of elite nations exploring the `terra incognita' of yesteryears. This paper presents a resume of the story of the initial commitment of the Geological Survey of India. It highlights the salient achievements acquired in the first 25th years of India's presence in the Antarctic region (1981-2006).

SCIENTIFIC IMPORTANCE OF ANTARCTICA

The continental drift theory suggests that Antarctica formed a major part of the supercontinent Gondwanaland from which India, Australia, Africa, and Madagascar have drifted apart. Therefore, from the point of view of Earth Sciences, Antarctica provides a vital link in the reconstruction of Gondwanaland. To understand the process and mechanism of drifting of continents, the geology of Antarctica plays an important role. The importance of studying the Geology of Antarctica from India perspective are following;

- The reconstructed models of the supercontinent suggest the proximity of the Eastern Ghats Mobile Belt (EGMB) of India to East Antarctica in
- Mineral-enriched zones of Gondwanaland continents were in contact with Antarctica in the geological past,

Antarctica, being the storehouse of 70% of the world's ice and 90% of freshwater locked in the form of ice, provides a unique platform to study continental glaciers and sea ice in a pristine environment. The thick continental ice sheet (~2 km wide on average) contains palaeoenvironment / paleoclimatic signatures.

Thus far, the major fields of studies undertaken by GSI in Antarctica comprise regional geological mapping, thematic mapping, continental ice sheet, dynamics of shelf ice, drilling for ice cores, and lake sediment studies.

For more than two decades, regional geological mapping was the core programme of GSI in Antarctica to prepare a database for a little-known area in the Central Dronning Maud Land (cDML), East Antarctica. Without creating this basic knowledge about the site's geology, any detailed study would have been infructuous and unsystematic, which may have led to erroneous conclusions.

A good understanding of regional geology helped target and chart out areas for detailed thematic studies. We have aimed to understand the formation of Gondwanaland deeply and the eventual drifting apart of the continents. Also, whenever the Antarctic Treaty nations agree to go for exploration-oriented studies, a thorough understanding of the tectonic processes involved, related magmatism and metamorphism will provide the constraints to look for mineral occurrence in the region. The well-known Pan-African Orogeny and Greenvellian Orogeny, which have modified the entire Gondwanaland, also constituted one of the primary objectives for a detailed study.

Antarctica's ever-dynamic continental ice sheet and the shelf ice surrounding the continent are useful natural sensors of change in the global climatic pattern. Therefore, monitoring the behaviour of these vast ice bodies, their growth or decay rate and the direction of the flow of their movement, etc., form an integral part of the long term studies in Antarctica. The database must be comprehensive in tandem with simultaneous recording of meteorological parameters.

The thick cover of ice over the Antarctic continent provides a wealth of information on climate records from the geological past. Trapped gases, soil, aerosol particles etc., in the ice at various depths are unadulterated indicators of the past (palaeo-) climate. Therefore, it is imperative that ice cores be retrieved through drilling and studied to decipher the record of the paleoclimate preserved in them. A collaborative ice core drilling programme of GSIs aims to recover roots from the selected area for isotopic and geochemical studies to build up the paleoclimatic history and, consequently, to understand the global climatic patterns over the ages.

The paleoclimate influence and imprints are also stored in the sediments of freshwater lakes found in abundance in Antarctica. Hence, recently GSI has also taken up the task of retrieving cores from the lakes of the Schirmacher Oasis. The collected sediment cores will be studied for multi-proxies of paleoclimates. The Thermal Lumicence (TL) Dating of these samples is also to be done along with the studies of environmental magnetism stored in the sediment cores.

Keeping in mind the geological importance of the Schirmacher Oasis region of East Antarctica, a detailed geological investigation was focussed on the initial twenty-five years.

GEOLOGICAL STUDIES

The central Dronning Maud land (cDML) that hosts the Indian Antarctic Base- Maitri, off the Princess Astrid Coast, has attracted many scientists in Geodynamic studies. Varied aspects of geology and geodynamics of cDML are addressed. During the early Cambrian period, the central Dronning Maud Land played a pivotal role in forming the supercontinent Gondwanaland. Thus, the area has attained importance for geological studies related to the evolution and growth of the continent.

GSI has actively participated in all the Antarctic Expeditions since its inception in the austral summer of 1981-82. During the 5th Indian Antarctic Expedition, GSI ventured into an ambitious programme of geologically mapping the linear Mountain belt exposed ~200 km inside from the coast of cDML. The area is an array of Antarctica's most dramatic mountain scenery, interspersed by glaciers and moraine fields. The mountains stand out as a unique spectacle in vertical rocky walls, pinnacle peaks and knife-edge ridges. Different segments in the mountain chain exposed off the cDML between 140 East Longitude and 0-degree meridian are the Wohlthat, Orvin, Muhlig-Hofmann and Gjelsvik ranges. The geological studies, which began mapping the Gruber mountains of Wohlthat range during the fifth expedition, continued until the 24th Indian Antarctic Expedition (2004-05), when geological studies of Muhlig Hofmann ranges were conducted at 4° East Longitude. The continued effort has enabled GSI to systematically evaluate outcrops exposed over an area of around 19000 sq. km, providing a unique opportunity to understand the geological processes associated with the development of this cratonic segment of East Antarctica. The expedition-wise detail of the study areas covered by GSI is given below in Table 1

Table 1. Area between longitudes, covered during different expeditions

EXPEDITION	YEAR	AREA COVERED	LONGITUDE
V	1985-86	Gruber Mountains and nunataks south of Schirmacher	13° to 14° E
VI	1986-87	North Petermann	12°15' to 13° E
VII	1987-88	North Humboldt	11° to 12° E
VIII	1988-89	South Petermann, Zweissel	11° 30' to 12° 45' E
IX	1989-90	Middle Humboldt	10° to 12°E
Weddel Sea Expedition	1989-90	Littlewood, Bertrab, Moltke nunataks, Weddell Sea	34°10' to 35°04' W
X	1990-91	Skeids area, Humboldt Mountains	11° to 11° 30' E
XI	1991-92	Payer-Weyprecht Mountains	13°09' to 14°49'E
XII	1992-93	Schi Oasis	
XIII	1993-94	Dallmannfjellet, Orvin Range	10° to 11°E
XIV	1994-95	Conrad Mountains	9°30' to 10°E
XV	1995-96	Conrad, Kurze	7°58' to 9°30'
XVI	1996-97	Filchnerfjella	7° to 8°E

Continued on following page

Table 1. Continued

EXPEDITION	YEAR	AREA COVERED	LONGITUDE
XVII	1997-98	Thematic studies in the central Humboldt Mountains	
XVIII	1998-99	South Zweissel Mountains	12°05' to 12°40'E
XIX	1999-00	Baalsrudfjellet, Sonstebynuten, Pevikhornet, Starheimtind nunataks south of Schirmacher	11° 30' to12°E
XX	2000-01	East Muhlig-Hofmann	7° to 6°30'E
XXI	2001-02	East-central Muhlig-Hofmann	5°45' to 6°20'E
XXII	2002-03	West-central Muhlig-Hofmann	5° to 5°45'E
XXIII	2003-04	Thematic Gruber Mountains, Schirmacher	13° to 14°E
XXIV	2004-05	East Muhlig-Hofmann	5° to 4°20'E

Many geoscientists have thoroughly studied the western DML, mainly from Germany, Norway, and South Africa. Its geological evolution is chronicled, yet its cDML lacks data, primarily due to its inaccessibility. The early initiative taken by GSI proved to be highly beneficial, and the department has continued its Geological Mapping Program in the cDML area on a 1:50000 Scale as its primary activity to date. During this period, GSI has published geological maps such as Geology of the Schirmacher-Wohlthat region central Dronning Maud Land, Antarctica, Geology of Muhlig Hofmannfjella, Geological map of Orvinfjella, and Geological map of Schirmacher Oasis. Also, to understand the processes involved in crustal evolution, GSI has planned and undertaken precise thematic studies in cDML.

Geology of Central Dronning Maud Land was first studied and described by Russian geologists in the early 1960s. The area's geology was primarily neglected until GSI took up the Regional Geological mapping Program of Wohlthat mountain Range in 1985. The programme has been systematically pursued to date. Thus, GSI is the only organisation in India that has developed a regional perspective on all the critical aspects of the geology of cDML. Countries like Germany, Norway and Japan have carried out some specific thematic studies through unique expeditions in the area. The involvement of Geoscientists from some other institutes from India has been sporadic but restricted accessible Schirmacher Oasis area only.

MAJOR FINDINGS/OUTCOME

GSI initiated its studies on an early assumption made by Russian Scientists that the area was part of an old crust dating back to the Archaean age. The Gruer area exposes massif anorthosite, metamorphosed ortho- and para-gneisses and associated undeformed magmatic rocks. The area being studied was, in fact, part of a poly-deformed and poly-metamorphosed mobile belt with the possibility of a geological evolutionary history much younger than initially thought of by the Russian geoscientists. The area possibly had evolved through two major tectono-thermal events, the older Grenvillian (1000Ma) event, involving the formation of the crust and its earlier metamorphism. It was followed by considerable reworking during later Pan-African (550Ma) orogeny, responsible for the building of Gondwanaland. Central Dronning Maud Land (cDML) soon attained importance with the realisation that it was a part of two accretionary tectonic events involving the amalgamation of continents through collision. The area was an ideal ground to study the processes involved in the growth of continents through collision and their fragmentation through break up termed as Wilson Cycle. Thus, the cDML area has an exciting history revealing its growth through two important tectono-thermal events and the break-up of the supercontinent during the Mesozoic period. A brief on the geology of cDML are enumerated below-

Regional Geological Setup

The geological studies in cDML have indicated that the exposed crust has a prominent Grenvillian (1000 Ma) component extensively modified by a later (600-500 Ma) Pan-African orogeny.

The Mesoproterozoic Grenvillian tectono-thermal event recognised as the first crust forming event in the region is represented by a thick sequence of metasedimentary and meta igneous rocks comprising metapelites, metapsammites, calc-silicates, amphibolites, pyroxene-granulites and orthogneisses depicting a bi-volcanic character. These Grenvillian supracrustals are extensively deformed and have suffered metamorphism up to granulite grade during the Grenvillian event. These rocks have been re-metamorphosed up to granulite grade and structurally overprinted during the last Pan–African activity. Juvenile crust associated with the Pan-African event has not been detected in the region. However, the Pan-African event is associated with the dominant magmatic activity, which started with the intrusion of massif anorthosite and AMCG suite of rocks in the Wohlthat Mountains and possibly culminated with the massive plutonic intrusion of A-type granitoid in different parts of cDML.

Magmatic rocks exposed in cDML are associated with Pan-African tectonic events form a significant component of the litho units. The dating of these magmatic rocks represents an early Pan-African phase and a post-collisional late phase. The early intrusives (600 Ma) are the anorthosite massif exposed in Gruber Mountain and Nordvestoya Mountains and possibly the charnockites, monzodiorites of Wohlthat ranges. The massive plutonic outcrops comprising granite, syenites, and charnockites exposed in Petermann Mountains, Conradfjella and Filchnerfjella have been dated from 510 to 530 Ma post-collisional A-type granitoid. The intrusion of A-type granitoids is followed by lamprophyres dated at 412 Ma, reflecting in the culmination of magmatism which started with the collision during the Pan-African orogeny. The Mesozoic olivine basalt dykes in the region may be associated with the break-up of Gondwanaland.

Regional Geological Data Acquisition

The GSI initially undertook geological studies in the Central Dronning Maud Land, East Antarctica, as a reconnoitre appraisal followed by regional systematic geological mapping. By the end of the year 2004-2005, a total of 19,000 sq. km (inclusive of intervening ice-covered areas) has been geologically mapped on a 1: 50,000 scale in CDML up to 4030'E longitude. The mapping was undertaken to create a geological database before taking thematic studies to understand the East Antarctic craton (Fig. 9. 1) .

Figure 1. Geological mapping in central Dronning Maud Land, East Antarctica

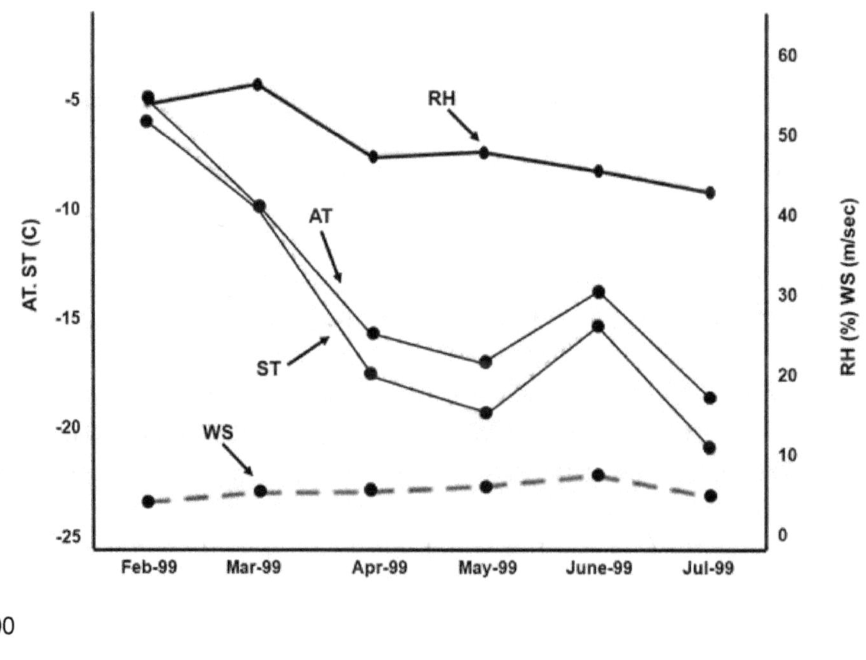

THEMATIC STUDIES

Based on the work carried out by regional systematic geological mapping, thematic studies covering the following areas are planned in CDML;

1. Nunataks South of Schirmacher Oasis
2. Petrogenesis, mineral chemistry, and geochemistry of Gruber Anorthosite Massif and adjoining areas where the AMCG suites of Zweissel, Petermann rocks are exposed.
3. Geochemistry and petrogenesis of alkaline and calc-alkaline lamprophyres and mafic-ultramafic intrusives in and around Schirmacher.
4. Evaluation of deformational, metamorphic and magmatic history of the Humboldt group of rocks.
5. Detailed studies on the charnockites of cDML.
6. Fluid inclusion studies to characterise the fluid regime, work out the genesis of charnockites and related suite of rocks of cDML.
7. Genetic classification of granites and charnockites of different ages.

Thematic studies have been initiated in the Gruber Mountains area to delineate different phases in the anorthosite complex and decipher their interrelationships and the surrounding rocks of the AMCG suite. A three-year programme in this area is underway, and the first phase of detailed mapping and sample collection has been accomplished. A similar program for the study of the AMCG rocks in the Zweissel area is envisaged. These studies will improve significantly our understanding of the geological evolution of cDML and crustal processes involved in the genesis of Precambrian anorthosites in general.

The outcome of the sustained geological program of GSI has resulted in the publication of geological maps of Schirmacher Oasis and a part of cDML, including Gruber, Petermann and Humboldt mountains, the Orvin Mountains and Muhlig Hofmann mountains. These maps are recognised and notified by the SCAR. Apart from this, geoscientists from GSI have published scientific papers in national/ international journals/periodicals.

The chronologic succession of rocks exposed in the cDML is given in Table 2

Table 2. The chronologic succession of rocks exposed in the cDML

Unit	Rock type	Area of Exposure
Late intrusive	Dolerite, basalt as dykes and lamprophyre, melasyenite and norite	Minor exposures in entire cDML, norites in south Humboldt, Schirmacher and Gruber.
Cambrian granitoids charnockites, syenites etc. (510-540Ma)	Porphyritic alkali granite and Syenite, Charnockites and its variants ferrogabbro	Muhlig-Hofmann mountains, Zweissel, north Humboldt and Kurze areas
Neoproterozoic Gruber Anorthosite (600 Ma)	Anorthosite with marginal noritic variants	Gruber Mountains
Mesoproterozoic foliated charnockite (Magmatic) (~1000Ma)	Coarse-grained charnockite, granulite, amphibolite, calc-silicates	Skeids, Dalmannfjellet, Conrad, Kurze and Schirmacher
Mesoproterozoic basement gneisses	Hbl-Bt-Qz-Kf gneiss, Px granulites, migmatite, amphibolite, metapelite, calc silicates, marbles, meta- psammite etc	Skeids, north Humboldt, Dallmannfjellet, south Conrad, Schirmacher

REGIONS COVERED IN ANTARCTICA

GSI has been participating continually in the Indian Antarctic Expeditions since the first expedition in 1981-82. Till the IV Indian Antarctic Expedition, geological and glaciological studies were confined to areas around the Dakshin Gangotri station, Princess Astrid Coast, the Lazarev ice shelf and Schirmacher Oasis. Regional systematic geological mapping was undertaken since the V Indian Antarctic Expedition in the eastern Wohlthat region encompassing Gruber Massif and the northern part of the Petermann range. A detailed account of the areas covered is given below in Table 3.

Table 3. Regions covered during different scientific expeditions

Expedition	Area	Co-ordinates
I to IV	Near Dakshin Gangotri station, Princess Astrid Coast, Queen Maud Land. Lazarev Ice Shelf.	S Lat 69^{0}59'12" E Long 11^{0}55'^{0}9"
	Schirmacher Oasis	S Lat 70^{0}44'30" to 70^{0}46'30" and E Long 11^{0}22'40" to 11^{0}54'00"
V	1. Eastern part of Wohlthat mountains encompassing Grubermassif and northern part of Petermannrange. 2. Nunataks south of Schirmacher-Tallaksenvarden, Stenersenknatten, Hauglandtoppen, Andersensata, Pevikhornet, Starheimtind, Baalsrudfjellet, Sonstebynuten.	
VI, VII, VIII	Petermann 1, 2, 3 ranges. Humboldt mts. (Wohlthat region)	
IX	Parts of Humboldt mountains (SW component of Wohlthat) Northern parts of Humboldt viz Nordvestoya Hjornehorna area	S Lat 71040' to 71^{0}58' and E Long 10^{0}45' to 12^{0}00' S Lat 71^{0}26' to 71^{0}33' and E Long 11^{0}17' to 11^{0}30'
Weddell Sea Expedition	Vahsel Bay of Weddell Sea Berkner Island Moltke nunataks Bertrab nunataks Littlewood nunataks	78^{0}S:35^{0}04'W . 77^{0}54'S:34040'W 77^{0}53'30" S:34^{0}10'W
X	Skeids area, Humboldt Mts (Wohlthat Region)	S Lat 70^{0}40' to 72^{0} 15' E Long 11^{0} to 15^{0}
XI	Payer-Weyprecht Mts (80 km SW of Gruber Massif, Wohlthat Region)	S Lat 71^{0}49'36" to"72^{0}05'48 and E Long 13° 09' 34" to 14049'27"
XII	Schirmacher area 2 km east of Veteheia	S Lat 70° 40' to 70° 47' and
		E Long 11° 28' to 11° 45.'
XIII	Dallmannfjellet (ORVIN I)	S Lat 71^{0}4^{0}' to 72^{0}1^{0}' and
		E Long 10^{0} to 11^{0}
XIV	Conrad (ORVIN II)	S Lat 71^{0}40' to 72^{0}3
		E Long 9^{0}30' to 10^{0}
XV	Kurze, Skorvestallen, Holtedahl, Fenris (ORVIN III)	S Lat 71^{0}25'to 72^{0}15' and E Long 07^{0}58' to 09^{0}30'
XVI	Filchnerfjella (ORVIN IV)	S Lat 71^{0}52' to 720^{0}7' E Long 7^{0} to 8^{0}

Continued on following page

Table 3. Continued

Expedition	Area	Co-ordinates
XVII	Muhlig Hofmann fjella	S Lat 71^{0}40’ to 72^{0}
	(Eastern Part)	E Long 6^{0}30’ to 7^{0}
	Central Humboldt	
XVIII	Zweissel area	S Lat 71^{0}51’ to 72^{0}05’
		E Long 12^{0}5’ to 12^{0}40’
XIX	Nunataks Southeast of Schirmacher Oasis – Baalsrudfjellet, Sonstebynuten, Pevikhornet, Starheimtind	S Lat 70^{0}55’ to 71^{0}05’E Long 11^{0}45’ to 12^{0}15’S
XX	Muhlig Hofmann fjella Sigurd-Svodene Nunataks	Lat 71040’ to 72005’E Long 06020’ to 0707’S Lat 71° 20’ 28” and E Long07° 37’ 56” E Long 9^{0} to 6030’S
XXI	Muhlig Hofmannfjella (central part)	Lat 71^{0}40’ to 72^{0}00’E Long 5^{0}40’ to 6^{0}20’
XXII	Muhlig Hofmannfjella (Western part)	S Lat 71^{0}45 to 72^{0}10’ and E Long 5^{0}0’to 5^{0}45’
XXIII	Thematic Mapping in Gruber area	S Lat 71^{0}10’ to 71^{0}30’ E Long 13^{0} to 140
XXIV	Muhlig Hofmannfjella (Western part)	S Lat 71^{0}47’00” to 72^{0}08’00” E Long 04^{0}30’00” to 05^{0}00’00”
XXV	Thematic Mapping in Zweissel area	

GLACIOLOGICAL STUDIES

GSI is engaged in studies on the dynamic processes of different snow and ice regimes under the programme of glaciological studies. These include:

1. Glacio-geomorphological mapping in Schirmacher Oasis on 1:25,000 Scale includes tracing the fluvial regime, neotectonic studies and sampling for

limnological studies. The Geomorphological map of Schirmacher Oasis has been published.

2. Snow accumulation/ablation studies by monitoring the network of stakes on shelf ice
3. Studies on polar ice front fluctuation and monitoring of ice shelf using digital satellite data
4. Iceberg monitoring
5. Ice dynamics (In collaboration with IIG.)
6. Thermal behaviour of different ground media
7. Ice core drilling

General glaciological observations were carried out in the 1st expedition. These include observation of significant concentration of icebergs, pack ice density during the voyage, melting rates of the ice shelf, augmentation of melting in polar daylight and finding an ionic composition of ice. The ions were found to be below the detection limits of the AAS used at that time.

The monitoring of icebergs during the voyage has been continued in subsequent expeditions. The first sighting of icebergs was generally at ~51°S latitude. The earliest sighting has been at 47°22'S latitude in the 13th expedition. Usually, two concentrations of icebergs are found in cold waters, the first zone near 56°S margin and the second one around 69°S latitude. The former is due to the Southern Ocean current, which moves in a clockwise direction. At the same time, the latter is attributed to Antarctic Coastal Current, which runs in an anti-clockwise direction. Giant icebergs are found mainly in the coastal current zone. Since the icebergs near the coastal areas are generally less eroded, these tend to be flat-topped, while the ones away from the Antarctic shelf are weathered, carved or upturned.

In the 2nd expedition, two long-term glaciological projects were identified, which have been continuing since 1983. The first observed a glacial snout's advance/retreat, and the second monitored accumulation/ablation patterns on the ice shelf.

For the first observation of monitoring fluctuations in the continental ice margin, a prominent glacial tongue was identified in 1983 in Schirmacher Oasis. It was named 'Dakshin Gangotri Glacier Snout.' This snout has been monitored on an annual basis every year. From 1983 to 1995, it was observed by surveying instruments from fixed survey points. The results indicated an average recession of 6.8 m per decade for this snout. In 1996, peripheral points were marked along the entire ice-margin of the snout for better quantification of observation of individual points. Twice, the snout was observed monthly for the whole year in 1996 and 2000 to understand the snout's advancing pattern. The snout surges (about 64 cm) from April to October before melting, and calving starts the recession from November. The melting reaches its peak in February, averaging about 70 cm every year. However, the northern wall of the glacier is receding at a faster rate than the eastern border, which possibly is because it is exposed to an increased amount of solar radiation. The observations in 2006 have completed the decadal monitoring for the 1996-2006 period. The overall average recession is 7.45 m per decade, but it is 10.33 m per decade for the northern wall. The maximum recession is shown by Obs-Pt.14, which is 18.40 m for this decade (Figure 2, Figure 3, Figure 4).

Figure 2. Snout of Dakshin Gangotri glacier, Schirmacher Oasis, East Antarctica

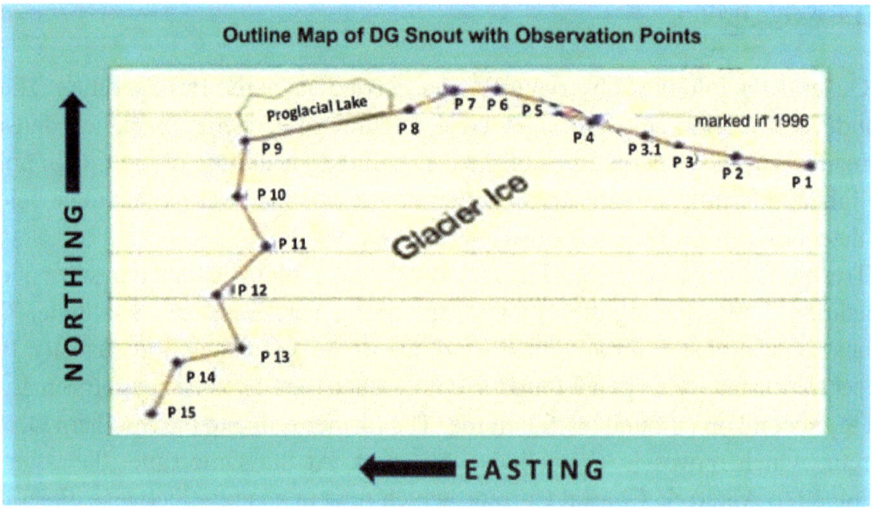

Figure 3. Recession of Dakshin Gangotri glacier's snout between 1996 and 2005

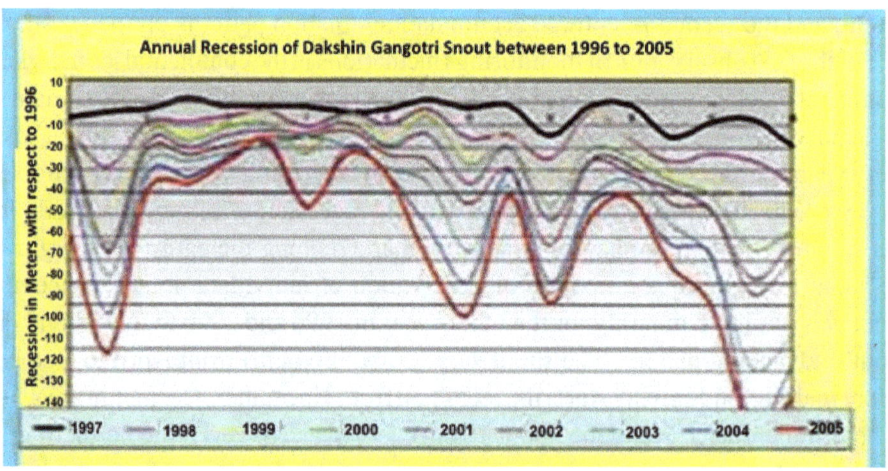

Figure 4. Average annual and overall recession of Dakshin Gangotri snout between 1996 and 2006

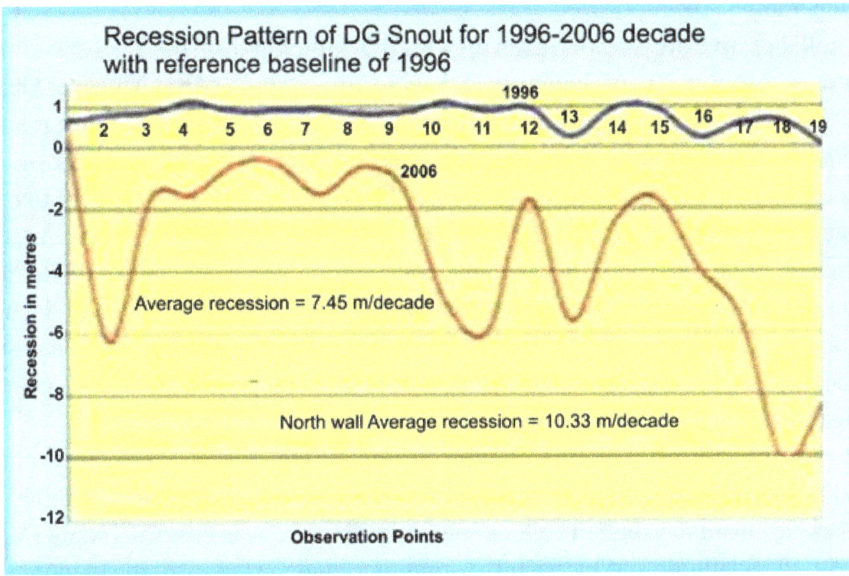

In the year 2001, the observation area was enlarged to confirm this recession pattern from other parts. For this, 20 additional observation points were marked over nine line-km along the ice wall margin of western Schirmacher Oasis. In 2006, the five-year observation of this ice wall margin showed an average recessional rate of 13.64 m per decade.

The second long-term project is about a km away from the old DG station and is about studying the accumulation pattern on the ice shelf, which has found a consistent trend for this part of the shelf. Every year, snow accumulation starts building up from March onwards and continues till September, depositing about 120 to 140 cm of snow. From October, the ablation process starts, and at its highest point in February, leaves a net balance of ~55 cm of snow. When it is buried and compacted to ice, this snow gives about 17 to 19 cm of ice-layer for each year.

In 1996, round-the-year observations were taken for thermal profiling of glacier ice, snow and permafrost by installing various thermistors at different depths. It indicated that the polar cold front does not penetrate below a depth of 24 m of ice. The response pattern to varying depths within the glacier shows a consistent time lag of up to 2 days. The average annual surface temperature was found in conformity with the temperature at a depth of 10-m. A temperature inversion was seen within the glacier, where the ice layers below 25 m depth became progressively warmer till the warmest temperature was found at the bottom of the glacier. The comparative study with snow and permafrost have established mathematical relationships by which the temperature of other media can be calculated if the data is available for air temperature.

A collaborative project was taken up with IIG-Mumbai from the 23rd expedition onwards to discover the ice dynamics and neotectonics south of Schirmacher Oasis. 21 GPS stations were installed on continental ice. These stations were observed in subsequent years. Initial studies indicate that the velocity of the ice sheet in this area ranges from 1.2 to 8.5 m per annum. Also, Schirmacher Oasis acts like an obstruction, and the ice sheet flows around it.

The Antarctic coastline is in dynamic equilibrium, and large chunks of floating ice shelves keep breaking off into the sea. A study has been undertaken by the GSI utilising successive satellite images for the past 30 years for studying the changing face of the coastline near the India Bay area. Aerial photographs and ground verification support these studies.

Ice Core Drilling: The first attempt for ice coring was made in the summer period of the 1st expedition (Jan 1982). A hand-auger was used for collecting ice core samples from the shelf. The location of the first site was near the Indian Camp at 69°59'12" S latitude and 11°55'09" E longitude, and the length of the ice core was 4 m. The second site was about 8 km away from the Camp, and the core recovered was 6 m long. In addition to these cores, two bulk ice samples of ~10 lt and ~20 lt volume were also collected. These samples were transported to India in frozen condition and were analysed at PRL-Ahmedabad. By studying radio isotopes ^{10}Be and ^{36}Cl, an attempt was made to detect any nuclear signature for natural radioactivity, but the result was negative. From the melt-water filter of ice samples, particles were separated under the microscope. From those particles, only 'spherules' were taken up for detailed examination, as spherules of extra-terrestrial origin are meteoroid ablation products. The size of most of the spherules ranged between 40 -m and 100 -m. Out of the 30 spherules studied, six appear to be of extra-terrestrial origin.

In the 2nd expedition (1982-83), five cores were collected using a hand-auger, four from shelf ice and one from inland polar ice. The depth of the cores ranged from 1.6 m to 5.65 m. Detailed optical and density profiling of the cores was carried out, detecting many lenses of firn and ice in the cores. The density ranged from 0.44 gm/cm^3 to 0.90 gm/ cm^3. One of these cores, a 3 m long section, was collected for isotope studies. The location of the sample was at 70^0 12' S latitude and 12^0 00' E longitude. As analysed in PRL-Ahmedabad, the oxygen isotope ratio suggests that this sample's mean annual surface air temperature (MASAT) is around –9°C.

In the 5th expedition (1985-86), three shallow ice cores were collected from the ice shelf, and the depth of these cores was 2 m, 5 m, and 9 m, respectively. In addition to the ice cores, 73 snow samples were also collected from Schirmacher Oasis, glaciers, and icebergs. Eighteen freshwater samples from the lakes of Schirmacher

Oasis were taken for comparative studies. These samples were analysed in PRL–Ahmedabad. Oxygen isotopes studies and a comparison of δ18O and MASAT were made for the ice cores, snow samples, and lake-waters. The results indicate that the lakes of Schirmacher Oasis receive 70% inflow from fresh snow-melt and only about 30% inflow from melt-water of old glacier ice. For January, the MASAT value for surface snow corresponds to –7°C, which agrees well with the observed meteorological value of –6.8°C. The MASAT value for the ice cores indicates it to be around –10°C at the sampling site. However, the MASAT value at the bottom of the glacier south of Schirmacher Oasis tends towards –40°C, which means that this ice is quite old; it was deposited at higher latitudes in the past and has moved to its present location.

The hand augers were used again in the 9th expedition (1989-90). The area of ice coring was shifted from the coastal shelf ice to the interior of the continent. But since the continental blue ice is severe (density ~0.9 gm/cm^3), the attempts with augers were not so successful.

The Ministry of Earth Sciences (Department of Ocean Development) procured one electromechanical ice core drilling machine in 1991, and it was taken to Antarctica by the 11th expedition team. Due to unforeseen circumstances and consequent constraints, test drilling could only take up during the next austral summer (Jan 1993). During the wintering period of the 12th expedition (April-Oct 1993), the first significant effort at motorised drilling was initiated. A site was selected on polar continental blue ice at 70°47'S latitude and 11°44'E longitude, about 3 km south of Maitri station. The first borehole (IND-1) was drilled, and 60.46 m of the ice core was recovered. It was transported to India in frozen condition and was analysed in the Isotope Lab of PRL-Ahmedabad. Due to some financial and technical constraints, only the top 30 m of the ice core could be explored in collaboration with Neils Bohr Institute of Astronomy, Physics, and Geophysics (NBI), Copenhagen-Denmark. Electrical conductance, $\delta^{18}O$, ^{210}Pb and cosmogenic radioisotopes ^{10}Be and ^{36}Cl were measured. The results indicate that the mean accumulation rate has been ~19 cm per annum during the past ~150 years. A significant marker horizon of mega-volcanic signatures of 1815 AD was identified ('Tambora' in Indonesia at 10°S latitude). Another sharp peak of a volcanic eruption of 1963 ('Augung' at 10°S latitude) was identified at 5 m depth in the ice core. The mean accumulation rate deducted from 'Tambora''s marker horizon is ~17 cm per annum. The MASAT data shows that the beginning of the 19th century was ~2°C cooler than in the recent past. Isotope studies of ^{36}Cl indicate that the cosmogenic fall-out in Greenland in the Arctic is higher than the present site in Antarctica by ~ 6. Also, the nuclear testing pulse (1940-1980) have not been detected at the current site.

In the wintering period of the the15th expedition (April-July 1996), the same machine was utilised with some modifications. This time, two shallow boreholes were made in polar continental ice. The location of the first borehole (IND-2) was at 70°46'51.9" S latitude and 11°43'5.3" E longitude, while the second borehole (IND-3) was about 15 km SW of Maitri station at 70°48'56.6" S latitude and 11°33'35.9" E longitude. In the IND-2 borehole, core recovery was good till rock was encountered at 76.23 m depth. Thermal profiling of the ice sheet from the surface to the depth at which rock was located was done. From the IND-3 borehole, an ice core of 84.10 m was retrieved. Detailed density profiling and extensive core photography were carried out in the field. The density variation was ~0.4 gm/cm^3 in the upper firn layers to ~0.9 gm/cm^3 in the lower blue ice layers. Cores from both these boreholes, IND-2 and IND-3, are preserved in a frozen condition in the ice core lab at NCAOR, Goa, to undertake detailed geochemistry, isotope studies and palaeoclimatic interpretation the

In the 22nd expedition, regular ice core drilling was tried for the first time during the summer period (Feb 2003). Borehole, IND-4, was drilled at 70°51'383' 'S latitude and 11°32'180''E longitude. An ice core of 62.19 m was recovered from this site. For Thermo-luminescence-dating studies, the last 11.44 m of core was retrieved and packed in dark conditions. This core was melted and shifted to the TL-dating lab in PRL-Ahmedabad. The remaining core has been analysed in detail by the Ice Core Lab at NCAOR, Goa and the results are ready for publication (April 2006).

In the 25th expedition, drilling was again taken up during the summer period (Feb 2006). The site selected was more in-depth into the continental ice, about 160 km from the coast. A surface convoy of vehicles was taken closer to the mouth of Humboldt Glacier. Borehole IND-5 was drilled at 71°20' 38.5" S latitude and 11°35'38.8" E longitude. An ice core of 64.97 m was recovered from this site. This core is being shifted to the Ice Core Lab of NCAOR, Goa, in frozen condition (April 2006).

PALAEOCLIMATE/PALAEOENVIRONMENT STUDIES

Drilling for ice cores to decipher the palaeoclimatic history of the polar continental ice has been successfully achieved at five different locations during XII, XV, XXII and XXV Indian Antarctic Expeditions, and 346 meters of ice core has been generated so far. Inferences on palaeoclimatic conditions can be obtained by subjecting the cores to O, Be, Pb and Cs isotopic analysis. Studies on the variation of a cold front in different seasons and media have yielded significant results. GSI has signed a memorandum of understanding on ice core drilling in Antarctica with NCAOR, Goa, which has a state-of-the-art laboratory established for ice core analysis with the following objectives;

- Sediment core sampling from the freshwater lakes of Schirmacher Oasis, East Antarctica
- Ice core drilling in polar continental ice.
- Monitoring of Antarctic Coastline in the India Bay region of Princess d) Astrid Coast, East Antarctica - (R & D item w.e.f. XXIV IAE)
- Exploring the two Antarctic lakes in Gruber and Schirmacher area of central Dronning Maud Land (cDML), East Antarctica using the Ground Penetrating Radar - (R & D item w.e.f. XXV IAE)

Geomorphology

Geomorphological mapping in Schirmacher Oasis was initiated during the 18th Indian Antarctic Expedition. In mapping, identification of palaeo-shorelines, fossil glacial valleys, and various depositional landforms attributed to glacial activities have been reported. A digital terrain model of Schirmacher Oasis with different layers of polar and peri-glacial geomorphological features has been prepared.

Due to the retreat of the ice cap and consequential uplifting of the landmass, the low-lying rocky mass of Schirmacher Oasis disposed of in the form of small hills. The existence of comparatively higher relief of the structural hills on the northern periphery of the landmass than the central corridor, the steep escarpment in the north margin and the indications of a fault running all along the north margin are some of the supporting pieces of evidence of the proposed possibility. The Schirmacher Oasis has evolved due to different depositional and erosional processes in a periglacial environment (Ravindra, 2001).

SEDIMENTOLOGICAL AND LIMNOLOGICAL STUDIES

A portable sediment coring device designed and fabricated at the Antarctica Division of GSI has been successfully used in the 22nd, 23rd and 24th Antarctic Expeditions to raise sediment cores from the freshwater lakes in Schirmacher Oasis and Larsemann Hills. The glacial lake sediments characteristically preserve various depositional episodes. Hence, they serve as valuable indicators of climatic fluctuations.

Terrace sediment samples (clay-rich portions sandwiched between glacial boulder horizons) collected from Priyadarshini (Zub) lake area for thermo-luminescence (TL) dating is expected to give a first-order estimate of the rate of deglaciation since the last glacial maxima.

Water samples from proglacial, land-locked and epi-shelf lakes of Schirmacher Oasis were collected for geochemical assessment. Water samples from lakes in the Amery ice shelf area, Prydz Bay Region, and East Antarctica were analysed for their major ion and trace element concentrations to establish base values for these constituents in the natural environment. These studies aimed to establish threshold/ base values for the physical and chemical parameters before the human occupation in the area. These would be useful in assessing the impact of anthropogenic activity at a later stage.

The most characteristic feature of the glacial environment is the deposition of debris supplied by ice-mass – i.e. settling of material from melting ice in the ablation zone (Asthana and Chaturvedi, 1998). The topographic undulations and irregularities from pools/lakes of variable sizes are characterised by almost stagnant water in which the finest sediment particles settle under a typical lacustrine environment. The sediments formed in such lakes are characteristically glacio-lacustrine deposits derived from melting ice around the vicinity of lakes, polar ice-sheet – south of Schirmacher Oasis through the melt-water channels in the lakes along with debris-bearing strong polar winds. The intense action of aeolian sedimentation in the present area remains a conspicuous feature. Wind action, coupled with snowstorms, also erodes and transports particles larger than sand, thus giving rise to coarse fractions occurring within the typical lacustrine sediment systems (Asthana et al., 2012). Most of the glacio-lacustrine sediments in the area are relatively fine-grained because the water column generally remains stagnant or has a low flow velocity so that even the finest particles get settled. Factors responsible for the final depositional process are settling out of suspension (windblown materials also) by the surficial currents, rolling and traction populations through insignificant bottom currents and little contribution from the loosely held thin veneer of glacial-debris occurring on the sub-aerial or sub-aqueous adjoining slopes (Asthana et al., 2012). The granulometric characteristics commonly show relatively coarse lake margin sediments and fine-grained bottom sets that originate from seasonal settling due to glacial melt-water in the spring seasons. The sediment cores contain mostly faintly laminated silts or sandy mud with frequent pebble-sized rock fragments, typical of deposits found in water bodies covered by an ice sheet (Asthana et al., 2012).

Furthermore, detailed limnological studies from the Antarctic Peninsula, McMurdo Dry Valley and East Antarctic lake-rich Oases: Vestfold Hills, Larsemann Hills, Bunger Hills, Rauer Islands and Windmill Islands have already contributed significantly in the various aspects of lacustrine systems. Earlier attempts of retrieving Lake Bottom sediment cores from Schirmacher Oasis succeeded in getting a few cores of about 50 to 70 cm lengths (Asthana et al., 2012). Threefold climatic oscillation (warm-dry, warm-humid and warm more humid) since last 8,000 BP (Bera and Khandelwal, 2003; Bera, 2004; Bera, 2006) are reported, besides the identification of the first record of some Arcellaceans (Mathur et al., 2006) are significant ecological indicators.

A total number of 39 lake bottom sediment cores (cumulative core length - 43.50m) have been retrieved using indigenously designed and fabricated sediment coring device during the austral winter of the 24th Indian Scientific Expedition to Antarctica (2004 – 06). From each coring site, the entire sedimentation column from top to bedrock has been raised with cores of more than 200 cm lengths (most extended core being 221 cm) that are likely to provide data beyond LGM (Asthana et al., 2012). Sedimentological, geochemical, biological, palynological, micropalaeontological and mineral magnetic susceptibility studies have been envisaged (Mathur et al., 2006; Asthana et al., 2009). The isotopic analyses and thermoluminescence (TL) dating were carried out. Such data will be helpful to understand the global climatic changes in the recent geologic past (Asthana et al., 2012).

Neotectonics

Evidence of neotectonic activity is observed in Schirmacher Oasis. It includes dislocation of glacial morainal trails, uplifting of deglaciated valleys, faults, exposed escarpments and palaeo-shorelines at higher topographic levels in the northern margin of the Schirmacher Oasis range. During XXI Antarctic Expedition, studies regarding these aspects were carried out in the north margin in the western part of Schirmacher Oasis.

SIGNIFICANT ACHIEVEMENTS

1. Compilation of an Interactive Digital Photographic database of work is done with Multiple Query Options, leading to the publication of an Illustrated Atlas of Antarctic Expeditions
2. Creation of theme-based relational database information system on Antarctic Rock Samples
3. Input for PPOD. Laboratory, AMSE- Fluid inclusion studies of gneiss-charnockite association in central Dronning Maud Land, East Antarctica
4. Input for Geochronology and Isotope Geology Division, CHQ-Geochronological studies in cDML, East Antarctica.
5. Compilation of geological map of the Orvin Mountains, central Dronning Maud Land
6. Compilation of geological map of Schirmacher Oasis
7. Compilation of geomorphological map of Schirmacher Oasis
8. Compilation of geological map of Schirmacher – Wohlthat region, cDML, East Antarctica

This chapter provided details of the geological, glaciological and limnological studies in the Antarctic region. It provided a firm ground to undertake futuristic geological investigations in the Schirmacher Oasis region.

REFERENCES

Asthana, R., & Chaturvedi, A. (1998). The grain size behaviour and morphoscopy of supraglacial sediments, south of Schirmacher Oasis, Antarctica. *Journal of the Geological Society of India*, *52*(5), 557–568.

Asthana, R., Shrivastava, P. K., Beg, M. J., Shome, S., & Kachroo, K. (2009). Surface microtexture of quartz grains from glaciolacustrine sediment core from Priyadarshini lake, Schirmacher Oasis, East Antarctica as revealed under Scanning electrom microscope. *Indian Journal of Geoscience*, *63*(2), 205–214.

Asthana, R., Shrivastava, P. K., & Ramar, K. (2012) *Lake Bottom Sediment Coring from Schirmacher Oasis During 24th Indian Antarctic Expedition.* Scientific Report Twenty Fourth Indian Antarctic Expedition 2003-2005 Ministry of Earth Sciences, Tech. Pub. No. 22.

Bera, S. K. (2004). Late Holocene Palaeo- winds and climatic changes in Eastern Antarctica as indicated by long-distance transported pollen-spores and local microbiota in polar lake core sediments. *Current Science*, *86*(11), 1485–1488.

Bera, S. K. (2006). *Pollen Analysis of Surface Deposits and Holocene Lake Sediment of Schirmacher Oasis, Central Dronning maud Land (CDML), East Antarctica.* Twentieth Indian Expedition to Antarctica, Scientific Report, MoES, Technical Publication No. 18.

Bera, S. K., & Khandelwal, A. (2003). Aerospora over the Southern Ocean and Schirmacher Oasis, East Antarctica. *Current Science*, *85*(2), 137–140.

Mathur, A. K., Asthana, R., & Ravindra, R. (2006). Arcellaceans (thecamoebians) from core sediments of Priyadarshini Lake, Schirmacher Oasis, Eastern Antarctica. *Current Science*, *90*(12).

Ravindra, R. (2001). Geomorphology of Schirmacher Oasis, *East Antarctica. Proceedings. Symposium on Snow, Ice and Glaciers, Geological Survey of India*, *53*, 379-390.

Chapter 10
An Overview of Survey and Mapping of the Antarctic Region Around Maitri Research Base

Nagarajan Balasubramanian
Indian Institute of Technology, Kanpur, India

ABSTRACT

Survey of India, the premier mapping agency of India, has been associated with Indian Antarctica Research Program since the 10th Indian Scientific Expedition to Antarctica (1991-92). It has been taking part in all the expeditions ever since, except for the 15th expedition. The focus till the 22nd expedition has remained to carry out the surveying and mapping of the entire Antarctica region of Indian interest and provide additional technical support to the various other participating agencies. During the first two expeditions, global positioning system (GPS) and conventional triangulation techniques were used for providing primary control work to facilitate detailed mapping of the area using conventional mapping techniques. In the successive expeditions, initially, mapping was done on the scale of 1:5000 with contour interval 5m and then on a scale of 1:1000 with a contour interval of 1m. 7.8 sq. km of Schirmacher Oasis has been surveyed, and mapping has been undertaken on a large scale. Analogue and digital maps have been prepared for most of the areas for scientific use.

DOI: 10.4018/978-1-6684-4078-0.ch010

INTRODUCTION

Antarctica has become an essential region of International Cooperation in scientific research governed by the norms of the International Treaty (Wikipedia, n.d.). India started research in Antarctica with the first scientific expedition to Antarctica in 1981-82. The initial days of Indian scientific efforts in Antarctic science were challenging and required herculean tasks to start from scratch. Therefore, the Indian mapping and survey in the Antarctic region during the first three decades are essential and historical. The present chapter details those significant scientific efforts made in the initial about 30 years.

The Survey of India began participating in the research at Antarctica during the 10th Indian Scientific Expedition. Survey of India has been continuously participating in research work in Antarctica, except for the 15th Indian Scientific Expedition to Antarctica (Survey of India, Annual Report 2011-12). Primary control points have been established in Schirmacher Oasis during the 10th Antarctica Expedition for mapping and other scientific studies in the region.

Survey of India has established Geodetic and Geophysical Control in Schirmacher Oasis and its surroundings for future scientific explorations in glaciology and global change. It has provided control points for the Dakshin Gangotri Glacier snout for the Geological Survey of India, control points for the observatory of the India Meteorological Department. It has also aided survey work related to the National Physical Laboratory, Defence Electronics Application Laboratory and National Environmental Engineering Research Institute.

It was felt that, before starting any mapping activities in a region, there is a need for height control for subsequent mapping to establish planimetric and location-based research activities. GPS (British Antarctic Survey, n.d.) was used during the 10th expedition to select 14 planimetric control points in Schirmacher Oasis and surrounding nunataks. The heights of the stations were established by connecting the actual sea level from two different locations in the sea through GPS observations. In addition to geomagnetic, gravity measurements were also taken on all control points established through GPS techniques because the GPS observations properly represent the ongoing crustal movement in the Antarctic Region (Ohzono et. al., 2006) All these activities were carried out given providing primary control and further studies for crustal movement or Glacier Movement. It was emphasized by the scientific community that there is a need for a large-scale map around the Maitri station for other developmental activities. During the 11th and 12th expeditions, a map was prepared on a scale of 1: 5,000 using Total Station along with conventional techniques. GPS control work was also extended up to that region. Mapping at 1:1000 for Schirmacher Oasis at a contour interval of 1m was undertaken during the 13th to 21st expedition. Due to specific reasons, the mapping could not be carried out during the 15th & 17th expeditions. The large-scale map was thought to be very useful for other scientific activities like Environmental studies, Soil studies, Crustal Movement Studies and other Atmospheric Parametric studies (Tropospheric and Ionospheric). In addition to mapping, the Survey of India team also carried out a Glacier Movement Study in Dakshin Gangotri glacier and glaciers located around Schirmacher Oasis and Geological Survey of India.

During the 16th expedition (Rawat & Mehta, 2000), the Survey of India team assisted the Norway Geodetic team in carrying out GPS observation at Maitri Station to provide a control point for their use.

During the 17th expedition (Ravichandran, 2000), a new GPS Bench Mark MAII was established near Maitri on the hilltop, adjoining its eastern side of the Priyadarshni (ZUB) lake. Survey of India, NGRI, and Institute of Planetary Geodesy University have jointly taken up the study of Planetary Geodesy in Antarctica to study the plate movement of Antarctica concerning other plates and to understand Ice-Shelf dynamics.

During the 22nd expedition (Pal & Shandilya, 2008), mapping was extended in the Schirmacher Oasis on a 1: 5,000 scale with a 5m Contour Interval, and mapping of 2.5 sq. km was completed. In this way, the total area covered by the 1: 5,000 scale was 7.8 sq. km.

From the 23rd expedition (2003-2004) (Jayaram, 2013), the Survey of India has started neo-tectonic and glacier movement studies in the Antarctica region. The aim is to undertake the following geodetic work:

- To establish 40 GPS control points (with permanent documentation) in the entire region of Indian interest and provide precise gravity values at these locations.
- Connect them by High Precision Levelling for accurate Vertical Control.
- Repeat levelling,
- GPS and gravity observations for at least three epochs.
- Carry out GPS observation campaign for seven days during each Expedition at Maitri station for interplate movement studies between Indian and Antarctica plates.

Twenty-eight stations have been established for neotectonics movement studies. There was a GPS observation campaign for seven days during the 23rd, 24th, 25th and 26th expedition at Maitri station for interplate movement studies between the Indian and the Antarctic plates. Details are provided below:

During the 23rd expedition, 13 new and one old GPS point were observed for neotectonics and glacier movement studies in the Antarctica region.

During the 24th Expedition (Singh & Tariyal, 2012), II epoch observations were carried out, in which 13 stations and 15 new stations were constructed and observed. Seven days of GPS observation at Maitri Station were also carried out for interplate movement studies between the Indian and Antarctic plates.

During the 25th Expedition, III epoch observation in 13 stations and II epoch observation in 15 stations were carried out. Seven days GPS campaign was carried out at Maitri Station for interplate movement studies. Mapping on a scale of 1:5,000 was also performed, covering an area of 2.2 sq. km.

MAPPING AT LARSEMANN HILLS REGION

During the 26th expedition, a Survey of India team consisting of 4 members went to 'Maitri' and the new Indian station site at 'Larsemann Hills'. Survey of India team established Primary control in WGS 84 Datum using GPS. Vertical control was provided using GPS and by applying the EGM96 geoid model. The team mapped on a 1:5000 scale with a contour interval of 5m at both destinations and completed the mapping job at 'Larsemann Hills' within a stipulated time frame (Figure 1).

The team carried out nine days of GPS observation at a new control point at Larsemann Hills, which will be used for inter-plate movement studies

Figure 1. Mapping work in Larsemann Hills

Simultaneously, the SOI team established seven ground control points well distributed and worked at 'Larsemann Hills' using a Trimble 5700 dual-frequency GPS receiver for future reference.

At 'Maitri', mapping in a 1:5000 scale is continued in the unmapped areas of 'Schirmacher Oasis'. Simultaneously, 7-days GPS observation was taken at Maitri Station' S' for Interplate Movement studies between the Indian and Antarctic plates.

Establishment of Horizontal Datum

A network of well-distributed precise control points is essential for carrying out mapping or other location-based research activities. To establish Horizontal Datum in Schirmacher Oasis, Survey of India provided Primary Control using Conventional Triangulation Techniques and state of the art Global Positioning System (GPS) Technology during the first two expeditions (Figure 2). Eighteen planimetric control points in the WGS84 datum were established in Schirmacher Oasis and surrounding Nunataks using Ashtech dual-frequency GPS Receiver.

These control points were provided with Broadcast Ephemerides in WGS84 Coordinate System with Post Processing Software POPS.

Figure 2. GPS Stations at Schirmacher Oasis

Establishment of Vertical Datum

Heights of the stations were provided in two different ways (a) Connecting to Instantaneous Mean Sea level, (b) Computing ellipsoidal height from GPS and then subtracting the Geoidal Undulation value from OSU86D Global Geopotential Model (Figure 3).

Also, gravity measurements were taken on all control points established through GPS techniques. All these activities were carried out given providing primary control and further studies for the Crustal Movement or Glacier Movement.

Figure 3. Establishment of vatical datum

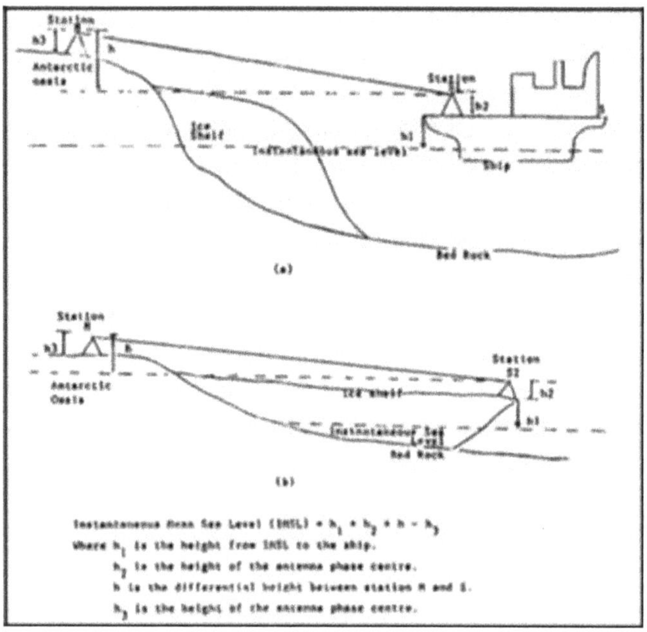

INTERPLATE MOVEMENT STUDIES

Since the 23rd expedition, the Survey of India started studying Interplate movement between the Indian and Antarctic plates. Global Positioning System can determine the distance between the thousands of km apart stations with considerable accuracy. If the same lengths (Baselines) are determined repeatedly at different epochs, the variation will give an idea of the relative movement of the stations. This method is applied for plate movement studies between the Indian plate and the Antarctic plate.

A seven-day GPS campaign has been performed at 'Maitri Station' in every expedition since the 23rd expedition. The data is processed concerning the IGS stations using precise ephemerides and the baselines to the permanent Indian GPS stations, which are compared for any variations. During the 26th expedition, the 4th epoch observations were carried out at Maitri (Figure 4).

Figure 4. GPS observation at Maitri station for inter plate Movement studies

During the 26th expedition, the Survey of India team carried out successive nine days of observation near the proposed new station site at Larsemann Hills. From then on, this control point was also to be used for inter-plate movement studies. Seven days of continuous GPS observation will be carried out on this point in each expedition.

The results obtained so far show that the baselines from Maitri Station to Indian Permanent Stations are contracting. However, considering the accuracy limit of the GPS, some baseline changes may be due to the errors. Hence, more epoch observations will be required to confirm this and for determining the rate of movement.

NEOTECTONICS AND CRUSTAL MOVEMENT STUDIES

Neotectonics and Crustal Movement Studies commenced since the 23rd expedition. Twenty-eight stations have been established in Schirmacher Oasis on (Stål, et. al, 2020) which regular GPS observations on campaign mode are planned. Until now, on 13 of these stations, three epoch observations have been completed, while on 15 stations, two epoch observations are over. The changes in the coordinates from one epoch to another epoch are monitored

Variation in the coordinates was noticed (within the accuracy limit of the instrument) from the observations taken so far. It will be inappropriate to come to any conclusion in just 2-3 epoch observations. More observations will be carried out in the subsequent few expeditions.

Surveying and Mapping

Surveying and mapping are the primary tasks of the Survey of India. In Schirmacher Oasis, mapping of 7.8 square km on scale 1:5000 with 5 meters counter interval has been completed. The area around Maitri has been mapped on a 1:1000 scale on demand of the team leaders of various expeditions. During initial trips, the conventional methods of triangulation were used to provide control work for mapping. For detailed surveying and contouring, the traditional way of plane tabling was used. These were tedious and time taking methods. A combination of various techniques needs to be used to complete the mapping work rapidly. During the 26th expedition, the Survey of India was asked to map the area around the proposed new station at the Larsemann Hills region. The available working days expected were less than ten days. Hence, it was planned to use a combination of techniques. Satellite imagery of the area was obtained to plan the work. For providing essential control work, two dual-frequency GPS receivers were used. One GPS receiver was continuously kept for nine days, while the other provided control points. (Figure 5)

Figure 5. Map of Larsemann Hills

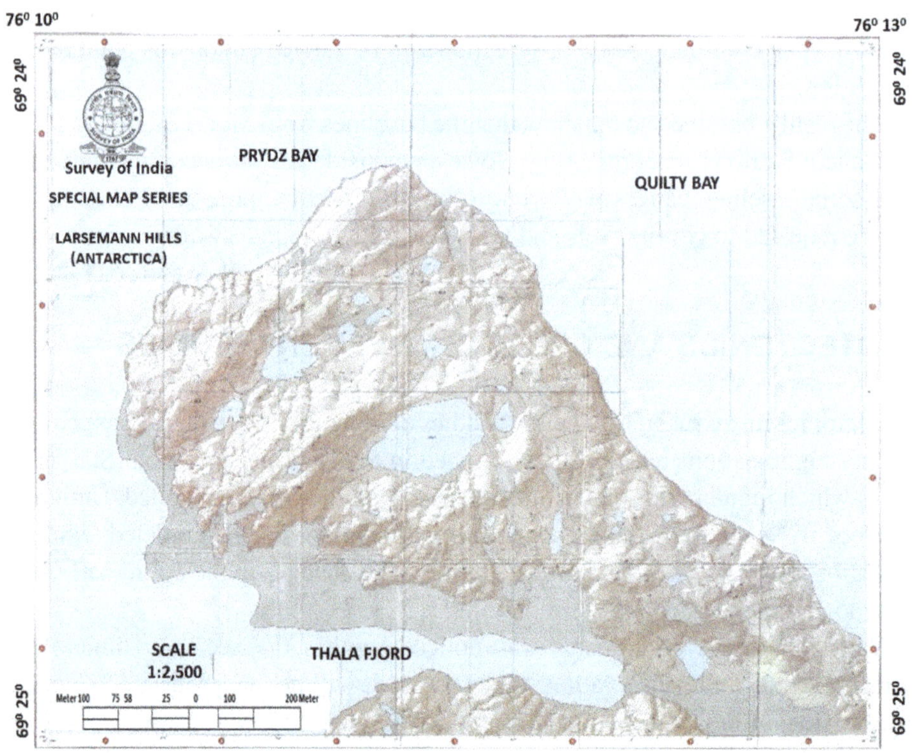

Simultaneously, mapping work was also done. Two single-frequency GPS receivers (Trimble Pathfinder with Recon controller) were used for detailed survey and contouring. The data of single-frequency receivers were processed with the data of dual-frequency receivers, and refined coordinates were obtained. The points collected by these receivers were plotted on a plane table, and contouring was done on the field to ensure that the best possible terrain representation was performed. Surveying was completed within nine days. A similar methodology was followed at Schirmacher Oasis also. At the Headquarters, the plane table section was scanned and digitized. Preliminary proof of the map of the area around the proposed site in the Larsemann Hills region has been given to NCAOR Goa.

Survey of India has contributed significantly to Antarctic science by way of;

- Establishment of Horizontal and Vertical Reference frame for Mapping of Schirmacher Oasis.
- Provision of Geodetic Control using Global Positioning System (GPS) technique for Geophysical and Glaciological Studies'.
- Mapping of Schirmacher Oasis on Scale 1:5000 with Contour Interval for 5 m for multi Organizational Studies and on Scale 1:1000 with Contour Interval for 1 m for specific detailed studies.
- Neo-tectonic and Crustal Movement Studies using GPS, Gravimetry and High Precision Levelling.
- Interplate movement studies between Antarctica and India.
- Mapping of new station site at Larsemann Hill in 1:2,500 scale with contour interval 5m.

For the benefit of Indian Scientists, the Survey of India has already published large scale maps (both in analogue and digital form) covering most of the areas of interest which can be effectively used for plotting research locations. The summary of the survey during the different Expeditions is provided in Table-10.1.

Table 1. Work done during Antarctica expeditions

SL. No.	Year of Expedition	Expedition	Work Done	Scale of survey	Contour Interval in metres	Locations	Area surveyed
1.	1990-91	X	Control Work	-------	-------	Maitri and	50 Sq. Km.
			By G.PS.			Surrounding	
						area	
2.	1991-92	XI	P.T.	1:5000	5	- do -	1.8 Sq.km.
3.	1992-93	XII	P.T.	1:5000	5	- do -	3.5 Sq.km.
4.	1993-94	XIII	P.T.	1:1000	1	- do -	0.25 Sq.km.
5.	1994-95	XIV	P.T.	1:1000	1	- do -	0.25 Sq.km.
6.	1995-96	XV	SOI	not	participated		
7.	1996-97	XVI	P.T.	1:1000	1	- do -	0.20 Sq. km.
8.	1997-98	XVII				- do -	
9.	1998-99	XVIII	P.T.	1:1000	1	- do -	0.20 Sq. km.
10.	1999-2000	XIX	P.T.	1:1000	1	- do -	0.20 Sq. km.
11.	2000-2001	XX	P.T.	1:1000	1	- do -	0.25 Sq. km.
12.	2001-2002	XXI	P.T.	1:1000	1	- do -	0.22 Sq. km.
13.	2002-2003	XXII	P.T.	1:5000	5	- do -	2.5 Sq. km.
14.	2003-2004	XXIII	GPS	------	------	- do -	13 New
			Work				1 Old
15.	2004-2005	XXIV	GPS	------	------	- do -	14 New
			Work				14 Old
16.	2005-2006	XXV	GPS	-------	-------	- do -	28 Old
			Work				
			P.T	1:5000	5	- do -	2.2 Sq. km.
17.	2006-2007	XXVI	GPS	-----	-----	1. Maitri	7 New
			Work			2. Larsemann	1 Old
			P.T	1:5000	1:5000	1. Maitri	2.3 Sq. km.
						2. Larsemann	4.2 Sq. Km.

In conclusion, the Survey of India has provided Ground Control Points in Schirmacher Oasis and surrounding nunataks. Gravity and geomagnetic observations were also taken in the region. These activities were carried out to provide primary control and further studies for crustal movement and Glacier Movement. Mapping on 1:5000 scale and 1:1000 scale was also undertaken around Maitri Station,

Schirmacher Oasis and Priyadarshini lake. Survey of India has published two maps on a 1:5000 scale, one on a 1:1000 scale under special map series, and other maps are in the printing stage.

The initial scientific mission of the department is to provide an accurate large-scale map on Scale 1:5000 with contour interval 5M for various scientific purposes. However, the team leader of the multiple expeditions has used the survey potential for large scale mapping on a scale of 1:1000 for their infrastructure installation needs. The mapping was also done for Larsemann Hills, where India planned to establish its new station, 'Bharti', which is now functional. This map was of great use in preparing the locations of various facilities of the station. As the map is in WGS 84 coordinate system, it will be helpful for scientists in location-based sample collection using Hand Held GPS.

In addition to the mapping, the Survey of India had also planned to establish about 40 Nos. of GPS Control Points with permanent documentation in the entire region of Indian interest for monitoring the glacial movement and Inter-plate movement studies between the Indian and the Antarctica Plate. Survey of India has also provided technical assistance to other participating organizations.

REFERENCES

British Antarctic Survey. (n.d.) *Global Positioning System (GPS).* National Environment Research Council. Retrieved from https://www.bas.ac.uk/polar-operations/engineering-and-technology/technology-tools-and-methods/global-positioning-system-gps/

Jayaram, S. (2013). *Technical Report of 23rd Indian Scientific- Expedition to Antarctica—An overview.* Twenty Third Indian Antarctic Expedition, Technical Report No. 21, pp 11-20 published by Ministry of Earth Sciences, Government of India. http://14.139.119.23:8080/dspace/bitstream/123456789/908/1/pages+11-20.pdf

Ohzono, M., Tabei, T., Doi, K., Shibuya, K., & Sagiya, T. (2006). Crustal movement of Antarctica and Syowa Station based on GPS measurements. *Earth Planets and Space.* https://www.researchgate.net/publication/44258528_Crustal_movement_of_Antarctica_and_Syowa_Station_based_on_GPS_measurements doi:10.1186/BF03351984

Pal, R. K., & Shandilya, S. K. (2008). *Large Scale Mapping in Schirmacher Oasis in Antarctica*. XXII Indian Scientific Expedition to Antarctica – Technical Publication No. 20. http://14.139.119.23:8080/dspace/bitstream/123456789/859/1/cover-pages.pdf

Ravichandran, V. (2000). *Scientific Activities Report of Survey of India during XVII IAE 1997-98*. XVII Indian Scientific Expedition to Antarctica – Technical Publication No. 15, 145-154 published by Department of Ocean Development, Government of India. http://14.139.119.23:8080/dspace/bitstream/123456789/499/3/17TH.pdf

Rawat, S. S., & Mehta, S. K. (2000). *A report on surveying in Antarctica during 16th Indian Scientific Expedition to Antarctica*. Scientific Report XVI Indian Expedition to Antarctica – Technical Publication No. 14, 335-340 published by Department of Ocean Development, Government of India. http://14.139.119.23:8080/dspace/bitstream/123456789/513/3/INTRODUCTION.pdf

Singh, M., & Tariyal, J. S. (2012). *Report on the work done by Survey of India team during XXIV Indian Scientific Expedition to Antarctica*. XXIV Indian Scientific Expedition to Antarctica – Technical Publication No. 22, 199-206 published by National Centre for Antarctic and Ocean Research/Ministry of Earth Sciences, Government of India. http://14.139.119.23:8080/dspace/bitstream/123456789/941/1/RRContent.pdf

Stål, T., Reading, A. M., Halpin, A., Phipps, S. J., & Whittaker, M. (2020). *The Antarctic Crust and Upper Mantle: A Flexible 3D Model and Software Framework for Interdisciplinary Research -Technology and Code Article*. Frontier Earth Science. doi:10.3389/feart.2020.577502

Survey of India. (n.d.). *Annual Report 2011-12*. https://surveyofindia.gov.in/documents/annual-reports/annual-report-2011-12.pdf

Wikipedia. (n.d.). Antarctic Treaty System. In *Wikipedia*. Retrieved from https://en.wikipedia.org/wiki/Antarctic_Treaty_System#cite_note-3

Chapter 11
Antarctic Laboratory Ground Truth for a Microwave Eye in the Sky

Mukund Kurtadikar
Department of Physics, JES College, India

ABSTRACT

A C-band microwave bench setup was carried out during the 22nd Indian Scientific Expedition to Antarctica (December 2002 to March 2003) to measure the dielectric properties of Antarctic geophysical materials like samples of ice, soil, and rocks around Indian Antarctic Station, Maitri. Laboratory validation of these properties of Antarctic frosts and soils was done during the summer period of the expedition, as they are significant for microwave remote sensing applications. Dielectric measurements of nine Antarctic rock samples were made after returning to India, which are substantial from the geological point of view. Here, the authors report the first attempt to measure the dielectric properties of these materials of Antarctica.

INTRODUCTION

Antarctica ice sheets play an essential role in influencing the climate system. Global climate change has a significant impact on the ice shelves. They are sensitive to changes in air and ocean temperature or circulation near Antarctica. Many Indian scientific institutions like Indian Space Research Organization (ISRO), Space Applications Centre (SAC) National Remote Sensing Centre (NRSC) participated during various Indian Scientific Expeditions to Antarctica in the current decade (2010-2020). They used equipment like Ground Penetrating Radar (GPR) to study

DOI: 10.4018/978-1-6684-4078-0.ch011

the ice terrines and used the data of RISAT 1, a microwave remote sensing satellite, launched in April 2012. It was the first indigenous satellite imaging mission of ISRO using an active radar sensor system, namely a C-band SAR(Synthetic Aperture Radar) imager. The RISAT mission aimed to use the all-weather as well as the day-and-night SARobservations.

It was also intended to capture its capability in various applications. Some significant implications were agriculture, forestry, soil moisture, geology, sea ice, coastal monitoring, object identification, flood monitoring, etc. Accordingly, the RISAT specifications have been drawn with the national requirements in mind. **The Sea Ice observations of RISAT 1 are significant for studying Antarctic Ice Sheets**. (https://www.isro.gov.in/isros-participation-antarctic-expedition).

In anticipation of the proposed launch of C-band microwave remote sensing satellite in future, ISRO had sanctioned a small project for laboratory validation of dielectric properties of soil, seawater, vegetation etc., at C-Band microwave frequency, under RESPOND scheme to Kurtadikar M L of JES College, Jalna, Maharashtra, India; way back in 1990. A C-band microwave bench setup was procured in this project. The set up was further modified and suitable methodology was developed for measuring dielectric properties of these materials. The modification was mainly about atomizing the manually operated microwave bench and interfacing it to a PC, achieved way back in the 1990s. With the meagre resources and several constraints, the facility was extensively used, even after completion of the project, for about a decade and a half leading to papers and doctoral dissertation of several researchers. (Navarkhele *et al.*, 1998; Agrawal *et al.*, 2004; Joshi and Kurtadikar, 2013; Kurtadikar *et al.* 2013; Deshpande *et al.*, 2015; Itolikar and Kurtadikar, 2017a ; Itolikar and Kurtadikar, 2017b; Itolikar *et al.* 2020; Navarkhele 1996; Agrawal 2000; Murugkar 2002; Kulkarni 2006; Joshi 2012; Deshpande 2012; Itolikar 2016).

The woek of experimental measurement of dielectric properties of Antarctic ice and other geophymsterials was carried out during the XXII Indian Scientific Expedition during December 2002-March 2003 (Kurtadikar, 2008). The work done during the expedition is presented in this chapter.

EXPERIMENTAL SETUP

In the C-band microwave bench, the detector probe vertically passes through the slot, senses the electromagnetic field inside the 'slotted section' of the bench setup and measures the strength of the electric field at that position manifested as detector current (Figure 1). Then, by traversing the probe along the slot-line, variation in the detector current (probe reading) at different locations in the waveguide could be measured. The dielectric Cell is a removable cavity of a rectangular waveguide with a perfect reflecting metal plate at the back (far) end. The microwave bench setup was interfaced with a PC through an add-on microcontroller card. The microwave power (or detector current) and the corresponding probe position are acquired and stored in a data file, later used in the source code to calculate dielectric properties.

Figure 1a. (A) The C-band microwave bench experimental setup; (B) C-band microwave bench at Maitri; (C) Indian Antarctic station (2003)

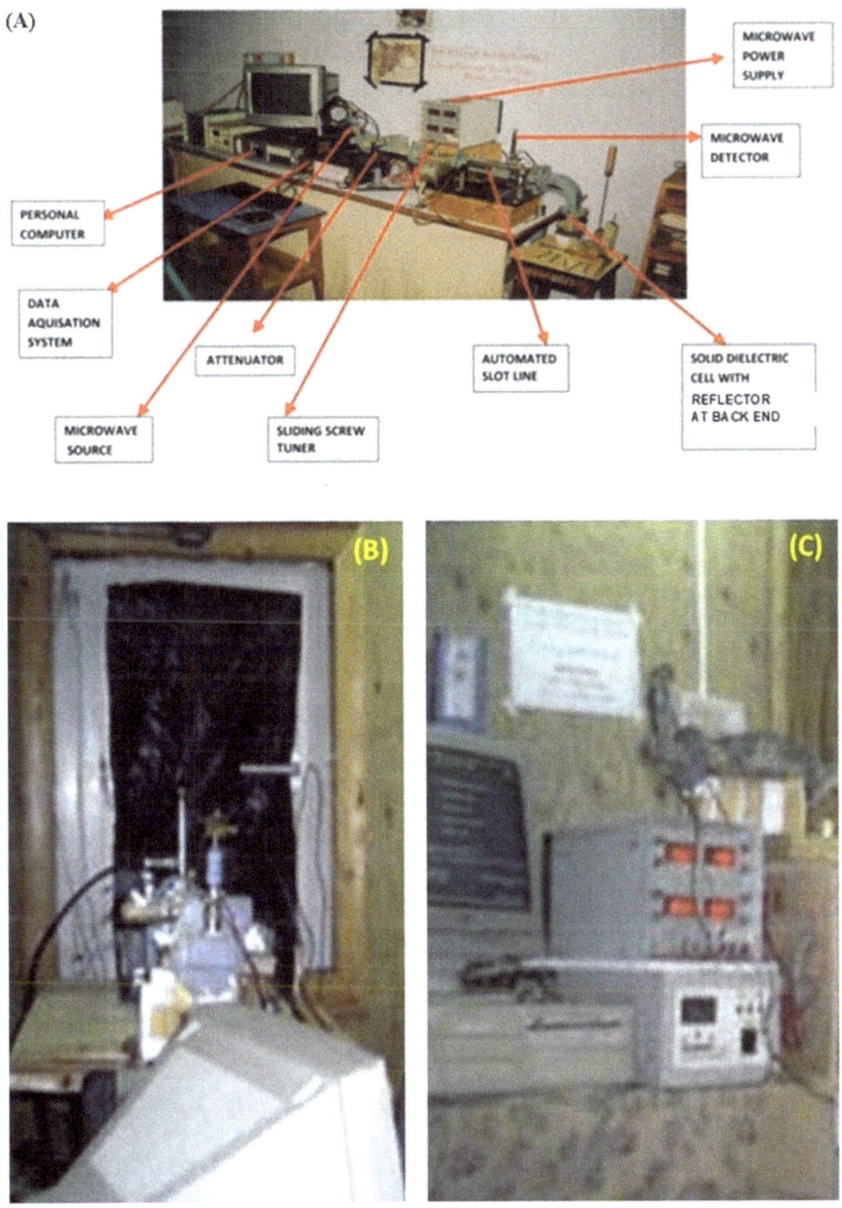

Prior to begin the experiment, the bench needs to be carefully and adequately tuned so that a perfect standing wave pattern of the microwaves in the slotted section was obtained with 'no sample' in the dielectric Cell (empty Cell), (Figure 2). Then similar wave patterns are obtained with the sample material (ice, soil, rock etc.) filled inside the Cell. Data is acquired and stored for samples of three different thicknesses (say 1cm (Figure 3), 2cm (Figure 4) and 3cm (Figure 5). All these four data files are clubbed together to make an 'input file' and used while running the source code developed to calculate the dielectric constant, dielectric loss and the errors in their measurement are stored in the 'output' file. The least-square fitting technique was used during the calculations. Standard formulae used by Von Hippal (Von Hippel, 1954) were used for analysis.

Figure 2. Standing wave pattern for empty cell (Without sample)

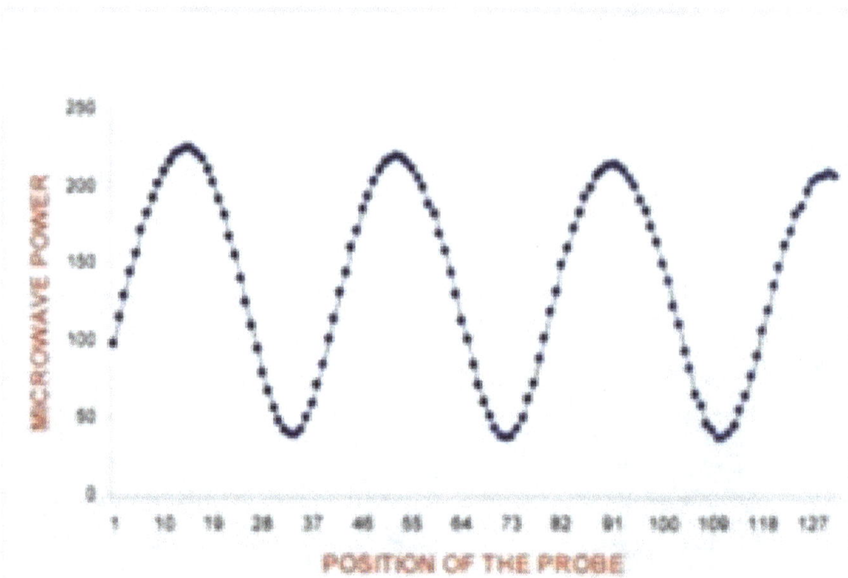

Figure 3. Standing wave pattern for sample in the cell (1 cm thickness)

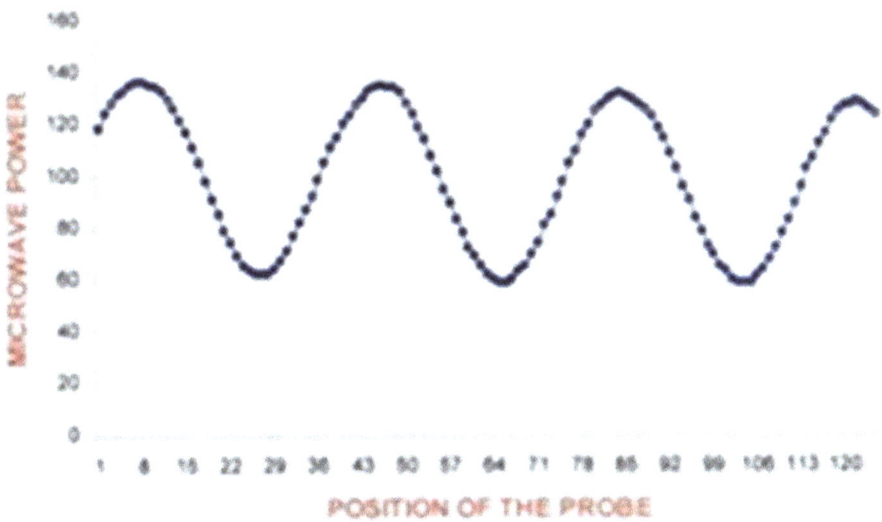

Figure 4. Standing wave pattern for sample in the cell (2 cm thickness)

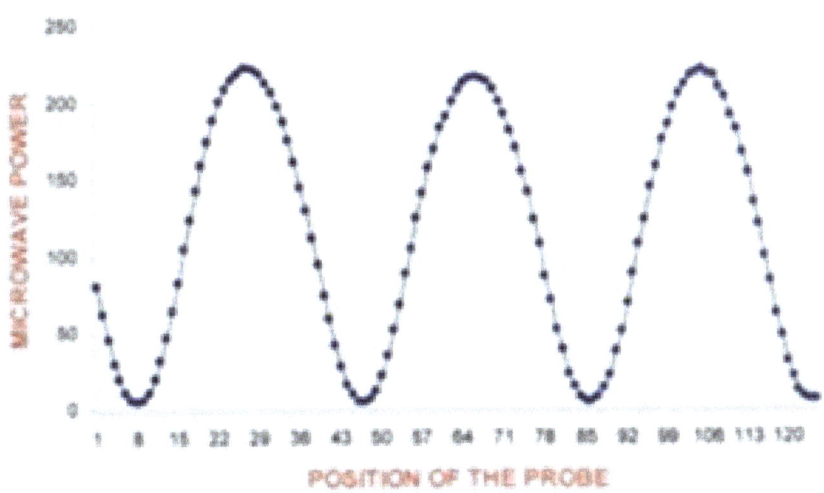

Figure 5. Standing wave pattern for sample in the cell (3 cm thickness)

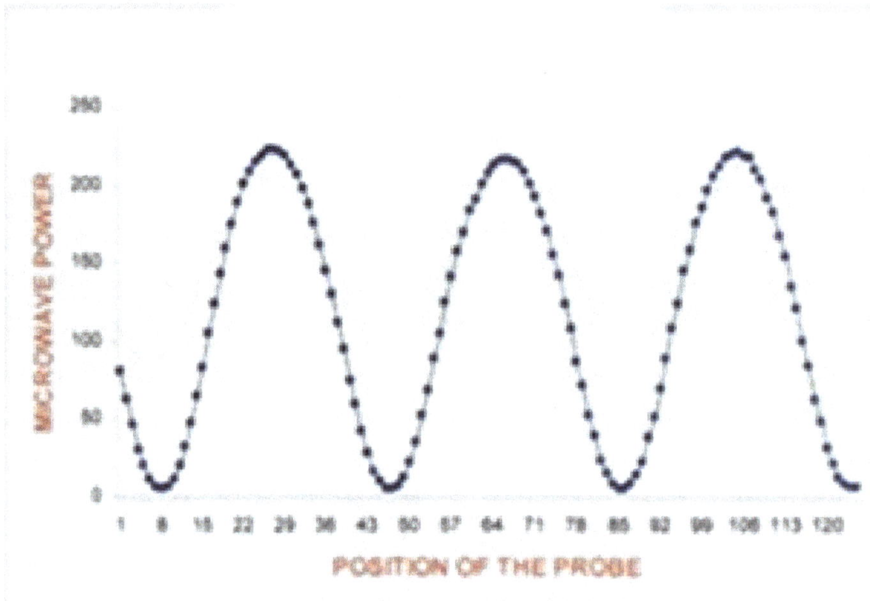

SAMPLE PREPARATION

The solid dielectric Cell is a rectangular cavity of C-band waveguide (of internal dimensions 1inch x 2 inches) and about 6cm depth. After tuning the bench for an empty Cell, the cavity is required to be filled by incerting the sample under test with depths 1cm, 2cm, 3cm as per requirement depending on the sample thickness to be used with a perfect reflector (highly polished metal plate) fixable to the flange at the far end of the Cell. However, the Cell needs to be kept vertical with a reflector plate downwards and then fill it with the ice, rock, soil or sand with uniform compaction. All the time to hold the sample vertically in the cavity, for keeping the reflecting plate downwards, we used an additional passive waveguide component of C-band, a plane E-bend, which diverts the microwave propagation through 90 degrees, without affecting the propagation characteristics. The plane E-bend conveniently connects the horizontal slotted section on one side and the vertical Cell on the other. (Figure 1) with the help of standard flanges. After filling the Cell cavity with the sample under test, no air gap should remain between the inner surfaces of sample and waveguide, in contact.

However, this problem is not very severe in case of the cold Antarctic ice samples. The ice samples of oversize i.e. of size more than the internal dimensions of the cavity, were kept ready in advance before starting the work. 'Hand plane' tool was used to cut the samples as per the exact dimensions of the Cell cavity and then by inserting it into the Cell, readings are immediately taken to obtain the standing wave pattern for that particular sample. Samples of three such thicknesses are needed for any specific material.

The sample preparation of soil and sand is not complex as it is a loose material and can be easily filled in the Cell cavity as per the required sample thickness.

Sample preparation of rock had another issue of cutting as the rocks.They are not easy for cutting as per the dimensions of the Cell. A high-speed hard cutter of precision quality is required for this purpose. Therefore, measurements for rock samples were done after reaching Jalna, India.

FORMULAE AND CALCULATIONS

A source code for computing the complex dielectric constant, dielectric loss and errors in their measurement was developed. The source code makes use of α and β as fitting parameters obtained from the input data file, where α = attenuation factor, β = phase shift constant. The guided wavelength λg is measured from the minima of the standing wave pattern.

$$\beta = \frac{2\pi}{\lambda_g} \tag{1}$$

The free space wavelength, λ_0 is determined using the relation refer equation (2)

$$\frac{1}{\lambda_0^2} = \frac{1}{\lambda_g^2} + \frac{1}{\lambda_c^2} \tag{2}$$

Where, $\lambda_c = 2\times a = 2\times 4.73 = 9.46 cm$, ‘a’ being the broader side of the C-band rectangular wave-guide.

The complex dielectric constant (The real and imaginary parts) are calculated using the relations given in equations (3) and (4)

$$\varepsilon' = \lambda_0^2 \left(\frac{1}{\lambda_c^2} + \frac{\left(\alpha^2 - \beta^2\right)}{4\pi^2} \right) \tag{3}$$

$$\varepsilon'' = \frac{\lambda_0^2 \alpha \beta}{2\pi^2} \tag{4}$$

The emissivity is a vital parameter of the material under consideration. It provides ground radiation characteristics for microwave remote sensing. It is the ratio of the microwave radiation from the surface of the material and blackbody thermal radiation with the same temperature as that of the material surface. It can be expressed as in equation.

$$e_{s(p)} = (1 - R_s) \tag{5}$$

Where R_s is the smooth- surface reflectivity.

SCIENTIFIC IMPORTANCE

The physical properties like dielectric constant, dielectric loss and emissivity of Antarctic Ice could not be measured in India in 2002-2003 due to the lack of a cold laboratory facility in National Centre for Antarctic and Ocean Research (NCAOR). Therefore, it was practically convenient to carry the equipment (C-band microwave bench) to Antarctica and do the measurements in the natural sub-zero temperature environment prevailing there. So far till the date, ours was the only research project regarding laboratory validation of these properties carried out in Antarctica. Our project "Study of Dielectric Properties of Antarctic Geophysical Materials at C-Band Microwave Frequencies" was conducted at Maitri during XXII Indian Antarctic expedition in 2003 by participating in the Summer Team of Indian Antarctic Expedition.

Most of the ice samples were collected from the region surrounding the Maitri Station. However, few samples were collected from the drilling site of the Geological Survey of India, about 25 Km away from Maitri. A couple of saamples of fast ice were collected from the India Bay site of Antarctica. The rock samples around Maitri were collected. During the expedition, the dielectric properties of about 20 ice and seven soil samples were measured at 5 GHz frequency. The Antarctic rock samples were brought to India and samples were cut to the size of the cavity of the dielectric Cell. Then the dielectric measurements were done in the Microwave Laboratory of JES College at Jalna Kurtadikar *et al.* (2009). Not much significant work on the laboratory measurements of the dielectric properties of Antarctic ices, soils and rocks has been done. There is perfect agreement between present values of ice dielectric constant which is is 3.19 (average of 8 samples) and those measured by Matzler and Wagmuller (2000) using radiometric method 3.17.

Also, there is a need for the laboratory measurement of dielectric properties of Antarctic ices at L and S-band microwave frequencies from the remote sensing application point of view. Any Indian laboratory does not have such facilities.

RESULTS

The dielectric constant (ε'), dielectric loss (ε'') and emissivity (e) are given for Antarctic ice samples in Table 1, soil samples in Table 2. The dielectric properties (ε' and ε'') of Antarctic rock samples are provided in Table 3. Comparison of the relative Permittivity (ε_r) of the rock samples at X and C-bands is given in Table 4.

Table 1. Dielectric properties of ice samples collected from Antarctica

Sr. No.	Sample Name	Dielectric Constant	Dielectric Loss	Emissivity	Sample type / Location
		ε'	ε''	e	
1	Snow1	3.1624	3.3387E-002	0.9215	new fern
2	Snow2	3.2138	1.1401E-002	0.9194	old fern
3	Snow3	2.9670	1.0438E-002	0.9295	old fern
4	Snow4	3.2904	7.6004E-003	0.9163	fresh snow(lake)
5	Snow6	3.1938	1.6217E-002	0.9202	Ice cored moraine
6	Blueice1	3.2750	2.0755E-002	0.9169	Near Maitri ice wall
7	Blueice2	3.2762	2.5372E-002	0.9513	GSI drilling site
8	Blueice4	3.1897	1.6516E-002	0.9204	GSI drilling site
9	Coreice1	3.1900	2.2012E-002	0.9203	GSI Camp 16.7m deep
10	Coreice2	3.1880	2.2870E-002	0.9204	GSI Camp 22.4m deep
11	Coreice3	3.1105	1.8681E-002	0.9236	GSI Camp 26.7m deep
12	Coreice4	3.1903	2.3254E-002	0.9203	GSI Camp 48.0m deep
13	Coreice5	3.2031	9.9326E-003	0.9198	GSI Camp 51.0m deep
14	Coreice6	3.1911	2.2200E-002	0.9203	GSI Camp 52.0m deep
15	Fastice1	3.0721	1.6028E-002	0.9251	New shelf site
16	Fastice2	3.0373	1.9613E-002	0.9266	New shelf site
17	Fern core	3.1989	2.5908E-002	0.9200	Fern core GSI site 0.3m 0.3 m deep Fern core GSI camp
18	Fern	3.1915	2.0886E-002	0.9203	Fresh fern (packed)
19	Loose fern	3.1884	1.9477E-002	0.9204	Fresh fern (loose)
20	Icy Soil 1	6.1911	4.2277E-002	0.8179	Near Maitri Ice Wall

Table 2. Dielectric properties of soil samples collected from land terrane near Maitri Station of Antarctica

Sr. No.	Sample Name	Dielectric Constant	Dielectric Loss	Emissivity	Location
		ε′	ε″	e	
1	Soil 1	6.35	1.05E-002	0.8135	Near Maitri Station
2	Soil 2	5.13	7.82E-003	0.8499	Near Lake L51
3	Soil 3	6.19	1.82E-002	0.8181	Near Priyadarshini Lake
4	Soil 4	3.36	2.48E-002	0.9136	10 m from the ice wall
5	Soil 5	3.19	6.81E-003	0.9203	Bottom of the ice wall
6	Sand 1	4.96	8.46E-003	0.8554	Aeolian sand
7	Sand 2	6.19	3.93E-002	0.8181	Sand near Russian Station - Nova

Table 3. Dielectric properties of Antarctic rock samples around Maitri at 5 GHz

Sr. No.	Type of Rock Sample	Dielectric Properties		Errors	
		ε'	ε"	Δε'	Δε"
1	Pink Granite	3.62	2.48E-002	1.77E-002	1.69E-002
2	Porphyritic Biolite Granite	3.85	2.94E-002	1.82E-002	1.72E-002
3	Norite	3.69	9.81 E-002	1.93E-002	1.93E-002
4	Chamockito	3.53	1.09E-002	1.36E-002	1.28E-002
5	Gneiss	3.69	1.38E-002	1.97E-002	1.91E-002
6	Mylonite	3.78	1.30E-002	1.54E-002	1.52E-002
7	Pyroxene Granulite	3.68	2.17E-002	2.13E-002	2.10E-002
8	Basic Dyke	3.79	1.51E-002	2.16E-002	2.00E-002
9	Pegmatite	3.66	9.98E-002	3.47E-002	3.59E-002
10	Banded Gneiss	2.59	1.29E-002	1.81E-002	1.68E-002

Table 4. Comparison of the relative permittivity (ε_r) of Antarctic rock samples at X-band and C-band

Sr. No.	Type of Rock Sample	Relative Permittivity ε_r	
		X -band	C-band
1	Pyroxene Granulite	3.50	3.65
2	Porphyritic Biolite Granite	3.31	3.83
3	Chamockito	3.92	3.51
4	Gneiss	3.74	3.67
5	Basic Dyke	3.70	3.76
6	Pink Granite	3.50	3.60
7	Mylonite	3.81	3.76

ACKNOWLEDGMENT

The author is thankful to ISRO for providing a C-Band microwave bench setup under RESPOND programme and NCAOR to support for participating in the XXII Indian Scientific Expedition to Antarctica. He also acknowledges Prof. S. C. Mehrotra and Dr P. C. Pandey for helpful discussions. Thanks are due to JES College and the University Grants Commission (UGC) of India for travel support.

REFERENCES

Agrawal, R. S. (2000). *Permittivity Measurement of DeccanTrap Soil of Microwave Frequencies* [Doctoral thesis, Dr Babasaheb Ambedkar Marathwada UniversityAurangabad]. Shodhganga: a reservoir of Indian theses at INFLIBNET.

Agrawal, R. S., Kurtadikar, M. L., & Murugkar, A. G. (2004). Dielectric Properties of Soil at 5 GHz: Microwaves and Optoelectronics. Anamaya Publishers.

Deshpande, S. S. (2012). *Dielectric properties of saline soils at c band microwave frequency* [Doctoral thesis, Dr Babasaheb Ambedkar Marathwada UniversityAurangabad]. Shodhganga: a reservoir of Indian theses at INFLIBNET.

Deshpande, S. S., Itolikar, A. B., Joshi, A. S., & Kurtadikar, M. L. (2015). Dielectric Properties of Gujarat and Uttar Pradesh Saline Soils At 5GHz. *Proceeding of 11th International Conference on Microwaves, Antenna, Propagation and Remote Sensing (ICMARS-2015)*, 87-91.

ISRO's Participation in Antarctic Expedition. (n.d.). https://www.isro.gov.in/isros-participation-antarctic-expedition

Itolikar, A. B. (2016). *Microwave Dielectric Properties of Vegetation at C-Band* [Doctoral thesis, Dr Babasaheb Ambedkar Marathwada UniversityAurangabad]. Shodhganga: a reservoir of Indian theses at INFLIBNET.

Itolikar, A. B., Joshi, A.S., Deshpande, S. S., Arole, V. M. & Kurtadikar, M. L. (2020). Dielectric and Emissive Properties of Sorghum (Jowar) Vegetation at C-Band Microwave Frequency. *Materials Today: Proceedings, 23P2,* 236-245.

Itolikar, A. B., & Kurtadikar, M. L. (2017a). Microwave dielectric properties and emissivity estimation of freshly cut Banana leaves at 5GHz. *International Journal of Advances in Remote Sensing and GIS*, *5*(1), 58–66.

Itolikar, A. B., & Kurtadikar, M. L. (2017b). Microwave measurements of dielectric properties of corn vegetation at C-Band and comparison with Debye Cole–Dual Dispersion model. *Journal of Microwaves, Optoelectronics and Electromagnetic Applications*, *16*(4), 954–965. doi:10.1590/2179-10742017v16i41087

Joshi, A. S. (2012). *Dielectric Properties of Pre-monsoon and Post monsoon Seawater at C-Band Microwave Frequency* [Unpublished doctoral thesis]. Dr Babasaheb Ambedkar Marathwada University, Aurangabad.

Joshi, A. S., & Kurtadikar, M. L.(2013). Study of Sea Water Permittivity Models and Laboratory and Validation at 5GHz. *Journal of Geomatics SAC-ISRO-The Indian society of Geomatics, 7*(1), 33-40.

Kulkarni, P. G. (2006). *Study of dielectric properties of different soil texture at C-Band microwave frequency* [Doctoral thesis, Dr Babasaheb Ambedkar Marathwada UniversityAurangabad]. Shodhganga: a reservoir of Indian theses at INFLIBNET.

Kurtadikar, M. L., Popalghat, S. K., Kulkarni, P. G., & Khare, N. (2009). Indian. *Journal of Geosciences (Prague)*, *63*(2), 241.

Kurtadikar, M. L., Popalghat, S. K., & Mehrotra, S. C. (2013). *Laboratory Validation of Dielectric Properties of Earth Resources* [Paper Presentation]. International Expert Meeting on Microwave Remote Sensing, Ahmadabad, India.

Matzler, C., & Wagmuller, U. (2000). Dielectric properties of fresh-water ice at microwave frequencies. *Journal of Applied Physics*, *20*, 1623.

Ministry of Earth Sciences Technical Publication. (2008) *Dielectric Properties of Antarctic Ices and Soils at 5 GHz Frequency* (Technical Publication No. 20).

Murugkar, A. G. (2002). *Dielectric Properties of Seawater* [Doctoral thesis, Dr Babasaheb Ambedkar Marathwada University Aurangabad]. Shodhganga: a reservoir of Indian theses at INFLIBNET.

Navarkhele, V., Agrawal, R., & Kurtadikar, M. L. (1998). Dielectric properties of electrolytic solutions. *Pramana, 51*(3-4), 511–518. doi:10.1007/BF02828943

Navarkhele, V. V. (1996). *Study of dielectric properties of liquids of geophysical interest* [Unpublished doctoral thesis]. Dr Babasaheb Ambedkar Marathwada University, Aurangabad.

Von Hippel, A. R. (1954). *Dielectrics and Waves.* Wiley.

Chapter 12

Ice Core Records for Paleo-Volcanism, Climate, and Snow Accumulation Rates Over the Past 150 Years

D. K. Rao
Physical Research Laboratory, Department of Space, India

R. A. Jani
Physical Research Laboratory, Department of Space, India

ABSTRACT

The annual fallout of radionuclides 32Si, 7Be, 210Pb, and 137Cs in the shelf snow samples collected near the Dakshin Gangotri, East Antarctica, has been estimated. The polar fallout of cosmic-ray produced 32Si estimated to be 2.34 X10-5 dpm cm-2 y-1. The fallout of 7Be and 210Pb is estimated to be 4.2 and 1.86 X 10-2 dpm cm-2 y-1. The depth profiles of electrical conductance, 210Pb, δ18O, and cosmogenic radioisotopes 10Be and 36Cl have been measured in a 60 m long ice core. Based on 210Pb and δ18O, the mean annual accumulation rates have been calculated. These rates are 0.20 and 0.23 m of ice equivalent per year during the past ~150 years. Based on electrical conductance measurements and using these accumulation rates, a volcanic event, 'Tambora' that occurred in 1815 AD, was identified. δ18O values suggested that the beginning of the 19th century was colder by about 2°C than the recent past and middle of the 18th century. The fallout of 36Cl reported here agrees well with the mean global production rate estimated earlier.

DOI: 10.4018/978-1-6684-4078-0.ch012

INTRODUCTION

Long ice cores from the Polar ice sheets provide information on continuous records of past volcanic eruptions and accumulation rates of ice, climatic and environmental changes, atmospheric and nuclear Fallout, and solar and terrestrial variability for more extended periods (Orheim et al., 1986; Delmas et al., 1992; Nijampurkar & Rao 1993a; Beer et al., 1994; Stuiver et al., 1995 and Hammer et al., 1997). Considerable interest is shown in studying high-resolution records during the last few centuries from different archives like glaciers, ice sheets, tree rings, and lake sediments (Orheim et al., 1986; Ramesh et al., 1989; Nijampurkar & Rao 1993a and Petterson et al., 1993). Such studies record interannual signals of variations in the worldwide surface air temperatures, sea surface temperatures, precipitation, which appear to correlate well with the 11-year solar cycle, heavy rainfall, and flood events (Seleshi et al., 1994 and Currie,1994).

Measurements of the D.C. (Electrical Conductivity[EC] of solid ice) is an essential tool for obtaining historical records of volcanic eruptions in terms of solid acids like H_2SO_4, HCI, and microparticles by estimation of actual accumulation rates of ice, giving accurate time index during past several thousands of years (Hammer et al., 1997). Stable isotopes of oxygen ($\delta^{18}O$) are excellent time markers and climatic indicators (Jouzel et al., 1987). Natural radioisotopes of different half-lives (^{210}Pb, ^{10}Be, and ^{36}CI) give information about their Fallout and depositional history during different time scales. ^{210}Pb ($t_{1/2}$=22.3a) provides accumulation rates of ice during the last century. In contrast, the cosmogenic radioisotope ^{10}Be ($t_{1/2}$ =1.6Ma) with a much longer half-life gives long term accumulation rates of ice ~ during millions of years and information on climatic changes, solar cycles, and sunspot activities during different time scales (Beer et al., 1991). Another cosmogenic isotope ^{36}CI ($t_{1/2}$=0.3Ma), also produced during nuclear weapons testing from 1940-1980, is a valuable tracer to study the transport process in the atmosphere and hydrology (Synal et al., 1990). These two cosmogenic isotopes are the most suitable proxies to reconstruct solar and terrestrial variability on scales longer than a few centuries. The ratio of $^{10}Be/^{36}CI$ is advantageous in dating older ice (Nishiizumi et al., 1983). With this introduction, our scientific study is to obtain short term high-resolution records of climatic changes, identifying the volcanic events and annual Fallout of radioisotopes during the past few centuries based on electrical conductance, ^{32}Si, ^{210}Pb, $\delta^{18}O$, ^{10}Bc, and ^{36}CI analysis in snow, ice and ice core samples from Antarctica near the Indian station "Dakshin Gangotri."

We have collected snow, ice, and ice core samples since the 1st Indian expedition to Antarctica. A few surface snow and ice samples were collected in 1982 for isotopic and related studies as a preliminary study. In continuation of these studies, Physical Research Laboratory [PRL] participated in the 5th scientific expedition in

1985 and collected many snow samples ranging from 25 to 300 Kg from the shelf ice at Dakshin Gangotri station. These were used to study the present-day Fallout of different natural (^{32}Si, ^{10}Be, and ^{210}Pb) and artificial (^{137}Cs) radioisotopes. These polar fallout values help estimate ice accumulation rates in remote regions where direct measurements pose several problems. With this background, a 60 m long ice core was raised by the Geological Survey of India [GSI] in 1992, and samples were given to PRL to study the snow accumulation rates and climatic changes and identify volcanic events in the last 150 years. Samples were also collected systematically from the ice wall near Maitri's southwest to study variations during the 21st expedition in 2002. Here we discuss the results obtained from the initial measurement of radioisotopes, electrical conductivity, and stable isotopes on a 60 m long ice core.

MATERIALS AND METHODS

Sample Collection

Snow samples ranging from 4-300 kg were collected from the shelf ice at Dakshin Gangotri Station near the Indian Bay in Antarctica during the 5th Indian Scientific Antarctic expedition. The top few inches of the surface snow were removed (to avoid any contamination), and the samples were collected up to 50 cms to represent the annual snow deposition. The snow samples were melted in clean plastic drums and acidified to pH-3 with 1:1 HNO_3. Stable carriers were added for different isotopes. After homogenization, the activities were scavenged by iron hydroxide in an ammoniacal medium following the procedures described earlier (Nijampurkar et al., 1982; Koide et al., 1982).The 60.46meter-long ice core was raised in the continental ice by the GSI, south of the Indian Research Station, Maitri, in Central Dronning Maud Land, East Antarctica, during the austral winter of the 12th Indian scientific expedition to Antarctica in 1992 (Chaturvedi et al., 1996). The drilling site was ~ 3 km east of the Nunatak Vateheia at location 70°47'S, 11°'44'E south of Schirmachar Oasis (figure 1). The entire core was cut into 302 sections of about 20 cm each at the sampling site and kept frozen in ultra-clean wide-mouth plastic bottles. The complete core was transported from Antarctica to Goa by ship in a frozen condition, preserved at −20°C in cold storage, and later hauled in a refrigerated vehicle to Ahmedabad and stored in a refrigerator until the analysis.

Figure 1. Location map of the site from where 60 m ice core was raised during Indian expedition to Antarctica in 1992

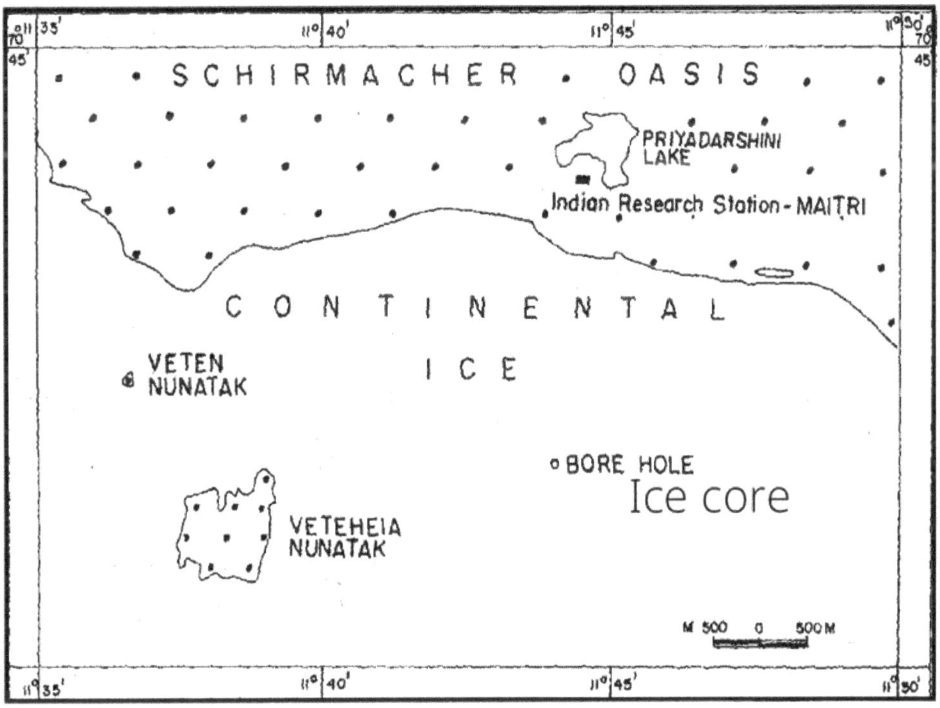

Chemical Procedures and Counting Methods

Various techniques and their applications used in the present study are given in Table 1, and a brief description is shown here. The chemical procedures for radiochemical separation and purification of radionuclides for the snow samples have been discussed elsewhere (Nijampurkar et al., 1982; Koide et al., 1982). The counting procedures required that the b activity of ^{32}P from ^{32}Si and ^{210}Bi from ^{210}Pb was followed for several weeks on a gas flow G.M. counter in anti-coincidence with Nal (Tl) crystal having a background of about 1.5 cpm and counting efficiency of 35%. The activity of ^{7}Be and ^{137}Cs by a non-destructive method was estimated on an HPGe detector system located in a 10 cm lead shield. Background in ^{7}Be was 0.14±0.006 cpm, whcrcas for ^{137}Cs was 0.11±0.01 cpm. The counting efficiency of the system is 3.8%. For the ice core samples, the following procedures were adopted.

Table 1. Various parameters, techniques used in the present study and their applications

S.No.	Parameter	Technique	Importance
1.	$\delta^{18}O$	Mass Spectrometry	Climatic changes Accumulation rates
2.	^{10}Be, ^{36}Cl	Accelerator Mass Spectrometry	Solar & Terrestrial Variability Atmospheric Transport Climatic variations Dating of ice
3.	^{210}Pb,^{32}Si Total β	Radiation Detectors (α, β)	Accumulation rates Polar Fallout
4.	Solid Conductance	Indigenous conductometer	Post volcanic records Accumulation rates
5.	Density	Conventional Technique	Study of densification

1. Electrical Conductivity (E.C.) measurements

About 20 cm cylindrical sections of the core from 3–30 m depth were cleaned by scraping off a few mm of the surface ice using an ultra-clean stainless-steel knife. The cleaned solid ice core section's electrical conductivity was measured at –18°C in a deep vertical freezer using the Electrical Conductivity Method. The core's acidity was monitored in terms of the current using an E.C. system developed at PRL on similar lines following those available (Hammer, 1980).

2. $\delta^{18}O$, ^{210}Pb, ^{10}Be, and ^{36}Cl measurements

After E.C. measurements, each 20 cm core section was subsampled in 2–3 cm intervals for $\delta^{18}O$ measures and preserved frozen until analysis. The remaining 20 cm section (~ 2L) was frozen before analysis for ^{210}Pb, ^{10}Be, and ^{36}Cl. The $\delta^{18}O$ measurements were made at the Department of Geophysics, Copenhagen, Denmark using an auto mass spectrometer as per standard procedures (Dansgaard 1964). ^{210}Pb activities in the meltwater samples (~1L) were measured as per the standard guidelines using α spectrometry (Sarin et al., 1992). In brief, these samples were acidified with 6N HCl to maintain the pH ~ 1–2 and were spiked with known activity of ^{209}Po. After equilibration of the spike, samples were evaporated to dryness, and the residue was dissolved in 50 ml of 0.6 N HCl. From this solution, polonium isotopes were plated on silver planchet, and their activities were assayed by alpha spectrometry using silicon surface barrier detectors. From the measured activity of ^{210}Po, the in situ ^{210}Pb activity in the samples was calculated. ^{10}Be and ^{36}Cl activity measurements were made at the AMS facility of ETH/AST in Zurich, Switzerland. Be and Cl spikes (supplied by Prof. J Beer) were added to bi-annual samples during the ice melting. The samples in the ultra-clean, leak-proof plastic bottles were sent to the AMS laboratory for ^{10}Be and ^{36}Cl analysis. The samples cover approximately the period of nuclear testing from 1980–1940 A.D. They were analyzed as per the standard procedures published earlier (Beer et al., 1994).

RESULTS AND DISCUSSION

Polar Fallouts of Different Radioisotopes

1. Polar Fallout of ^{32}Si

The concentration of ^{32}Si in the snow accumulated over the shelf ice (0-30 cm) at Dakshin Gangotri station, which represents annual precipitation, has been observed to be $(0.78 \pm 0.4) \times 10^{-3}$ dpm/L (Table-12.2). Based on this available single measurement at 70°S and the average precipitation of 40 cm [mean of estimated precipitation, 30 cm, (Nijampurkar et al., 1988) and the average precipitation of 50 cm in the coastal Antarctica regions], the Fallout of ^{32}Si has been estimated to be 3 X 10^{-5} dpm/cm^2. Yr (Table 2), which may be an average fallout of ^{32}Si in the latitude belt, of 30-90° in the absence of data at other latitudes in this belt, and the Fallout of ^{32}Si being independent of the latitude > 30° (Lal and Peters, 1967; Clausen, 1974). Approximately 40 and 60 per cent of ^{32}Si activities will be deposited in the 0-30° and 30-90° cells, respectively. Hence the Fallout of ^{32}Si should be lower in the 0-30°S cell by a factor of 1.5, i.e., 2 X 10^{-5} dpm/cm^2 Yr than that observed in 30-90°S cell. The global mean Fallout of ^{32}Si is estimated to be 2.5 X 10^{-5} dpm/cm^2.yr based on the measured value of 3 X 10^{-5} dpm/cm^2 yr for the 30-90°S cell and the calculated value of 2 X 10^{-5} dpm/cm^2.yr for 0-30° tropical cell (the surface areas are same for the two cells). The average Fallout in the 0-30°S and 30-90°S should be measured by collecting samples at every 10°latitude in the Southern hemisphere, which will give the correct picture of the mean global Fallout of ^{32}Si in the Southern hemisphere. The mean global fallout of 2.5 X 10^{-5} dpm/cm^2.yr of ^{32}Si corresponds to a mean global production rate of 0.8 X 10^{-4} atoms/cm^2. sec. (calculated using 140 yr as the half-life of ^{32}Si). This estimate is lower than the calculated value of 1.6 X 10-4 atoms/cm^2.sec (Lal and Peters, 1962) due to the uncertainty in the cross-sections used for calculating the ^{32}Si production rate or in the half-life of ^{32}Si. If it is due to the latter, to obtain agreement with the measured production rate, the half-life of ^{32}Si may be closer to 270 yr, similar to the values reported by Clausen (1973) and Demaster (1980). More direct measurements in this region might help to settle the discrepancy. Additionally, the higher values of ^{32}Si observed in the circumpolar waters of the Atlantic waters (Somayajulu et al., 1987) match well with the high values of ^{32}Si observed in the present work compared to that of 0.5 X 10^{-3} dpm/L in the same latitude belt in the northern hemisphere and the subsequent contribution of meltwaters of the Antarctic ice sheet to coastal Antarctic waters and their mixing with the mid-latitude ocean waters in the Southern hemisphere.

Table 2. Concentration and annual fallout of radioisotopes ^{7}Be, ^{210}Pb, ^{32}Si and ^{137}Cs at Dakshin Gangotri Station, Antarctica

Radio Isotope	Half-life	Mode of decay and Energy (MeV)	The volume of water processed (L)	Concentration (dpm/L)	Annual Fallout (dpm/cm^2.yr)
^{7}Be	53 days	$E_\gamma = 0.477$	25	(8.8±0.2)	3.0
^{210}Pb	22.3 yrs	$E_\beta = 0.061$	25	(0.62±0.10 0.12	0.12
^{32}Si	140 yrs	$E_\beta = 0.21$	300	(0.78±0.4) X 10^{-3}	3.0 X 10^{-5}
^{137}Cs	30 yrs	$E_\gamma = 0.66$	4.5	(0.2±0.2)	8.0

2. Fallout of ^{7}Be and ^{210}Pb

The observed concentration of 8.8 ± 0.2 dpm/L (Table-13.2) for ^{7}Be is consistent with the earlier estimates (Bhandari et al., 1984) and comparable to that observed at tropical latitudes in the northern hemisphere from 15° to 31°N (Lal et al., 1979). It confirms that the snow samples collected for analysis were about 6-8 months old and would have been deposited in the shelf ice from June to September 1985. The ^{210}Pb activity in the fresh snow sample is observed to be 0.62±0.1 dpm/L (Table-13.2), consistent with the earlier findings (Nijampurkar et al., 1984) estimates a fallout value of 0.15 dpm/cm^2-yr. The shallow ice cores from Antarctica near the South Pole, far away from the coastal areas, have been dated by ^{210}Pb using the surface ice value of 1.8 dpm/L (Picciotto et al., 1968). The fallout values of ^{210}Pb in the northern hemisphere in Greenland, Alps, and the Himalayas lie in the range of 3 to 9 dpm/L (Crozaz & Langway, 1966; Nijampurkar et al.,1982 and Gaggler, 1983). The ^{210}Pb values observed at Dakshin Gangotri at shelf ice near the coast are slightly lower than those observed near the pole. Still, they are consistent with the statistical errors, which partial contributions can explain from the coastal seawater moisture.

^{210}Pb activity condensing over the shelf ice in summer does not occur, thus lowering the ^{210}Pb concentration in the snow deposited over the immediate coastal areas. However, assuming the total Fallout of ^{210}Pb remains constant for a few hundred years, the deep-seated ice can be safely dated using the ^{210}Pb value from Antarctica's average surface ice values.

3. Fallout of ^{137}Cs

The artificial radioactivity of ^{137}Cs in surface snow meltwaters at Dakshin Gangotri is found to be 0.2 X 10^{-5} dpm/L (Table-12.2), the very close background order of magnitudes lower than the peak fallout observed during the last four decades. The peak activity of ^{137}Cs (15 dpm/L) produced as a result of testing nuclear devices in the early '60s at the Dye-3 station of Greenland is reported by Koide *et al.* (1982). This work suggests that there has been no testing of nuclear devices in the atmosphere during the late '80s and that the activity produced in the late '70s by French tests in the Southern Hemisphere has been thoroughly washed out during the last decade. Even though no activity of ^{137}Cs has been observed in surface snow samples, the systematic nuclear records preserved in the Antarctic ice sheet can be investigated from a shallow ice core of about 20 meters.

Accumulation Rates of Ice from Ice Core Studies

1. 1.Electrical Conductivity (E.C.) measurements

The E.C. profile shown in figure 2 indicates that the instrument's high background (2 ± 2 μamp) is probably due to non-ideal laboratory conditions (like non-availability of cold room facility generally maintained at –18° to –20° C for such measurements) during E.C. measurements of all the samples. The sharp peaks of magnitude >10 μamp (saturated values of currents due to limitations of the instrument) at few depths (time) could be related to historical volcanic eruptions during the past 150 years. They can be subsequently used for dating older ice. A sharp peak of >10 μamp occurring at a depth of about 30 m could be related to the significant eruption '''TAMBORA' in Indonesia (10°S) in 1815 AD. The other sharp peak was observed at 5 m depth around 1963–64, attributed to the volcanic eruption at 1°S (AUGUNG) in 1963. Four unnamed peaks of similar magnitudes were observed at depths between 3 and 4 m, probably due to significant eruptions that might have occurred recently (i.e., after 1963) in the southern hemisphere. They, however, could not be related to any specific known volcanic events. The volcanic eruption of '''TAMBORA' that erupted in 1815AD has been used to calculate an accumulation rate of ~ 17 ± 0.08 cm/yr, which agrees reasonably well with those obtained ^{210}Pb and $\delta^{18}O$ (see discussions). Constraints in the logistic facilities in the ice core storage and transportation, the data are interpreted cautiously.

Figure 2. Electrical conductivity measurements of the solid ice core samples from 0-30 m depth at an interval of 20 cm. The saturated peak values of currents (μ amp) at about 5 & 30m depths identify the past volcanic episodes Augung (1963 A.D.) and Tambora (1815 AD), respectively

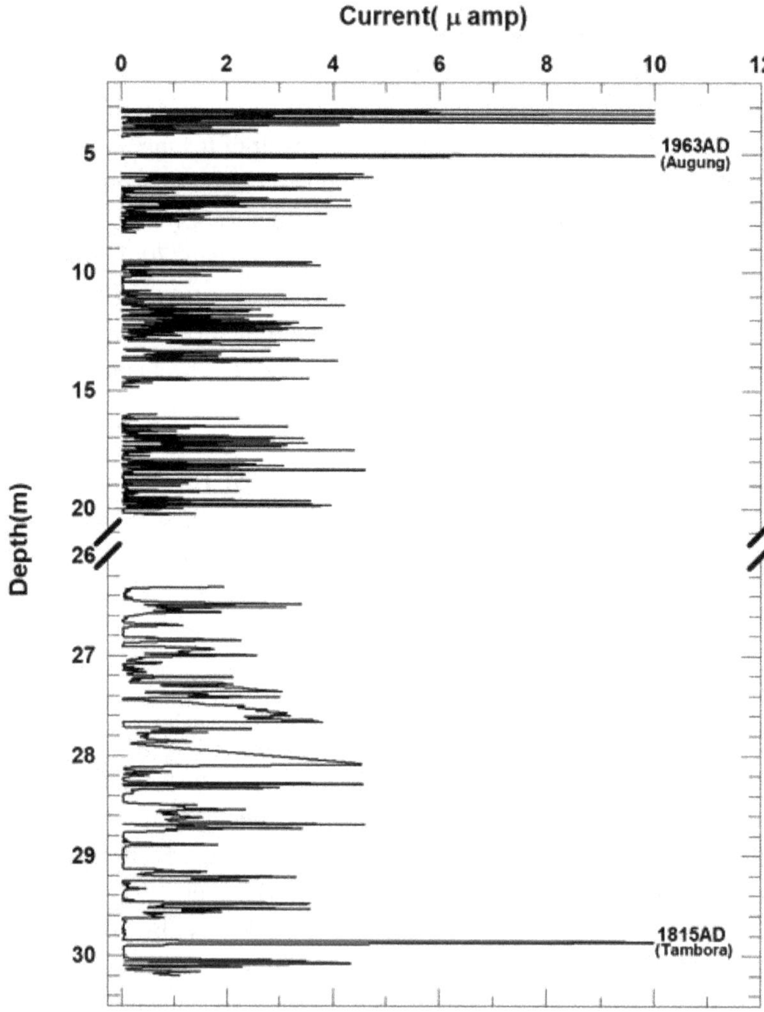

2. ^{210}Pb activities

The ^{210}Pb activities at a depth of 20 m have been measured (by combining eight sections of the core, each of 160 cm to obtain easily measurable signals of ^{210}Pb activities)—the results of 7 samples being plotted in figure 3. The data's best-fit line gives an average accumulation rate of 20 ± 2 cm/yr during the past 100 years. It appears from figure-13.3 that the accumulation rate of ice has not remained constant during the last century and varies from 12.5 to 26.6 cm/yr with a mean value of 20 ± 2 cm/yr. ^{210}Pb activity is four times decreased (about two half-lives of ^{210}Pb) at a depth of ~10 m compared to 3 m deep (Figure 3). It Indicated that the depth interval from 3–10 m covers ~ 50 years (1990–1940), a period of nuclear tests, and an age of 150 years at a depth of 30 m. The Fallout of ^{210}Pb calculated from the present work (2.6×10^{-2} dpm/cm^2 yr) agrees well with the mean Fallout of 3×10^{-2} dpm/cm^2 yr for a different region (Nijampurkar et al., 1993b). However, the mean value of ^{210}Pb Fallout estimated in the 60°–90° belt in the northern hemisphere is higher by ~ 2 (5×10^{-2} dpm/cm^2 yr (Crozaz & Langway, 1966).

Table 3. Mean annual δ ^{18}O(‰), average MASAT (°C) values at different periods derived from 60meters ice core raised at Dakshin Gangotri from East Antarctica

Sr.No	Depth (m)	Mean annual δ ^{18}O (‰)	Mean MASAT Temp (°C)	Age (yr)	Time (AD)
1	3.1-5.8	-24	-15	22	1970
2	9.8-10.0	-24.6	-15.8	49.5	1942
3	13.1-13.2	-24.5	-15.7	65.8	1926
4	17.8-18.0	-25.8	-17.6	89.6	1902
5	27.7-27.9	-24.6	-15.8	139	1853

Figure 3. Depth profile of ^{210}Pb specific activities in the ice core up to 18m depth at different depth intervals. The past average accumulation rate of ice (that has not remained constant during the last century) has been calculated to be 20 cm/yr using the best-fit line of these data

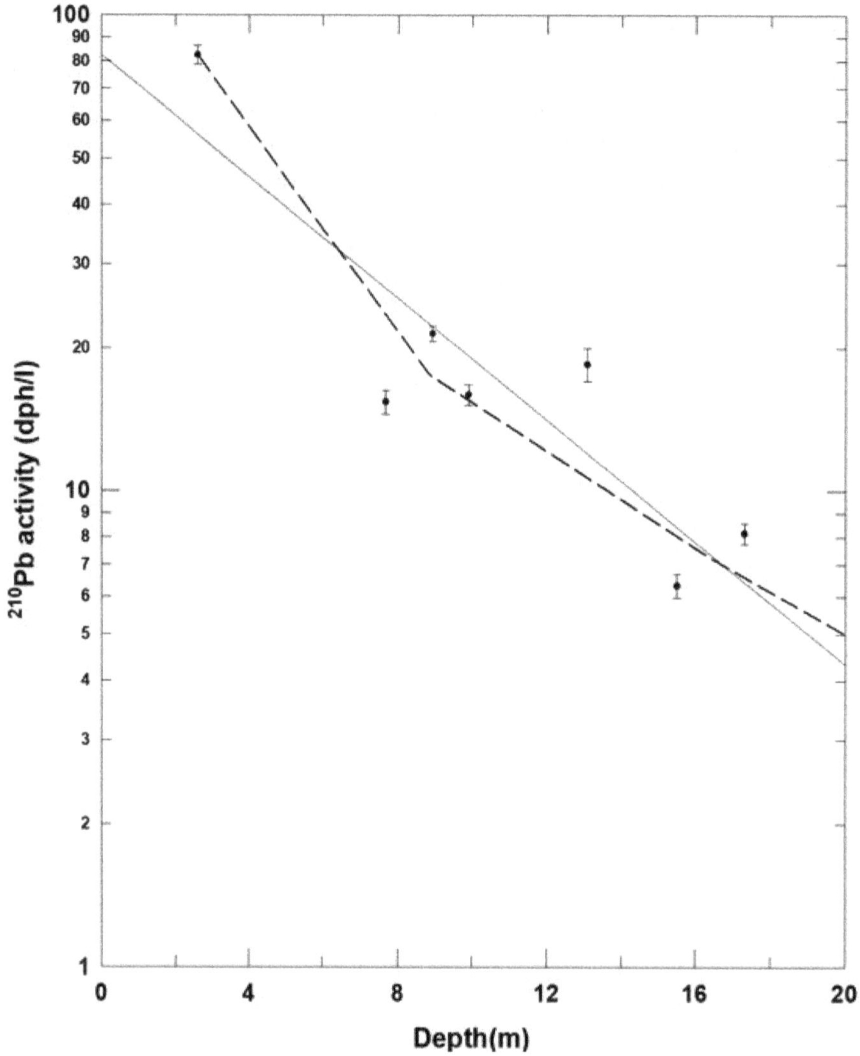

3. δ¹⁸O measurements

The figure-12.4 shows continuous and discontinuous records of depth profiles of ^{18}O concentrations from 3–30m. The δ ^{18}O values range from –21 to –27‰. $\delta^{18}O$ records in the polar ice cores have often been used to determine ice accumulation rates and the mean annual surface air temperatures (MASAT) at the precipitation site (Dansgaard 1964). $\delta^{18}O$ measurements carried out in the present work are discussed in light of these parameters. The depth profile of the continuous record of $\delta^{18}O$ measurements for 3–6 m depth of the core is given in figure-12.5. The constant high-resolution $\delta^{18}O$ record shows evident cyclic variations and annual cycles with summer and winter peaks. These data have been used for the calculation of (i) accumulation rates of ice and (ii) MASAT at the site of precipitation using an empirical linear relationship established between mean annual $\delta^{18}O$ and MASAT (Dansgaard, 1964). In the absence of any other reasonable time marker (e.g., ^{137}Cs), ^{210}Pb has been used to calculate and compare the ages of ice samples and used as a time index to study the depositional history of the ice core and climatic variations at the site of precipitation. The $\delta^{18}O$ values for the depth 3–6 m (figure 5), equivalent to 1977–1962 A.D., vary from −21.4 to −26.5‰. During this period, the accumulation rate of ice is not uniform and varies from 8 to 46 cm/yr, and the mean has been calculated to be 21 ± 0.10 cm/yr. The accumulation rate of ice during the first two years (corresponding to 3–4 m depth) is higher by a factor of two than the calculated mean value.

Similarly, the discontinuous cyclic record of $\delta^{18}O$ concentrations shows that ice's accumulation rate varies from 16 to 26 cm/yr during the period equivalent to 6-30 m (1960–1840AD). The average MASAT value for the period comparable to 3–6 m has been calculated to be −15°C (table 3). Figure 6 shows the average MASAT values versus a depth of 3–30 m. The individual MASAT values for the different periods have been calculated, varying from −15 to −17.6oC and showing the lowest MASAT of −17.6 ± 0.2°C at ~ 1900AD. Many similar measurements are needed to confirm the trend and minima at ~1900 A.D. Table 2 shows the mean annual $\delta^{18}O$, average MASAT, age (yr), and time (A.D.) for the different periods corresponding to varying depths from 3–30 m. It is clear from table-12.2 and figure 5 that the MASAT for the period equivalent to 17.8 to 18.0 m depth (i.e., beginning of the 19th century) was lower by ~ 2°C than other times. The data indicate that the average accumulation rate at this site during the last 150 years is about 20 cm/yr and that a few decades at the beginning of the century (1920–1890 AD) were cooler by ~ 2°C than the other periods during the recent past (1977–1920 A.D.) and middle of 18th century (1890–1840 AD). The accumulation, however, is not uniform during this period and shows significant interannual and decadal variations. $\delta^{18}O$ values from DMI (74°S 12°W), East Antarctica closer to our station vary from −14 to −26‰ during the last 50 years (Aldahan et al., 1998). The accumulation rate is double that of the accumulation rate than estimated from the present work.

Figure 4. Continuous and discontinuous records of depth profiles of δ ^{18}O concentrations from 3–30m. The δ ^{18}O values range from –21 to –27‰. The highly depleted δ ^{18}O values occur at about 18 m depth

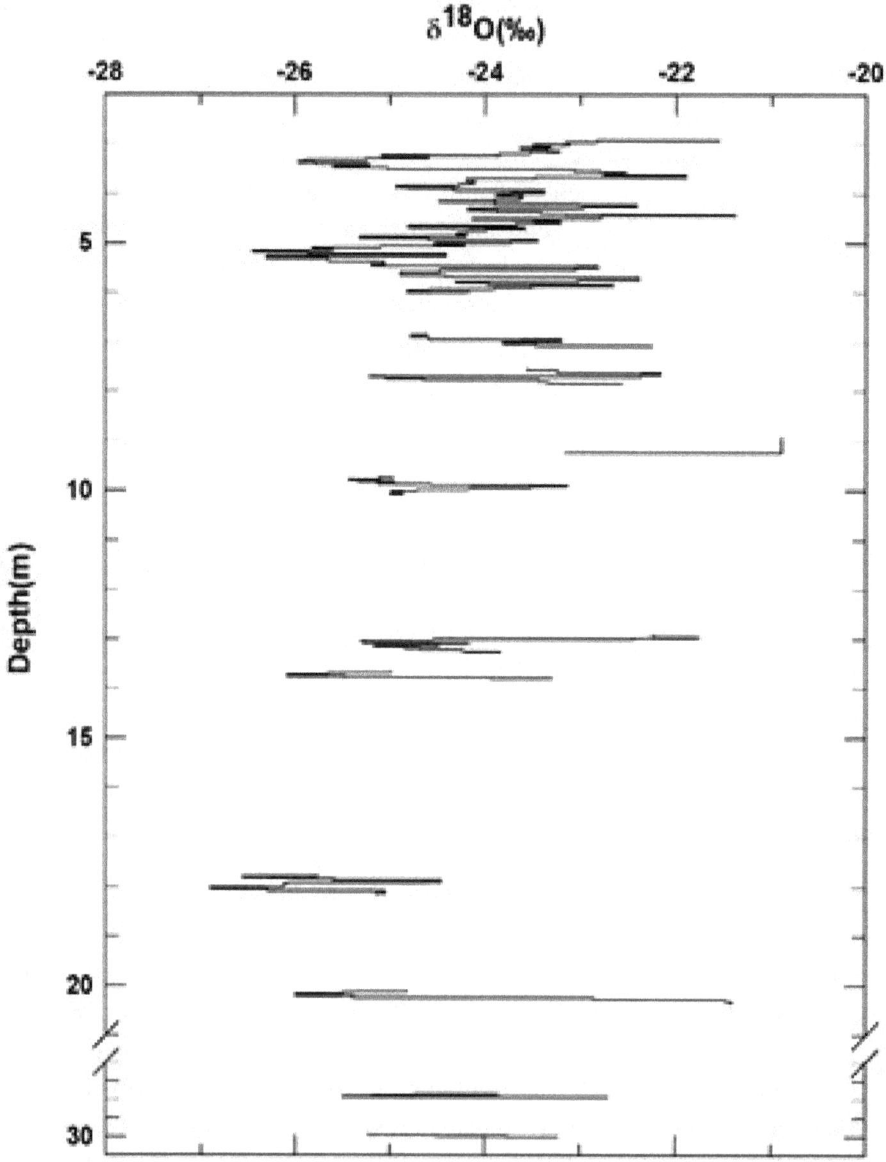

Figure 5. Variation for δ ^{18}O concentrations between 3-6m depths is drawn on an expanded scale. The ice accumulation rate has been estimated, which varies from 8 to 46 cm during that period with an average value of 20 cm/yr

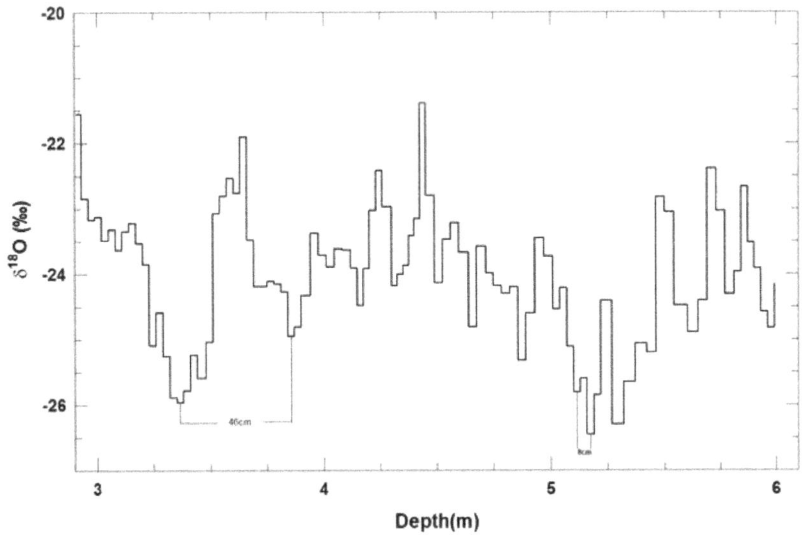

Figure 6. Average MASAT (°C) values calculated during different periods are plotted against depth (m) and Time (A.D.). At the beginning of the century, the lowest value indicates temperatures cooler by ~ 2 cC than in recent times

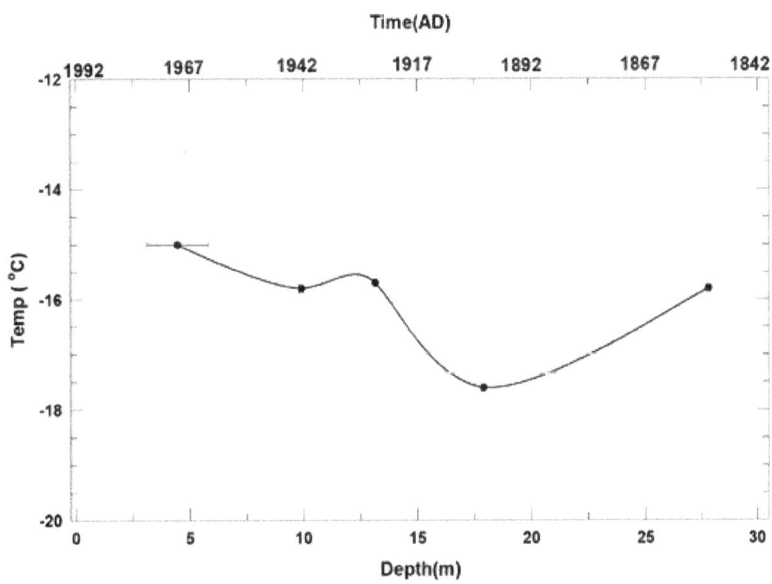

4. ^{10}Be and ^{36}Cl studies

The importance of polar Fallout of cosmogenic isotopes ^{10}Be and ^{36}Cl (also produced artificially during nuclear detonations) has been discussed briefly in the introduction. ^{10}Be and ^{36}Cl measurements have been made for the ice samples, representing 1940–1975AD. The ^{10}Be values (Table-13.3) range from 0.56 to 2.58 $\times 10^4$ atoms/g with a mean value of 1.16×10^4 atoms/g, which agrees well with those calculated at the Renland ice core from Greenland and DML ice core from Antarctica (Aldahan et al., 1998). Among the ^{10}Be concentrations, only one sample had ^{10}Be more than 1.5×10^4 atoms/g, at 500 cm depth with a value of 2.58×10^4 atoms/g. The deposition of ^{10}Be (2×10^5 atoms/cm^2.yr) determined in the present work agrees reasonably well with those derived for other regions at higher latitudes in Antarctica (73°S to 90°S). However, the 10Be Fallout at locations in the northern hemisphere is higher by factors of 2–3. The ^{10}Be deposition values have a smaller variance and lower annual average in Antarctica than in Greenland (Raisbeck & Yiou, 1985; Steig et al., 1996 and Aldahan et al., 1998). The ^{10}Be deposition flux calculated for Antarctic ice core by Aldahan et al. (1998) and the estimates of ^{10}Be made by Lal et al. (1987) are higher by a factor of 2 than that measured in the present work and that on Greenland ice core by a factor of 1.5 (Baumgartner et al., 1997) probably due to the difference in the annual precipitation. ^{36}Cl measurements have been made for the period (1980–40), which cover nuclear weapon tests. The ^{36}Cl values vary from 0.1 to 0.25×10^4 atoms/g (Table-13.3) with a mean value of 0.2×10^4 atoms/g. The mean deposition of ^{36}Cl has been estimated to be 3.91×10^4 atoms/cm^2 yr, which agrees reasonably well with the global production rate of 4.09×10^4 atoms/cm^2 yr (i.e., 1.3×10^{-3} atoms/cm^2 sec) by Lal and Peters (1967). The Fallout of ^{36}Cl (based on a single measurement available before this work) in Antarctica is higher by a factor of ~10 (Finkel et al., 1980) and that in Greenland for the cosmic ray component is higher by a factor of 2–6 compared to that in the present work. Figure-12.7 shows variations in bi-annual concentrations of ^{10}Be, ^{36}Cl, their ratios (^{10}Be/ ^{36}CI) and δ^{18}O for the period 1980–1940, which is a period of atmospheric nuclear tests. The ^{36}Cl concentrations unexpectedly do not show any variation throughout bomb tests, and the bomb pulse appears to be missing or diluted to cosmic ray background ^{36}Cl level at this location. The cause for the "missing" ^{36}Cl is unclear. It probably could be due to its dilution during transport from the northern hemisphere to Antarctica. Earlier work in Greenland shows a bomb pulse of ^{36}Cl concentration during 1980–1940, with concentrations by magnitude higher than its cosmogenic contribution (Beer, personnel communication, 1998; Synal et al., 1990; Elmore et al., 1982 and Conard et al., 1989). ^{10}Be/ ^{36}CI ratios measured in the present work for the last ~ 40 years though show a wide range 3.5–13, the bulk of the samples have values 5 ± 1 (Table-13.4) with one sample > 8. The mean ^{10}Be/

^{36}CI in all samples is 7, which agrees well with that calculated to be 6.7 (Nishiizumi et al., 1983) in surface ice from Yamato hills in Antarctica. We suggest that the mean ^{10}Be/ ^{36}CI ratio of 7 measured in the present work for the last ~ 40 years can be used as the initial ratio to date older ice from central Dronning Maud Land, East Antarctica (as indicated by Nishiizumi et al., 1983 to date older ice from the Allan hills). The biannual data on ^{10}Be and ^{36}CI for a short period in the present work are insufficient to determine their correlations with solar activity, sunspot numbers and cosmic ray intensities, etc., to draw meaningful conclusions. The limited data set does not explain the strange observation of the missing ^{36}Cl pulse at this location. More systematic measurements need to be carried out simultaneously or at neighbouring locations to justify the present observation.

However, the bomb pulse of ^{36}Cl appears to be missing during the nuclear testing period from 1940-1980. These data are also plotted for δ ^{18}O, which are good climatic indicators.

CONCLUSION

The annual Fallout of cosmic-ray produced ^{32}Si, which is the first measurement in Antarctica, has been estimated to be 2.34 X 10^{-5} dpm cm-^{2} y-^{1}, which can be used for dating the older ice from Antarctica in the time scales up to 1000 years. The mean accumulation rate of ice based on δ^{18}O (20 cm/yr), ^{210}Pb (19.8 cm/yr), Total β activities (21.5 cm/yr), and conductance (17cm/yr) has been estimated to be ~20 cm/yr in the polar ice near the Indian station, East Antarctica for the past few centuries. As reflected in the mean annual surface air temperatures (MASAT) data observed during the last 150 years, the beginning of the 19th century was more relaxed by about 2°C than the recent past and middle of the 18th century. The mean ^{10}Be/ ^{36}CI ratio of ~7 for the last 40 years may be used as the initial value to date older ice from East Antarctica.

ACKNOWLEDGMENT

We are grateful to the Secretary, Department of Ocean Development (now MOES), New Delhi, for providing financial support and logistics for raising the long ice core from East Antarctica. We also thank Late M.K.Kaul of GSI, Antarctic division, India, to collect the ice core samples. We kindly acknowledge Prof. J. Beer, P. Kubik, and H. A. Synal for their dedicated help in carrying out the ^{10}Be and ^{36}Cl measurements using the AMS facility at ETH/AST in Zurich, Switzerland, and for valuable discussions during the work. We also thank Prof. H. B. Clausen, Department of Geophysics, Niels Bohr Institute of Astronomy, Physics and Geophysics, DK 2200 Copenhagen O, Denmark, for help in $\delta^{18}O$ analysis of the ice core samples. We also thank the Director of PRL, Ahmedabad, for encouraging and financial support for taking up this project.

REFERENCES

Aldahan, A., Possnert, G., Johnson, S. J., & Clausen, H. B., Lsaksson, E., Karlen, W., & Hansson, M. (1998). Sixty-year ^{10}Be record from Greenland and Antarctica; Proceedings. *Indian Academy of Science*, *107*, 139–147.

Baumgartner, S., Beer, J., Suter, M., Hannen, B. D., Synal, H. A., Kubik, P. W., Hammer, C. U., & Johnsen, S. (1997). Chlorine 36 fallout in the summit Greenland ice core. *Journal of Geophysical Research*, *102*(C12), 26659–26662. doi:10.1029/97JC00166

Beer, J., Joos, Ch. F., Lukasczyk, Ch., Mende, W., Siegenthaler, U., & Stellmacher, R. (1994). 10Be as an indicator of solar variability and climate; In: The solar engine and its influence on the terrestrial atmosphere and climate. E NATO ASI Series, 25, 221–233.

Beer, J., Raisbeck, G. M., & Yiou, F. (1991). Time variations of ^{10}Be and solar activity. In C. P. Sonnett, M. S. Giampapa, & M. S. Matthews (Eds.), *The sun in time* (pp. 343–359). University of Arizona Press.

Bhandari, N., Bhattacharya, S. K., Nijampurkar, V. N., Sengupta, D., Shah, V., Singhvi, A.K., Shukla, P. N., Suthar, K. M. & Vohra, C. P. (1984). Isotopic and related studies of Antarctic ice samples, Proceedings. *Indian Academy of Science (Earth and Planetary Science)*, *93*(2), 135.

Bhandari, N., & Rama. (1963). Study of atmospheric washout processes by means of radon products. *Journal of Geophysical Research*, *68*(13), 3823–3826. doi:10.1029/JZ068i013p03823

Chaturvedi, A., & Asthana, R. (1996). *An attempt at shallow ice-core drilling in polar continental ice, central Dronning Maud Land.* 12th Indian Expedition to Antarctica DOD Tech. *Publ.* 10.

Clausen, H. B. (1973). Dating of polar ice by ^{32}Si. *Journal of Glaciology*, *12*(66), 411.

Conard, N. J., Kubik, P. W., Gove, H. E., & David, E. (1989). A ^{36}CI profile in Greenland ice from 1265 to 1865. *Radiocarbon*, *31*(03), 585–591. doi:10.1017/S0033822200012170

Crozaz, G., & Langway, C. C. Jr. (1966). Dating Greenland firn-ice cores with Pb-210. *Earth and Planetary Science Letters*, *1*(4), 194–196. doi:10.1016/0012-821X(66)90067-7

Crozaz, G., Picciotto, E., & De, W. (1964). Antarctic snow chronology with ^{210}Pb. *Journal of Geophysical Research*, *69*(12), 2597–2604. doi:10.1029/JZ069i012p02597

Currie, R. (1994). Variance contribution to Iuni-solar and solar cycle signals in St. Lawrence and Nile river records. *International Journal of Climatology*, *14*(8), 843–852. doi:10.1002/joc.3370140803

Dansgaard, W. (1964). Stable isotopes in precipitation. *Tellus*, *16*(4), 436–468. doi:10.3402/tellusa.v16i4.8993

Delmas, R. J., Krichner, S., Palis, J. M., & Petit, J. R. (1992). 1000 years of explosive volcanism records at the south pole. *Tellus*, *44B*(4), 335–350. doi:10.3402/tellusb.v44i4.15461

Demaster, D. J. (1980). The half-life of ^{32}Si was determined from a varved Gulf of Mexico sediment core. *Earth and Planetary Science Letters*, *48*(1), 209–217. doi:10.1016/0012-821X(80)90182-X

Elmore, D., Tubbs, L. E., Newman, D., Ma, X. Z., Finkel, R., Nishiizumi, K., Beer, J., Oeschger, H., & Andree, M. (1982). ^{36}CI bomb pulse measured in a shallow ice core from Dye 3, Greenland. *Nature*, *300*(5894), 735–737. doi:10.1038/300735a0

Finkel, R. C., Nishiizumi, K., Elmore, D., Ferraro, R. D., & Gove, H. E. (1980). ^{36}CI in polar ice, rainwater, and seawater. *Geophysical Research Letters*, *7*(11), 983–986. doi:10.1029/GL007i011p00983

Gaggeler, H., Von Gunten, H. R., Rossler, E., Oeschger, H., & Schotterer, U. (1983). ^{210}Pb dating of cold alpine firn/ice cores from Colle Gnifetti, Switzerland. *Journal of Glaciology*, *101*(101), 165–177. doi:10.1017/S0022143000005220

Hammer, C. U. (1980). The acidity of polar ice cores in relation to absolute dating, past volcanism and radio-echos. *Journal of Glaciology*, *93*(93), 359–372. doi:10.1017/S0022143000015227

Hammer, C. U., Clausen, H. B., & Langway, C. C. Jr. (1997). 50,000 years of recorded global volcanism. *Climatic Change*, *35*(1), 1–15. doi:10.1023/A:1005344225434

Jouzel, J., Lorius, C., Petit, J. R., Genthon, C., Barkov, N. I., Kotlyakov, V. M., & Petrov, V. M. (1987). Vostok ice core: A continuous isotope temperature record over the last climatic cycle (160,000 yrs). *Nature*, *329*(6138), 1–6. doi:10.1038/329403a0

Koide, M., Michel, R., Goldberg, E. D., Herron, M. M., & Langway, C. C. Jr. (1982). Characterization of radioactive Fallout from pre-and post-moratorium tests to polar ice caps. *Nature, 296*(5857), 544–547. doi:10.1038/296544a0

Kutschera, W., Paul, M., Smithers, R.K., Stephenson, E. J., Yntema, J. L., Alburger, D. E., Cumming, J. B., & Harbottle, G. (1980). Measurement of the ^{32}Si half-life via accelerator mass spectrometry. *Physics Review Letters, 45*(8), 592.

Lal, D. (1987). ^{10}Be in polar ice: Data reflect changes in cosmic ray flux or polar meteorology. *Geophysical Research Letters, 14*(8), 785–788. doi:10.1029/GL014i008p00785

Lal, D., Nijampurkar, V. N., Rajagopalan, G., & Somayajulu, B. L. K. (1979). Annual Fallout of 32Si, 2l0Pb, 22Na, 35S, and 7Be in rains in India. Proceedings Indian Academy of Science, 88(1), 29.

Lal, D., & Peters, B. (1967). Cosmic ray produced radioactivity on the Earth. In *Handbuch der Physik, 46* (pp. 551–612). Springer-Verlag. doi:10.1007/978-3-642-46079-1_7

Nijampurkar, V. N., Bhandari, N., Bhattacharya, S. K., Rao, D. K., Sengupta, D., Raina, V. K., & Kaul, M. K. (1985). *Isotopic and T.L. studies of Antarctic ice Samples.* Scientific Report of 2nd Indian Expedition to Antarctica, Technical *Publication* 2.

Nijampurkar, V. N., Bhandari, N., Vohra C P & Krishnan, V. (1982). Radiometric chronology of Nehnar glacier, Kashmir. *Journal of Glaciology, 28*(98), 91.

Nijampurkar, V. N., & Rao, D. K. (1993a). Ice dynamics and climatic studies on Himalayan glaciers based on stable and radioactive isotopes. *Snow and Glacier Hydrology, 218*, 355–369.

Nijampurkar, V. N., & Rao, D. K. (1993b). Polar Fallout of radionuclides ^{32}Si, ^{7}Be and ^{210}Pb and past accumulation rate of ice at Indian station Dakshin Gangotri, east Antarctica. *Journal of Environmental Radioactivity, 21*(2), 107–117. doi:10.1016/0265-931X(93)90048-C

Nishiizumi, K., Arnold, J. R., Elmore, D., Ma, K., Newman, D., & Gove, H. E. (1983). ^{36}CI and ^{53}Mn in Antarctic meteorites and ^{10}Be- ^{36}CI dating of Antarctic ice. *Earth and Planetary Science Letters, 62*(3), 407–417. doi:10.1016/0012-821X(83)90011-0

Orheim, O., Gjessing, Y., Lunde, T., Repp, K., Wold, B., Clausen, H. B., & Liestol, O. (1986). Oxygen isotopes and accumulation rates at Riiser-Larsenisen, Antarctica. *Norsk. Polarinstitutt. Skr., 187*, 33–37.

Petterson, G., Renberg, I., Geladi, P., Lindberg, A., & Lindgren, F. (1993). Spatial uniformity of sediment accumulation in varved lake sediments in Northern Sweden. *Journal of Paleolimnology*, *9*(3), 195–208. doi:10.1007/BF00677213

Picciotto, E., Cameron, R., Crozaz, G., Deutsch, S., & Wilgain, S. (1968). Determination of the rate of snow accumulation at the pole of relative inaccessibility, Eastern Antarctica: A comparison of glaciological and isotopic methods. *Journal of Glaciology, 7*(50), 273.

Picciotto, E., Crozaz, G., & De Breuk, W. (1964). Rate of accumulation of snow at the south pole as determined by radioactive measurements. *Nature*, *203*(4943), 393–394. doi:10.1038/203393a0

Raisbeck, G. M., & Yiou, F. (1985). ^{10}Be in polar ice and atmospheres. *Annals of Glaciology*, *7*, 138–140. doi:10.3189/S0260305500006054

Ramesh, R., Bhattacharya, S. K., & Pant, G. B. (1989). Climatic significance of δD variations in tropical tree species from India. *Nature*, *337*(6203), 149–150. doi:10.1038/337149a0

Sarin, M. M., Bhushan, R., Rengarajan, R., & Yadav, D. N. (1992). Simultaneous determination of ^{238}U series nuclides in the water of Arabian Sea and Bay of Bengal. *Indian Journal of Geo-Marine Sciences*, *21*, 121–127.

Seleshi, Y., Demaree, G. R., & Delleur, J. W. (1994). Sunspot numbers as a possible indicator of annual rainfall at Addis Ababa, Ethiopia. *International Journal of Climatology*, *14*(8), 911–923. doi:10.1002/joc.3370140807

Steig, E. J., Polissar, P. J., Stuiver, M., Grootes, P. M., & Finkel, R. C. (1996). Large amplitude solar modulation cycles of ^{10}Be in Antarctica: Implication for atmospheric mixing processes and interpretation of the ice core record. *Geophysical Research Letters*, *23*(5), 523–526. doi:10.1029/96GL00255

Stuiver, M., Grootes, P. M., & Braziunas, T. F. (1995). The GISP2 δ^{18}O climate record of the past 16,500 years and the role of sun, ocean, and volcanoes. *Quaternary Research*, *44*(3), 341–354. doi:10.1006/qres.1995.1079

Synal, H. A., Beer, J., Bonani, G., Sute, M., & Wolfli, W. (1990). Atmospheric transport of bomb produced ^{36}Cl. *Nuclear Instruments and Methods*, *B52*(3-4), 483–488. doi:10.1016/0168-583X(90)90462-4

Chapter 13
India's GeoScience Pursuit in the Antarctica:
Initial Attempts

Neloy Khare
Ministry of Earth Sciences, India

ABSTRACT

India made her maiden entry in the Antarctic realm way back in 1981 by launching her first scientific expedition to the icy continent. Previous chapters dealt with many significant geoscientific studies of the past Indian scientific expeditions to Antarctica. However, few geoscientists participated for a short duration but carried out essential Antarctic studies. This chapter highlights and briefly collates such short geoscientific investigations that are otherwise significant and add to the knowledge base about the icy continent.

INTRODUCTION

The Schirmacher Oasis, Baalsrudfjellet, Veteheia, and Veten nunataks are the prominent northernmost exposures in the cDML area and control the ice flow in this region, resulting in glacial erosional and depositional landforms (Shrivastava *et al.*, 2019). The Schirmacher Oasis (or Schirmacher Lake Plateau) is 25 km long and 3 km wide. With 34 km^2, the Schirmacher Oasis ranks among the smallest Antarctic oases and is a typical polar desert. Based on the ages and geologic setting of klippe in the Lurio Belt Foreland, the Schirmacher Oasis appeared to be a part of a klippe of the Lurio Belt southeast Mozambique. It can be considered an extension of the East African Orogen into Antarctica (Ravikant et al., 2004). Several geoscientific

DOI: 10.4018/978-1-6684-4078-0.ch013

investigations by Indian researchers have addressed various scientific issues of this region. Here we provide the main findings of other significant geoscientific studies in and around the Antarctic area.

GEOLOGICAL AND GEOCHEMICAL INVESTIGATIONS IN SCHIRMACHER OASIS

The Atomic Mineral Directorate for Exploration and Research (AMD) participated in the IX scientific expedition to Antarctica to study the distribution of radio elements (U, Th, K, and their ratios) in different lithologies evaluate the uranium potential of the terrain. Geoscientists from the Atomic Minerals Division, Department of Atomic Energy, sampled 140 stations spread over 50 sq. km in Schirmacher Oasis, using a four-channel gamma-ray spectrometer scintillometer and Geiger counter to carry out Geo-radiometric studies. They studied one hundred fifty rock samples, samples from freshwaters, polar ice caps, six pits around the 'Maitri' station to obtain traces of radon-radium content in the soil with the SSNTD method.

The AMD's primary aim was to search, identify, explore, and evaluate the atomic mineral potentials, mainly U, Th, REEs, Nb, Ta, and Be, required for the Atomic Power Programme of the country. With this objective, the geology, radiometry, and geochemistry of the Schirmacher oasis become important and addressed. The Schirmacher range forms a part of the East Antarctica Charnockite Province and represents the most extensive granulite facies terrain on the Earth. The area falls in the iron-oxide sub-province of the Iron-metallogenic province of East Antarctica. It is significantly crucial for the metallic mineral deposits of the Precambrian. As the granulitic terrains usually do not contain primary uranium mineralisation, these rock's geochemical studies were carried out. The frequency plots and relative distribution of radioelement data were found comparable to the granulites of South-East India.

In brief, the results of this study revealed that the Schirmacher rock Oasis area exposes granulites and high-grade amphibolite facies rocks. These quartzo-feldspathic rocks have characteristically low uranium, and the low order radioactivity in them is solely due to potassium (K) and rarely thorium (Th).

Later processes of migmatisation, pegmatite intrusions, and shear fractures' development did not result in anomalous uranium enrichment. However, local thorium enrichment occurs along with certain fractures, sub-parallel to major NE-SW and WNW-ESE trending shears.

Litho-geological mapping using gamma-ray spectrometric data was carried out over an area of 35 sq. km. The Regional geological map on a 1:25,000 scale shows the distribution of various gneisses/granulites prepared using U, Th, and K distribution data using the four-channel gamma-ray spectrometer. The Streaky gneiss, garnet-biotite-gneiss, and leucocratic gneiss are identifiable from the total count map (i.e., U + Th + K). The Augen gneiss, leucocratic gneiss, and garnet-biotite gneiss are identifiable in the thorium geochemical map, and likewise, Augen gneiss and leucocratic gneiss are identifiable in the potassium distribution map. The radio-elemental (U, Th, and K) distribution characterised low U, Th, and Th/U ratios. Overall variations in U, Th, and K contents can be due to the metamorphic redistribution of mixed crustal source rock, with a dominant acid igneous component.

Five representative gneiss samples were subjected to quartz-biotite-gneiss, garnet-biotite gneiss, sheared banded gneiss, leucocratic gneiss, and banded Instrumental Neutron Activation analysis (INAA) using neutron flux of 10-13ncm^{-2} sec^{-1} and suitable standards. Determination of major elements like Na, Ca, Fe; REEs, La, Ce, Nd, Sm, Eu, Tb, Tm, Yb, Lu; minor elements like U, Th and trace elements like Sc, Cr, Co, Rb, Hf, and Ta were made on these samples. The gneisses have low to moderate total REEs (LREE = 19-218 ppm), and the chondrite normalised REE patterns show weak to moderate fractionation trends with medium Eu anomaly.

In addition to these, radon (Rn) gas measurements were made on freshwater samples of Schirmacher landmass and ice samples' waters from the polar ice sheet to the south of Maitri Station. The analytical data indicated the absence of any anomalous radon (Rn) in both sample types. Cellulose Nitrate (C.N.) film studies in the vicinity of the same camping area also indicate the absence of high radon (Rn) in the moraines. Detailed Petrograph i.e. studies on samples indicated the three most common mineral assemblages (a) quartz - orthoclase-plagioclase garnet, (b) quartz-orthoclase-plagioclase-hypersthene, and (c) plagioclase-quartz-diopside.

The mafic granulites of the Schirmacher region, East Antarctica, occur more or less as concordant sills, lenses, or badinage structures within the felsic rocks, charnockites or metapelites of the area. They show variation from garnet-bearing two-pyroxene granulites and garnet free pyroxene granulites to transitional amphibolite-pyroxene granulites (Rao et al., 2000).

The rocks are characterised by enriched large-ion lithophile elemental concentrations than that of mid-oceanic ridge basalts. Their high-field strength elements and heavy rare-earth elemental concentrations are that of mid-oceanic ridge basalts. Such an inference can also be drawn from the ratios of their large-ion lithophile elements against high-field strength elements and heavy rare-earth elements (Rao et al., 2000),

The isotopic studies carried out on these samples show that the Sm-Nd and Rb-Sr dating did not yield much spread but suggested an Sm-Nd metamorphic age of ~960 Ma. Rb-Sr dating gave age's ~886 Ma, suggesting the Rb and Sr elements' reworking during subsequent tectonothermal overprinting. These rocks' Nd model ages (TDMNd) show a relatively restricted range of 1120 to 1357 Ma, suggesting mafic magmatism ~1200 Ma. These rocks' positive eNd values (+4.22 to +6.07) represent a juvenile crustal fragment derived from mantle precursors' melting without a significant reworking of older crustal material. It proposed that a mantle source's partial melting produced these rocks (Rao et al., 2000).

MARINE GEOPHYSICAL STUDIES

The Naval Physical and Oceanographic Laboratory (NPOL), Kochi, participated in the Second Scientific Expedition to Antarctica to undertake geophysical studies. The seismic studies were carried out at a base camp in Princess Astrid Kyst in the Dronning Maudland (69^0 59' S latitude and 11^0 55' E longitude) regions by exploding plastic explosives of 1 Kg capacity at an interval of 175 m at selected shot points along a track of length 1.925 Kms. The seismic signals picked up by geophones (placed at the other end of the same path) were recorded on a tape recorder. The seismic signals were replayed and interpreted for obtaining the strata below the track. The studies revealed two layers: one sedimentary layer thickness of 486 m, a P wave velocity of 2500m/s and a granitic layer of P wave velocity of 1700m/s. It confirmed the concept that the Antarctic shelf is of the continent's origin.

ANTARCTIC OCEANOGRAPHY AND LIMNOLOGICAL STUDIES

The National Institute of Oceanography (NIO), Goa regularly participated in earlier Antarctic expeditions and carried out significant oceanographic and limnological investigations around the Antarctic region, including the Southern Ocean (NIO – Report, 1988-89). Here we enumerate some of the critical findings of the studies undertaken by NIO during the earlier expeditions.

Petroleum Hydrocarbons: Surface seawater in the southernmost Indian Ocean showed dissolved petroleum hydrocarbons in the range 7.76-22.44 µg.I-1.

Trace Metals: Trace metals concentration in the lake sediments range Ni 29-31 µg. g^{-1}; Pb 10-12 µg. g^{-1}; Co 23-28 µg. g^{-1}; Cu 50-75 µg. g^{-1}; Cd 1-2 µg. g^{-1} The concentration of Pb and Cd in the Penguin feathers were below the detection limit. The samples of water, zooplankton, and krill are being for metals.

Major Ions: Water samples were also analysed for major ions Ca, Mg and F. Their concentration ranges are as follows Ca 416-448 mg.kg^{-1}; Mg 1249-1376 mg.kg^{-1}; F 0.85-0.93mg.kg^{-1}. The fluoride concentration in the Antarctic waters is comparatively lower than that in other seas/ocean waters.

SURFACE HEAT BUDGET OF THE POLYNYA IN THE COASTAL WATERS OFF QUEEN MAUD LAND, ANTARCTICA FOR AUSTRAL SUMMER, 1987

The incoming solar radiation was the essential component in the surface heat budget and its mean value for the study period was 209 w/m^2. The latent and sensible heat fluxes were found in opposition, nearly balancing each other out. The average net heat gain over the polynya for the study period was 141 w/m^2. Due to the Heat budget of shore polynya off Droning Maud Land Coast, Antarctica and hydrographical regime from the Droning Maud Land Coast to 30° S, during 1988 austral summer, the heat gain in the upper 50 m of the water column is lost.

One of the experiments carried out during the 7th Indian Antarctic expedition was the mooring of an array of current meters in a shore polynya off Droning Maud Land coast of Antarctica for the first time. The mooring of the current meters continued for about a fortnight from 9 to 23 January 1988. The current velocity field during January 1988 indicated a dominant mean northward flow. The heat budget of the shore polynya using the time series of the surface meteorology and hydrography was also obtained.

The salient results show that over the polynya, the net surface heat exchange was steady with an average heat gain of about 150 w/m^2 during the earlier part of austral summer, while in its later part, the sea loses heat. The higher rate of heat loss in the last period of austral summer compensated for the steady heat gain of the preceding period resulting in a meagre seasonal heat gain of about 20 w/m^2, impeded the faster growth of polynya. The seasonal heat gain orders and the heat spent for the melting of ice around polynya were comparable. The water body lost heat at an average rate of about 45 W/m^2 suggesting a net heat advection out of the region exporting heat around the same order as the change in the polynya's heat content.

The hydrographic regime from Droning Maud Land coast to 30°S shows the presence of three distinct water masses, viz., (i) a cold and fresh surface water (<34.5 ppt) with sinking nature around 50°S, (ii) a warm saline surface water between 40°S and 30°S with enhanced vertical stability, and (iii) a warm and saline (<34.5 ppt) deep-water mass rising near 66°S. It infers intense vertical mixing in the upper 300 m of the sub-Antarctic region. A sizeable warm saline eddy in the upper layer with a 4° north-south dimension was conspicuously present around 46°S latitude.

CARBOHYDRATE DISTRIBUTION

During the summer, paniculate carbohydrates (PHCO) in the coastal ice edge and offshore waters in Antarctica varied from 67.0 to 234.2 µg. I-1 and from 43.9 to

334.5 μg. I-1 respectively. The ratio of PHCO/Chlorophyll a and PHCO/ Particulate Organic Carbon from the surface up to 50 m depth decreased in offshore waters, but the same did not apply to the coastal waters. The dissolved carbohydrate (DHCO) in the Antarctic waters varied from 0.5 to 9.6 mg. l-1. In the Central Bay of Bengal's tropical region, the PHCO and DHCO values ranged from 18.4 to 147.7 μg I-1 and from 0.9 to 1.55 mg. $l^{-1,}$ much lower than those of the Antarctic waters. PHCO showed a high correlation with the chlorophyll in Antarctic waters, indicating that living phytoplankton contributed a significant part of the carbohydrates

Marine Biomass: A comparative study on krill production concerning the distribution and abundance of consumer species of seabirds and mammals revealed the following.

The annual variability in sea-ice coverage, as observed during the winter of 1985 and 1986 rather than the standing crop of krill, is a limiting factor for 38 species of seabirds and nine marine mammals.

Primary Production: Diurnal variation in chl-a and primary production at a station in the polynya. Three observations made for 24 hours showed that the surface chl-a varied from 0.53 to 0.92 mgm^{-2} (average 0.71 mgm^{-2}) and column values from 10.23 to 27.47 mgm^{-3} (average 18.0 mgm^{-3}). These values are almost equal to those observed in coastal waters of continents. The surface primary production showed a variation from 25.4 to 54.9 $mgCm^{-3}d^{-1}$(average 41.7 $mgCm^{-3}d^{-1}$). The column production varied from 650 to 914 $mgCm^{-2}d^{-1}$ (average 786 $mgCm^{-2}d^{-1}$).

Studies carried out on variations in chl-a and primary production between lat. 70°S and 30°S showed that between 69°S chl-a concentration was low which varied between 0.015 and 0.134 mg/m^3 (average 0.066 mg/m^3) at the surface and between 1.22 and 2.90 mg/m^2 (average 2.32 mg/m^2) in the column. In the same area primary production varied from 0.58 to 1.8 $mgCm^{-3}d$-^{1} (average 0.88 $mgCm^{-3}d^{-1}$) at the surface and in the column from 49 to 156 mgCm-2d^{-1} (average 91 mgCm-2d^{-1}). Between 50°S and 40°S chl a value varied from 0.075 to 0.159 mgm-3 (average 0.122 mgm^{-3}) at the surface and from 6.39 to 12.14 mgm^{-2} (average 9.78 in the column. The corresponding primary production values were between 0.84 and 1.24 $mgCm^{-3}d^{-1}$ (average 0.10 $mgCm^{-3}d^{-1}$) and 65 and 106 $mgCm^{-2}d^{-1}$ (average 84 $mgCm^{-2}d^{-1}$) at the surface and column respectively. Similarly, between 4°S and 30°S chl a at the surface varied from 0.009 to 0.052 mg/m^3 (average 0.028 mg/m^3) and in the column from 1 to 4.68 mg/m^2 (average 2.16 mgm^{-2}) Primary production in the same area varied from 0.46 to 1.32 $mgCm^{-3}d^{-1}$(average 0.1 $mgCm^{-3}d^{-1}$) and in the column from 94 to 149 $mgCm^{-2}d^{-1}$ (average 120 $mgCm^{-2}d^{-1}$).

The studies on the Antarctic lakes in general and the Priyadarshini lake by the Indian Institute of Technology, Kanpur, during the Antarctic summer period (1996-97) of XVI Indian Scientific Expedition Antarctica revealed the following.

Apart from hydrological and thermal stratification studies (Sinha et al., 1999, 2000a), studies also aimed to retrieve sediment cores from the Priyadarshini Lake and use sedimentological, isotopic, and palynological data to unravel the paleoclimatic fluctuation during the Quaternary period as recorded in the lacustrine sediments. Mineralogical study of the lacustrine sediment cores from lakes and varve deposits from the lakeshore showed quartz and feldspar as principal allogenic phases derived from local crystalline rocks surrounding the lake basin (Sinha and Chatterjee, 2000).

Illite, chlorite, and a minor amount of mixed-layered mineral in the lacustrine sediments could be weathering products in the lake catchment. In general, the clay mineralogy of sediment cores and varve sediments indicates low-hydrolysing depositional environments under high latitudes and subtle weathering. Further studies (Sinha et al., 2003) on elemental distributions in vertical profiles were correlated with the redox cycles and provide information about the depositional environment. A low sedimentation rate was inferred in the periglacial environment of Antarctica with a minimal sediment supply. A ^{14}C age of ~ eight kyr indicated that the lake formed in the early Holocene because of major deglaciation after the last glacial period. The palynological investigation of samples from the Priyadarshini Lake core (Sinha et al., 2000b) yielded low pollen compared to moss spores and other lower plants' remains. The Antarctica continent is floristically almost barren and restricted to only two occurring vascular plant species, i.e., *Deschamsia antarctica* (Poaceae) and *Colobanthes quitensis* (Caryophyllaceae). Their distribution confines to the western fringe or Antarctic peninsula region. Apart from the local mosses- *Polytrichum alpinum*, *Drepanocladus uncinatus*, the preponderance of algae and often seen thick growth of various lichens on exposed rocky substratum was reported. Based on the palynomorphs' overall representation fluctuations, three distinct phases were recognised in the present pollen sequence, indicating palaeoclimatic oscillations from the base towards the top. Based on the frequent encounter of *Cosmarium*, which is a freshwater alga, showing moderate values, coupled with preponderance of *Acritarch*, an indicator of a shallow freshwater/marine environment, the bottommost Phase I was representing a shallow and small lake. Phase covers the major part of the sequence and demonstrates the expansion of terrestrial flora, which varies in composition.

Improvement in the pollen depositional assessment is indicative of a change in the local environment, therefore to the retreat of ice and probably the inception of a warm climate in the region. During this Phase, the lake probably had a wider spread than seen earlier. The proliferation of *Cosmarium* and a decline in Acritarch occurred during this Phase. The uppermost parts of the sequence represented Phase III when climate once again seems to have deteriorated, as inferred from the sharp decrease in Poaceae values and the disappearance of most herbaceous elements. Mosses, along with ferns, are represented very well- the former is known to withstand too harsh climatic conditions. During this Phase, the lake became shallower, as indicated by the marked reduction in the frequencies of *Cosmarium*. It ends with the improvement in the values of *Cosmarium,* suggesting the expansion of the lake due to the amelioration of climatic conditions, which continued until recent times.

OXYGEN ISOTOPE STUDIES ON SURFACE ICE, SNOW, AND WATER SAMPLES AROUND MAITRI STATION

Geology Department of M.S. University of Baroda participated in XXI Scientific Expedition to Antarctica to carry out detailed sampling of the ice wall (at 20 cm interval) behind the Maitri station top to the base to study the vertical variations of delta ^{18}O. Similarly, the lateral variations in delta ^{18}O and modification (if any) in the isotopic signatures due to post-depositional processes in the ice and snow samples were studied. This study was also conducted on the water samples from Priyadarshini Lake.

REFERENCES

NIO-Annual Report. (1988-89). *Antarctic Lake Studies for Palaeoclimatic Interpretations*. Indian Institute of Technology, Kanpur.

Rao, D. R., Rashid, S. A., & Panthulu, G. V. C. (2000). Origin of Mg-Metatholeiites of the Schirmacher Region, East Antarctica: Constraints from Trace Elements and Nd-Sr Isotopic Systematics. *Gondwana Research, 3*(1), 91–104. doi:10.1016/S1342-937X(05)70060-5

Ravikant, V., Bhaskar Rao, Y. J., & Gopalan, K. (2004). Schirmacher Oasis as an extension of the Neoproterozoic East African Orogen into Antarctica: New Sm-Nd Isochron Age Constraints. *The Journal of Geology, 112*(5), 607–616. doi:10.1086/422669

Shrivastava, P. K., Roy, S. K., Srivastava, H. B., & Dharwadkar, A. (2019). Estimation of Paleo-ice Sheet Thickness and Evolution of Landforms in Schirmacher Oasis and Adjoining Area, cDML, East Antarctica. *Journal of the Geological Society of India, 93*(6), 638–644. doi:10.100712594-019-1242-5

Sinha, R., Sharma, C., & Chauhan, M.S. (2000b). Article. *Paleobotanist, 49*, 1-8.

Sinha, R., & Chatterjee, A. (2000). Article. *Journal of the Geological Society of India, 56*, 39–45.

Sinha, R., Chatterjee, A., Panda, A. K., & Mitra, A. (1999). Article. *Current Science*, *76*, 680–683.

Sinha, R., Navada, S. V., Chatterjee, A., Kumar, S., Mitra, A., & Nair, A. R. (2000a). Article. *Current Science*, *78*, 992–995.

Sinha, R., Sastry, V. N., & Rajagopalan, G. (2003). Article. *Journal of the Geological Society of India*, *61*, 717–723.

Compilation of References

Adamson, D., & Pickard, J. (1986). Cenozoic history of the Vestfold Hills. In J. Pickard (Ed.), *Antarctic Oasis* (pp. 63–98). Academic Press.

Agrawal, R. S. (2000). *Permittivity Measurement of DeccanTrap Soil of Microwave Frequencies* [Doctoral thesis, Dr Babasaheb Ambedkar Marathwada UniversityAurangabad]. Shodhganga: a reservoir of Indian theses at INFLIBNET.

Agrawal, R. S., Kurtadikar, M. L., & Murugkar, A. G. (2004). Dielectric Properties of Soil at 5 GHz: Microwaves and Optoelectronics. Anamaya Publishers.

Akilan, A., Abdul Azeez, K. K., Schuh, H., & Kumar, S. (2019). Perturbations in atmospheric gaseous components over coastal Antarctica detected in GPS signals and its natural origin to volcanic eruption. *Polar Science*, *19*, 69–76.

Akilan, A., Abdul Azeez, K. K., Schuh, H., Padhy, S., & Subhadra, N. (2016). Changes in atmospheric water content associated to an unusual high snowfall during June 2004 at Maitri station (Schirmacher Oasis, East Antarctica) and the role of south west Indian ridge geodynamics. *Natural Hazards*, *83*, 563–574.

Akilan, A., & Balaji, S. (2013). Crustal deformation around East-Antarctica and Southern Ocean using GPS-Geodesy. *Current Development in Oceanography*, *6*, 41–49.

Akilan, A., Balaji, S., Srinivas, Y., & Yuvaraj, N. (2013). Plate motion predictability using Hurst exponent applied to Maitri-Antarctica GPS network. *Journal of the Geological Society of India*, *82*, 613–620.

Aldahan, A., Possnert, G., Johnson, S. J., & Clausen, H. B., Lsaksson, E., Karlen, W., & Hansson, M. (1998). Sixty-year ^{10}Be record from Greenland and Antarctica; Proceedings. *Indian Academy of Science*, *107*, 139–147.

Andersen, T., Austrheim, H., & Buke, E. A. J. (1991). Fluid induced rctrogression of granulites in the Bergen Arcs, Caledonides of W Norway: Fluid inclusion evidence from amphibolites facies shear zones. *Lithosphere*, *27*, 29–42.

Anderson, J. B., Shipp, S. S., Lowe, A. L., Wellner, J. S., & Mosola, A. B. (2002). The Antarctic Ice Sheet during the Last Glacial Maximum and its subsequent retreat history: A review. *Quaternary Science Reviews, 21*(1-3), 49–70. doi:10.1016/S0277-3791(01)00083-X

Argus, D. F., & Dorden, R. G. (1991). No Net-rotation model of current plate velocities incorporating plate motion model NUVEL-1. *GRL, 18*(11), 2039–2042. doi:10.1029/91GL01532

Armijo, R., Tapponnier, P., Mercier, J. L., & Tanglin, H. (1986). Quaternary extension in southern Tibet: Field observations and tectonic implications. *Journal of Geophysical Research, 91*(B14), 803–13,872. doi:10.1029/JB091iB14p13803

Asthana, R., Shrivastava, P. K., & Ramar, K. (2012) *Lake Bottom Sediment Coring from Schirmacher Oasis During 24th Indian Antarctic Expedition.* Scientific Report Twenty Fourth Indian Antarctic Expedition 2003-2005 Ministry of Earth Sciences, Tech. Pub. No. 22.

Asthana, R., & Chaturvedi, A. (1998). The grain size behaviour and morphoscopy of supraglacial sediments, south of Schirmacher Oasis, Antarctica. *Journal of the Geological Society of India, 52*(5), 557–568.

Asthana, R., & Chaturvedi, A. (1998). The grain size behaviour and morphoscopy of supraglacial sediments, south of Schirmacher Oasis, E. Antarctica. *Journal of the Geological Society of India, 52*, 557–568.

Asthana, R., Shrivastava, P. K., Beg, M. J., Shome, S., & Kachroo, K. (2009). Surface microtexture of quartz grains from glaciolacustrine sediment core from Priyadarshini lake, Schirmacher Oasis, East Antarctica as revealed under Scanning electrom microscope. *Indian Journal of Geoscience, 63*(2), 205–214.

Bardin, V. I. (1982). Composition of East Antarctic moraines and some problems of Cenozoic history. In C. Craddock (Ed.), Antarctic Geoscience (pp. 1069-1076). Univ. Wisconsin Press.

Bardin, V. I., Bubnova, M. I., & Gerasimova, V. M. (1979). Clay minerals in unconsolidated deposits of the Prince Charles Mountains. *Inf. Bull. Sov. Antarct. Exped., 71*, 120–128.

Baumgartner, S., Beer, J., Suter, M., Hannen, B. D., Synal, H. A., Kubik, P. W., Hammer, C. U., & Johnsen, S. (1997). Chlorine 36 fallout in the summit Greenland ice core. *Journal of Geophysical Research, 102*(C12), 26659–26662. doi:10.1029/97JC00166

Beer, J., Joos, Ch. F., Lukasczyk, Ch., Mende, W., Siegenthaler, U., & Stellmacher, R. (1994). 10Be as an indicator of solar variability and climate; In: The solar engine and its influence on the terrestrial atmosphere and climate. E NATO ASI Series, 25, 221–233.

Beer, J., Raisbeck, G. M., & Yiou, F. (1991). Time variations of ^{10}Be and solar activity. In C. P. Sonnett, M. S. Giampapa, & M. S. Matthews (Eds.), *The sun in time* (pp. 343–359). University of Arizona Press.

Bendick, R., & Bilham, R. (1999). Search for buckling of the southwest Indian coast related to Himalayan collision. *Geological Society of America. Special Paper, 328*, 313–321.

Bera, S. K. (2006). *Pollen Analysis of Surface Deposits and Holocene Lake Sediment of Schirmacher Oasis, Central Dronning maud Land (CDML), East Antarctica*. Twentieth Indian Expedition to Antarctica, Scientific Report, MoES, Technical Publication No. 18.

Bera, S. K. (2004). Late Holocene Palaeo- winds and climatic changes in Eastern Antarctica as indicated by long-distance transported pollen-spores and local microbiota in polar lake core sediments. *Current Science*, *86*(11), 1485–1488.

Bera, S. K. (2004). Late Holocene palaeo-winds and climatic changes in Eastern Antarctica as indicated by long-distance transported pollen-spores and local microbiota in polar lake core sediments. *Current Science*, *86*(11), 1485–1488.

Bera, S. K. (2005). Recovery of airborne palyno debris from continental ice sheet qasis, East Antarctica. *Current Science*, *88*(10), 1550–1552.

Bera, S. K., & Khandelwal, A. (2003). Modern pollen – spore rain in Schirmacher Oasis, East Antarctica. *Current Science*, *85*(2), 137–140.

Bera, S. K., & Khandelwal, A. (2003). Aerospora over the Southern Ocean and Schirmacher Oasis, East Antarctica. *Current Science*, *85*(2), 137–140.

Bhandari, N., Bhattacharya, S. K., Nijampurkar, V. N., Sengupta, D., Shah, V., Singhvi, A.K., Shukla, P. N., Suthar, K. M. & Vohra, C. P. (1984). Isotopic and related studies of Antarctic ice samples, Proceedings. *Indian Academy of Science (Earth and Planetary Science)*, *93*(2), 135.

Bhandari, N., & Rama. (1963). Study of atmospheric washout processes by means of radon products. *Journal of Geophysical Research*, *68*(13), 3823–3826. doi:10.1029/JZ068i013p03823

Bhattacharya, A., Krishna Kumar, K. R., Raith, M., & Sen, S. K. (1991). An improved set of a-X parameters for Fe-Mg-Ca garnets and refinements of the orthopyroxene-garnet thermometer and orthopyroxene-garnet-plagioclase-quartz barometer. *Journal of Petrology*, *32*(3), 629–656. doi:10.1093/petrology/32.3.629

Bhattacharya, A., Mohanty, L., Maji, A., Sen, S. K., & Raith, M. (1992). Non-ideal mixing in the phlogophite-annite binary: Constraints from experimental data on Mg-Fe partitioning and a reformulation of the biotite-garnet geothermometers. *Contributions to Mineralogy and Petrology*, *111*(1), 87–93. doi:10.1007/BF00296580

Billham, R., & Gaur, V. K. (2000). Geodetic contributions to the study of seismotectonics in India. *Current Science*, *79*, 1259–1269.

Binns, R. A. (1968). Hydrothermal investigation of the amphibolite-granulite facies boundary. Special Publication in. *Geological Society of America*, *2*, 341–344.

Biscaye, P. E. (1965). Mineralogy and sedimentation of recent deep-sea clay in the Atlantic Ocean and adjacent seas and oceans. *Bulletin of the Geological Society of America*, *76*(7), 803–832. doi:10.1130/0016-7606(1965)76[803:MASORD]2.0.CO;2

Björc, K., Hjort, S., Ingólfsson, C., & Skog, G. (1991). Radiocarbon dates from the Antarctic Peninsula region - problems and potential. *Quaternary Proceedings*, *1*, 55–65.

Bohlen, S. R. (1987). Pressure-temperature-time paths and a tectonic model for the evolution of granulites. *The Journal of Geology*, *95*(5), 617–632. doi:10.1086/629159

Bohlen, S. R., & Mezger, K. (1989). Origin of granulite terranes and the formation of the lowermost continental crust. *Science*, *244*(4902), 326–329. doi:10.1126cience.244.4902.326 PMID:17738304

Bohlen, S. R., Wall, V. J., & Boettcher, A. L. (1983). Experimental investigation and application to garnet granulite equilibria. *Contributions to Mineralogy and Petrology*, *83*(1-2), 52–56. doi:10.1007/BF00373079

Borsi, L., Petrini, R., Talarica, R. F., & Palmeri, R. (1995). Geochemistry and Sr-Nd isotopes of amphibolite dykes of northern Vitoria Land, Antarctica. *Lithosphere*, *35*, 245–260.

Bose, S., & Sengupta, S. (2003). High-temperature mylonitization of quartzofeldspathic gneiss: An example from the Schirmacher Hills, East Antarctica. *Gondwana Research*, *6*(4), 805–816. doi:10.1016/S1342-937X(05)71026-1

Bouin, M.F., & Vigny, C. (2000). New constraints on Antarctic plate motion and deformation from GPS data. *J.Geophys.Res.*, *105*, 279-293.

Boyer, S. J. (1975). Chemical weathering of the rocks on the Lassiter Coast, Antarctic Peninsula, Antarctica. *New Zealand Journal of Geology and Geophysics*, *18*(4), 623–628. doi:10.1080/00288306.1975.10421561

British Antarctic Survey. (n.d.) *Global Positioning System (GPS)*. National Environment Research Council. Retrieved from https://www.bas.ac.uk/polar-operations/engineering-and-technology/technology-tools-and-methods/global-positioning-system-gps/

Brown, P. E. (1989). Flincor: A microcomputer programme for the reduction and investigation of fluid inclusion data. *American Journal of Mineralogy*, *74*, 1390–1393.

Brown, P. E., & Lamb, W. M. (1986). Mixing of H_2O-CO_2 in fluid inclusions: Geobarometry and Archean gold deposits. *Geochimica et Cosmochimica Acta*, *50*(5), 847–852. doi:10.1016/0016-7037(86)90360-1

Cabanis, B., Chantraine, J., & Rabu, D. (1987). Geochemistry study of the Brioverian (late Proterozoic) volcanic rocks in the northern Armorican Massif (France), Implications for geodynamic evolution during the Cadomian. In T. C. Beckinsale & D. Rickard (Eds.), Geological Society of London Special Publications: Vol. 33. Geochemistry and Mineralisation of Proterozoic Volcanic Suites' (pp. 525–539). Academic Press.

Campbell, I. B., & Claridge, G. G. C. (1987). *Antarctica: Soil, Weathering Processes, and Environment*. Elsevier.

Carswell, D. A., & Harley, S. L. (1990). Mineral barometry and thermometry. In Eclogite Facies Rocks. Blackie Press.

Carver, R. E. (1971). *Procedures in Sedimentary Petrology*. Wiley.

Chakraborty, S. K., Kaul, M. K., & Raina, V. K. (1985). Mineral variation in the gneisses from Dakshin Gangotri, Antarctica. *Department of Ocean Development. Technical Publication*, (2), 23–28.

Chamley, H. (1989). *Clay Sedimentology*. Springer. doi:10.1007/978-3-642-85916-8

Chandrashekharan, S., & Sreedharan, A. (2000). *Report on hydrographic survey of Antarctica by XVI Indian scientific expedition. Sixteenth Indian Expendilion to Antarctica, Scientific Report, 2000.* Department of Ocean Development, Technical Publication No. 14. http://14.139.119.23:8080/dspace/bitstream/123456789/622/3/ARTICLE+17-2.pdf

Chappell, B. W., & Stephens, W. E. (1988). Origin of infracrustal (I-type) granite magmas. *Transactions of the Royal Society of Edinburgh. Earth Sciences*, *79*(2-3), 71–86. doi:10.1017/S0263593300014139

Chappell, B. W., & White, A. J. R. (1974). Two contrasting granite types. *Pacific Geology*, *8*, 173–174.

Chaturvedi, A., & Singh, A. (2008). *First Attempt at Motorised Ice Core Drilling during the Antarctic Summer Season.* 22nd Indian Expedition to Antarctica, Scientific Report, 2008, Ministry of Earth Science, Technical Publication No.20.

Chaturvedi, A., & Asthana, R. (1996). *An attempt at shallow ice-core drilling in polar continental ice, central Dronning Maud Land.* 12th Indian Expedition to Antarctica DOD Tech. *Publ.* 10.

Claridge, G. G. C., & Campbell, J. B. (1989). Clay Mineralogy. In P. J. Barrett (Ed.), Antarctic Cenozoic History from the CIROS-1 Drillhole, McMurdo Sound (pp. 185-193). DSIR Bulletin.

Clausen, H. B. (1973). Dating of polar ice by ^{32}Si. *Journal of Glaciology, 12*(66), 411.

Cocherie, A., Rossi, P. L., Fouillac, A. M., & Vidal, Ph. (1994). Crust and mantle contribution to granite genesis - an example from the Variscan batholith of Corsica, France, studied by trace element and Nd-Sr-O-isotope systematics. *Chemical Geology*, *115*(3-4), 173–211. doi:10.1016/0009-2541(94)90186-4

Conard, N. J., Kubik, P. W., Gove, H. E., & David, E. (1989). A ^{36}CI profile in Greenland ice from 1265 to 1865. *Radiocarbon*, *31*(03), 585–591. doi:10.1017/S0033822200012170

Crozaz, G., & Langway, C. C. Jr. (1966). Dating Greenland firn-ice cores with Pb-210. *Earth and Planetary Science Letters*, *1*(4), 194–196. doi:10.1016/0012-821X(66)90067-7

Crozaz, G., Picciotto, E., & De, W. (1964). Antarctic snow chronology with ^{210}Pb. *Journal of Geophysical Research*, *69*(12), 2597–2604. doi:10.1029/JZ069i012p02597

Currie, R. (1994). Variance contribution to Iuni-solar and solar cycle signals in St. Lawrence and Nile river records. *International Journal of Climatology*, *14*(8), 843–852. doi:10.1002/joc.3370140803

Dahl, P. S. (1980). The thermal-compositional dependence of Fe^{+2}-Mg^{+2} distribution between coexisting garnet and pyroxene: Application to geothermometry. *American Journal of Mineralogy*, *65*, 852–866.

Dansgaard, W. (1964). Stable isotopes in precipitation. *Tellus*, *16*(4), 436–468. doi:10.3402/tellusa.v16i4.8993

Dasgupta, S., Sengupta, S., Bose, S., Fukuoka, M., & Dasgupta, S. (2001). Polymetanorphism in the Schirmacher Hills granulites, East Antarctica: Implication for the tectono-thermal reworking of an isobarically cooled deep continental crust. *Gondwana Research*, *4*(3), 337–357. doi:10.1016/S1342-937X(05)70334-8

Dayal, A. M., & Hussain, T. S. M. (1997). Rb-Sr ages of lamprophyre dykes from Schirmacher Oasis, Queen Mound Land, East Antarctica. *Journal of the Geological Society of India*, *50*, 457–460.

De Paolo, D. J. (1981). Neodymium isotopes in the Colorado Front Range and crust-mantle evolution in the Proterozoic. *Nature*, *291*(5812), 193–196. doi:10.1038/291193a0

Deer, W. A., Howie, R. A., & Zussman, J. (1992). *An Introduction to the Rock-Forming Minerals* (2nd ed.). Longman Sc. Technic.

Delmas, R. J., Krichner, S., Palis, J. M., & Petit, J. R. (1992). 1000 years of explosive volcanism records at the south pole. *Tellus*, *44B*(4), 335–350. doi:10.3402/tellusb.v44i4.15461

Demaster, D. J. (1980). The half-life of ^{32}Si was determined from a varved Gulf of Mexico sediment core. *Earth and Planetary Science Letters*, *48*(1), 209–217. doi:10.1016/0012-821X(80)90182-X

DeMets, Gorden, Argus, & Stein. (1994). Effect of recent revisions to geo magnetic time scale on estimates of current plate motions. *Geophysical Research Letters, 21*, 2191-2194.

DeMets, C., Gordon, R. G., Argus, D. F., & Stein, S. (1990). Current Plate Motions. *Geophysical Journal International*, *101*(2), 425–478. doi:10.1111/j.1365-246X.1990.tb06579.x

Deshpande, S. S. (2012). *Dielectric properties of saline soils at c band microwave frequency* [Doctoral thesis, Dr Babasaheb Ambedkar Marathwada UniversityAurangabad]. Shodhganga: a reservoir of Indian theses at INFLIBNET.

Deshpande, S. S., Itolikar, A. B., Joshi, A. S., & Kurtadikar, M. L. (2015). Dielectric Properties of Gujarat and Uttar Pradesh Saline Soils At 5GHz. *Proceeding of 11th International Conference on Microwaves, Antenna, Propagation and Remote Sensing (ICMARS-2015)*, 87-91.

Eckert, J.O. Jr, Newton, R.C., & Klepp, A, O.J. (1991). The ΔH of reaction and recalibration of garnet-pyroxene-plagioclase-quartz geobarometer in CMAS system in solution calorimetry. *American Journal of Mineralogy*, *76*, 148–160.

Ehrmann, W., Bloemendal, J., Hambrey, J. M., Mckelvey, B., & Whitehead, J. (2003). Variations in the composition of the clay fraction of the Cenozoic Pagodrama Group, East Antarctica: Implications for determining provenance. *Sedimentary Geology*, *161*(1-2), 131–152. doi:10.1016/S0037-0738(03)00069-1

Ehrmann, W., & Mackensen, A. (1992). Sedimentological evidence from the formation of an East Antarctic ice sheet in Eocene/Oligocene time. *Palaeogeography, Palaeoclimatology, Palaeoecology*, *93*(1-2), 85–112. doi:10.1016/0031-0182(92)90185-8

Ehrmann, W., & Polozek, K. (1999). The heavy mineral record in the Pliocene to Quaternary sediments of the CIROS-2 drill core, McMurdo Sound, Antarctica. *Sedimentary Geology*, *128*(3-4), 223–244. doi:10.1016/S0037-0738(99)00071-8

Ehrmann, W., Setti, M., & Marinoni, L. (2005). Clay minerals in Cenozoic sediments of Cape Roberts (McMurdo Sound, Antarctica) reveal palaeoclimatic history. *Palaeogeography, Palaeoclimatology, Palaeoecology*, *229*(3), 187–211. doi:10.1016/j.palaeo.2005.06.022

Elliot, D. H., Hoffman, S. M., & Rieske, D. E. (1992). The provenance of Paleocene strata, Seymour Island. In Recent Progress in Antarctic Earth Science (pp. 347-355). Terra Sci. Pub. Comp. (TERRAPUB).

Ellis, D. J. (1987). Origin and evolution of granulites in the normal and thickened crust. *Geology*, *15*(2), 167–170. doi:10.1130/0091-7613(1987)15<167:OAEOGI>2.0.CO;2

Ellis, D. J., & Green, D. (1979). An experimental study of the effect of Ca upon garnet-clinopyroxene Fe-Mg exchange equilibria. *Contributions to Mineralogy and Petrology*, *71*(1), 13–32. doi:10.1007/BF00371878

Elmore, D., Tubbs, L. E., Newman, D., Ma, X. Z., Finkel, R., Nishiizumi, K., Beer, J., Oeschger, H., & Andree, M. (1982). ^{36}CI bomb pulse measured in a shallow ice core from Dye 3, Greenland. *Nature*, *300*(5894), 735–737. doi:10.1038/300735a0

Erdtman, G. (1943). An introduction to pollen analysis. Academic Press.

Ewart, A. (1982). The mineralogy and petrology of Tertiary-Recent orogenic volcanic rocks with special reference to the andesite-basaltic compositional range. In R. S. Thorpe (Ed.), *Andesite: Orogenic andesites and related rocks* (pp. 25–95). Wiley.

Farmer, G. L., & De Paolo, D. J. (1983). Origin of Mesozoic and Tertiary granite in the Western United States and implications for pre-Mesozoic crustal structure. Nd and Sr isotopic studies in the Geoclue of the northern Great Basin. *Journal of Geophysical Research*, *88*(B4), 3379–3401. doi:10.1029/JB088iB04p03379

Ferry, J. M., & Spear, F. S. (1978). Experimental calibration of the partitioning of Fe and Mg between biotite and garnet. *Contributions to Mineralogy and Petrology*, *66*(2), 113–117. doi:10.1007/BF00372150

Finkel, R. C., Nishiizumi, K., Elmore, D., Ferraro, R. D., & Gove, H. E. (1980). ^{36}CI in polar ice, rainwater, and seawater. *Geophysical Research Letters*, *7*(11), 983–986. doi:10.1029/GL007i011p00983

Folk, R. L. (1980). *Petrology of Sedimentary Rocks*. Hemphill Austin.

Folk, R. L., & Ward, W. (1957). Brazos river bar: A study in the significance of grains size parameters. *Journal of Sedimentary Petrology*, *27*(1), 3–26. doi:10.1306/74D70646-2B21-11D7-8648000102C1865D

Force, E. R. (1980). The provenance of rutile. *Sedimentary Petrology*, *50*, 485–488.

Friedman, G. M. (1961). The distinction between dune, beach, and river sands from their textural characteristics. *Journal of Sedimentary Petrology*, *31*(4), 514–529.

Friedman, G. M. (1967). Dynamic processes and statistical parameters compared to the size-frequency distribution of beach and river sands. *Journal of Sedimentary Petrology*, *37*(2), 327–354.

Gaggeler, H., Von Gunten, H. R., Rossler, E., Oeschger, H., & Schotterer, U. (1983). ^{210}Pb dating of cold alpine firn/ice cores from Colle Gnifetti, Switzerland. *Journal of Glaciology*, *101*(101), 165–177. doi:10.1017/S0022143000005220

Ganguly, J. (1979). Garnet and clinopyroxene solid solutions, and geothermometry based on Fe-Mg distribution coefficient. *Geochimica et Cosmochimica Acta*, *43*(7), 1021–1029. doi:10.1016/0016-7037(79)90091-7

Ganguly, J., & Saxena, S. K. (1984). Mixing properties of aluminosilicate garnets: Constraints from natural and experimental data, and applications to geothermobarometry. *American Journal of Mineralogy*, *69*, 88–97.

Geoscience, M. (2002). *RAMAC GPR operating manual*. Mala Geoscience.

Gessmann, C. K., Spiering, B., & Raith, M. (1997). Experimental study of the Fe-Mg exchange between garnet and biotite: Constraints on the mixing behaviour and analysis of the cation exchange mechanism. *American Journal of Mineralogy*, *82*(11-12), 1225–1240. doi:10.2138/am-1997-11-1218

Ghent, E. D., & Stout, M. Z. (1981). Geobarometry and geothermometry of plagioclase-biotite-garnet-muscovite assemblages. *Contributions to Mineralogy and Petrology*, *76*(1), 92–97. doi:10.1007/BF00373688

Gill, J. B. (1981). *Orogenic andesites and plate tectonics*. Springer. doi:10.1007/978-3-642-68012-0

Gorden, R. G., DeMets, C., & Royer, J.-Y. (1998). Evidence of for long-term diffuse deformation of the lithosphere of equatorial Indian Ocean. *Nature*, *395*(6700), 370–374. doi:10.1038/26463

Gorden, R. G., & Stein, S. (1992). Global tectonics and space geodesy. *Science*, *256*(5055), 333–342. doi:10.1126cience.256.5055.333 PMID:17743109

Graham, C. M., & Powell, R. (1984). A garnet-hornblende geothermometer: Calibration testing and application to the Pelona Schist, Southern California. *Journal of Meteorology and Geology*, *2*(1), 13–31. doi:10.1111/j.1525-1314.1984.tb00282.x

Green, T. H. (1976). Experimental generation of cordierite - or garnet-bearing granitic liquids from a pelitic composition. *Geology*, *4*(2), 85–88. doi:10.1130/0091-7613(1976)4<85:EGOCGG>2.0.CO;2

Green, T. H. (1980). Island arc and continent-building magmatism - a review of petrogenic models based on experimental petrology and geochemistry. *Tectonophysics*, *63*(1-4), 367–385. doi:10.1016/0040-1951(80)90121-3

Grew, E. S. (1983). Saphirine-garnet and associated paragneisses in Antarctica. In R. L. Oliver, P. R. James, & J. B. Jago (Eds.), *Antarctic Earth Science* (pp. 40–43). Cambridge University Press.

Grew, E. S., & Manton, W. I. (1983). Geochronologic studies in east Antarctica: Reconnaissance uranium-thorium-lead data from rocks in Schirmacher Hills and Mount Steiner. *Antarctic Journal of the United States*, *18*, 6–8.

Griffin, J. J., Windon, H., & Goldberg, E. D. (1968). The distribution of clay minerals in the World Ocean. *Deep-Sea Research*, *15*, 433–459.

Groenewald, P. B., Moyes, A. B., Grantham, G. H., & Krynauw, J. R. (1995). East Antarctic crustal evolution: Geological constraints and modelling in Western Dronning Maud Land. *Precambrian Research*, *75*(3-4), 231–250. doi:10.1016/0301-9268(95)80008-6

Ground Penetrating Radar. (1992). Finnish Geotechnical Society.

Gupta, H. K., & Verma, A. K. (1986). Magnetic characteristics in Antarctica over geological contacts in Schirmacher Hill Region and Ice Shelf near Dakshin Gangotri (70° 05' 37" S, 12° 00' 00" E). In *Antarctica. Sci. Rep. 3rd Ind. Sci. Exp. Antarctica*. DOD, Govt. of Ind. Pub. Technical Publication.

Gupta, H. K., & Verma, S. K. (1995). *India's contribution to geophysical investigations in Antarctica during Precambrian* (Vol. 34). Mem. Geol. Soc.

Gupta, S., Nagaraju, K., & Akilan, A. (2017). Volcanic passive continental margin beneath the Maitri station in central DML, East Antarctica. constrains from the crustal shear velocity structure from receiver function modeling. *Polar Research*. Advance online publication. doi:10 .1080/17518369.2017.1332947

Güven, N. (1988). Smectites. In S. W. Bailey (Ed.), *Hydrous Phyllosilicates* (pp. 497–552). Mineralogical Society of America.

Hammer, C. U. (1980). The acidity of polar ice cores in relation to absolute dating, past volcanism and radio-echos. *Journal of Glaciology*, *93*(93), 359–372. doi:10.1017/S0022143000015227

Hammer, C. U., Clausen, H. B., & Langway, C. C. Jr. (1997). 50,000 years of recorded global volcanism. *Climatic Change*, *35*(1), 1–15. doi:10.1023/A:1005344225434

Harley, S. L. (1984). An experimental study of the partitioning of Fe and Mg between garnet and orthopyroxene. *Contributions to Mineralogy and Petrology*, *86*(4), 359–373. doi:10.1007/BF01187140

Harley, S. L. (1989). The origin of granulites: A metamorphic perspective. *Geological Magazine*, *126*(3), 215–247. doi:10.1017/S0016756800022330

Harris, N. B. W., Pearce, J. A., & Tindle, A. G. (1986). Geochemical characteristics of collision zone magmatism. In M. P. Coward & A. C. Ries (Eds.), Geological Society of London Special Publication: Vol. 19. Collision Tectonics (pp. 67–81). Academic Press.

Haselton, H. T. Jr, Hovis, G. L., Hemingway, B. S., & Robie, R. A. (1983). Calorimetric investigation of the excess entropy of mixing in an albite-sanidine solid solution: Lack of evidence for Na, K short-range order and implications for two-feldspar thermometry. *American Journal of Mineralogy*, *68*, 398–413.

Henry, D. J., & Guidottic, C. V. (1985). Tourmaline: As a petrogenetic indicator in abyssal and Alpine-type peridotites and spatially associated lavas. *Contributions to Mineralogy and Petrology*, *86*, 54–76.

Hillenbrand, C. D., & Ehrmann, W. (2001). Distribution of clay minerals in drift sediments on the continental rise west of Antarctic Peninsula, ODP leg 178, sites 1095 and 1096. *Proc. ODP, Science., Results.*, *178*, 1-29. 10.2973/odp.proc.sr.178.224.2001

Hillenbrand, C. D., & Ehrmann, W. (2003). Palaeoenvironmental implications of Tertiary sediments from Kainan Maru Seamount and northern Gunnerous Ridge. *Antarctic Science*, *15*(4), 522–536. doi:10.1017/S0954102003001640

Hillenbrand, C. D., & Ehrmann, W. (2005). Late Neogene to Quaternary environmental changes in the Antarctic Peninsula region: Evidence from drift sediments. *Global and Planetary Change*, *45*(1-3), 165–191. doi:10.1016/j.gloplacha.2004.09.006

Hoch, M. (1999). Geochemistry and petrology of ultramafic lamprophyres from Schirmacher Oasis, East Antarctica. *Mineralogy and Petrology*, *65*(1-2), 51–67. doi:10.1007/BF01161576

Hoch, M., Rehkamper, M., & Tobschall, H. J. (2001). Sr, Nd, Pb, and O Isotopes of minettes from Schirmacher Oasis, East Antarctica: A case of mantle metasomatism involving subducted continental material. *Journal of Petrology*, *42*(7), 1387–1400. doi:10.1093/petrology/42.7.1387

Hoch, M., & Tobschall, H. J. (1998). Minettes from Schirmacher Oasis, East Antarctica - indicators of an enriched mantle source. *Antarctic Science*, *10*(4), 476–486. doi:10.1017/S0954102098000571

Hodges, K. V., & Crowley, P. D. (1985). Error estimation and empirical geothermobarometry for pelitic systems. *American Journal of Mineralogy*, *70*, 702–709.

Hodges, K. V., & Spear, F. S. (1982). Geothermometry, geobarometry and the Al_2SiO_5 triple point at Mt. Moosilauke, New Hampshire. *American Journal of Mineralogy*, *67*, 1118–1134.

Hofmann-Wellenhof, B., Lichtenegger, H., & Collins, J. (1992). *Textbook on Global Positioning System Theory and Practice*. Academic Press.

Hoisch, T. D. (1990). Empirical calibration of six geobarometers for the mineral assemblage quartz+muscovite+biotite+plagioclase+garnet. *Contributions to Mineralogy and Petrology*, *104*(2), 225–234. doi:10.1007/BF00306445

Holdaway, M. J. (1971). Stability of andalusite and the aluminous silicate phase diagram. *American Journal of Science*, *271*(2), 97–131. doi:10.2475/ajs.271.2.97

Holdaway, M. J., Mukhopadhyay, B., Dyar, M. D., Guidotti, C. V., & Dutrow, B. L. (1997). Garnet-biotite geothermometry revised: New margules parameters and a natural specimen data set from Maine. *American Journal of Mineralogy*, *82*(5-6), 582–595. doi:10.2138/am-1997-5-618

Holmes, M. A. (2000). Clay mineral composition of glacial erratics, McMurdo Sound. In J. D. Stilwell & R. M. Feldmann (Eds.), *Palaeobiology and Palaeoenvironments of Eocene Rocks, McMurdo Sound East Antarctic, Antarctic. Research*. doi:10.1029/AR076p0063

Holm, P. E. (1985). The geochemical fingerprints of different tectonomagnetic environments using hygromagmatophile element abundances of tholeiitic basalts and basaltic andesites. *Chemical Geology*, *51*(3-4), 303–323. doi:10.1016/0009-2541(85)90139-1

Holt, W. E., Chamot-Rooke, N., Pichon, X., & Le. (2000). The velocity field in Asia inferred from Quaternary fault slip rates and Global Positioning System observations. *Journal of Geophysical Research*, *105*(B8), 19185–19209. doi:10.1029/2000JB900045

Hubert, J. F. (1962). A zircon-tourmaline-rutile maturity index and the interdependence of the composition of heavy mineral assemblages with the gross composition and texture of sandstones. *Sedimentary Petrology*, *32*, 440–450.

Hugentobler, U., Cchaer, S., & Fridez, P. (2001). *Bernese GPS software version 4.2 manual*. Astronomical Institute, University of Berne.

Hussain, T. S. M., & Diwakar, Rao, V. (1996). Geochemistry of polyphase gneisses from Schirmacher Oasis, East Antarctica. *Journal of the Geological Society of India*, *47*, 303–312.

Indares, A., & Martignole, J. (1985). Biotite-garnet thermometry in the granulite facies: The influence of Ti and Al in biotite. *American Journal of Mineralogy*, *70*, 272–278.

Ingole, B., & Dhargalkar, V. (1998). Eco-biological assessment of a freshwater lake at Schirmacher Oasis, East Antarctica, with reference to human activities. *Current Science*, *74*(6), 529–534.

Ingram, R. L. (1971). Sieve analysis. In R. E. Carver (Ed.), *Procedures in Sedimentary Petrology* (pp. 49–68). Wilson Interscience.

International Hydrographic Organization (IHO). (n.d.). https://web.archive.org/web/20140724100423/http://www.iho.int/srv1/index.php?option=com_content&view=article&id=299&Itemid=289

ISRO's Participation in Antarctic Expedition. (n.d.). https://www.isro.gov.in/isros-participation-antarctic-expedition

Itolikar, A. B. (2016). *Microwave Dielectric Properties of Vegetation at C-Band* [Doctoral thesis, Dr Babasaheb Ambedkar Marathwada UniversityAurangabad]. Shodhganga: a reservoir of Indian theses at INFLIBNET.

Itolikar, A. B., Joshi, A.S., Deshpande, S. S., Arole, V. M. & Kurtadikar, M. L. (2020). Dielectric and Emissive Properties of Sorghum (Jowar) Vegetation at C-Band Microwave Frequency. *Materials Today: Proceedings, 23P2,* 236-245.

Itolikar, A. B., & Kurtadikar, M. L. (2017a). Microwave dielectric properties and emissivity estimation of freshly cut Banana leaves at 5GHz. *International Journal of Advances in Remote Sensing and GIS*, *5*(1), 58–66.

Itolikar, A. B., & Kurtadikar, M. L. (2017b). Microwave measurements of dielectric properties of corn vegetation at C-Band and comparison with Debye Cole–Dual Dispersion model. *Journal of Microwaves, Optoelectronics and Electromagnetic Applications*, *16*(4), 954–965. doi:10.1590/2179-10742017v16i41087

Jackson, M. L. (1979). *Soil Chemical Analysis-Advance Course* (2nd ed.). Department of Soil Science, Wisconsin University.

Jacobs, J., & Thomas, R. J. (1994). Oblique collision at about 1.1 Ga along the southern margin of the Kaapval continent, south-east Africa. *Geologische Rundschau*, *83*, 322–333.

Jafri, S. H., Moeen, S., Charan, S. N., Narayan, B. L., & Diwakar Rao, V. (2002). The occurrence of Quench olivine crystals in a basaltic dyke from Schirmacher Oasis, Queen Mound Land, East Antarctica. *Journal of the Geological Society of India*, *60*, 67–74.

Jayapaul, D., Kundu, A., Gaur, M. P., & Krishnamurthy, K. V. (2005). Petrology of A-Type hornblende granite and fayalite granite enclaves from cDML, East Antarctica: Some constraints on their origin. *Geological Society of India*, *65*, 562–576.

Jayaram, S. (2013). *Technical Report of 23rd Indian Scientific- Expedition to Antarctica—An overview.* Twenty Third Indian Antarctic Expedition, Technical Report No. 21, pp 11-20 published by Ministry of Earth Sciences, Government of India. http://14.139.119.23:8080/dspace/bitstream/123456789/908/1/pages+11-20.pdf

Jeong, G. Y., & Yeon, H. I. (2001). The origin of clay minerals in soils of King George Island, South Shetland Islands, West Antarctica, and its implication for the clay mineral composition of marine sediments. *Journal of Sedimentary Petrology*, *71*(5), 833–847. doi:10.1306/2DC4096C-0E47-11D7-8643000102C1865D

Jones-Couture, S. (2020). *Increasing our knowledge of Antarctica with hydrography Cooperation and expeditions.* IHO Hydrographic Commission on Antarctica. https://storymaps.arcgis.com/stories/e51256ee28504092accf031864ea65c2

Joshi, A. S. (2012). *Dielectric Properties of Pre-monsoon and Post monsoon Seawater at C-Band Microwave Frequency* [Unpublished doctoral thesis]. Dr Babasaheb Ambedkar Marathwada University, Aurangabad.

Joshi, A. S., & Kurtadikar, M. L.(2013). Study of Sea Water Permittivity Models and Laboratory and Validation at 5GHz. *Journal of Geomatics SAC-ISRO-The Indian society of Geomatics, 7*(1), 33-40.

Jouzel, J., Lorius, C., Petit, J. R., Genthon, C., Barkov, N. I., Kotlyakov, V. M., & Petrov, V. M. (1987). Vostok ice core: A continuous isotope temperature record over the last climatic cycle (160,000 yrs). *Nature*, *329*(6138), 1–6. doi:10.1038/329403a0

Kappen, L., & Straka, H. (1988). Pollen and spores Transport into the Antarctic. *Polar Biology*, *8*(3), 173–180. doi:10.1007/BF00443450

Kaul, M. K., Chakraborty, S. K., & Raina, V. K. (1985). Petrography of the biotite-hornblende bearing quartzo-feldspathic gneisses from Dakshin Gangotri, Schirmacher Hill, Antarctica. *Scientific Report, Department of Ocean Development. Technical Publication*, *2*, 29–38.

Keller, W. D. (1970). Environmental aspects of clay minerals. *Journal of Sedimentary Petrology*, *40*, 788–813. doi:10.1306/74D720A4-2B21-11D7-8648000102C1865D

Kersting, A. B., & Arculus, R. J. (1994). Klyuchevskoy volcano, Kamchatka, Russia: The role of high-flux recharged, tapped, and fractionated magma chamber(s) in the genesis of high Al_2O_3 from high-MgO basalt. *Journal of Petrology*, *35*(1), 1–41. doi:10.1093/petrology/35.1.1

Khare, N., Chaturvedi, S. K., Saraswat, R., Srivastava, R., Raina, R., & Wanganeo, A. (2008). Some morphometric characteristics of Priyadarshini water body at Schirmacher Oasis, Central Dronning Maud Land with special reference to its bathymetry. *Indian Journal of Geo-Marine Sciences*, *37*(4), 435–438.

Kohn, M. J., & Spear, F. S. (1990). Two new geobarometers for garnet amphibolites with application to southeastern Vermont. *American Journal of Mineralogy*, *75*, 89–96.

Koide, M., Michel, R., Goldberg, E. D., Herron, M. M., & Langway, C. C. Jr. (1982). Characterization of radioactive Fallout from pre-and post-moratorium tests to polar ice caps. *Nature*, *296*(5857), 544–547. doi:10.1038/296544a0

Kotoky, P., Bezbarruah, D., Baruah, J., Borah, G. C., & Sharma, J. N. (2006). Characterisation of clay minerals in the Brahmaputra river sediments, Assam, India. *Current Science*, *91*(9), 1247–1250.

Kreemer, C., Holt, W. E., & Haines, A. J. (2003). An integrated global model of present-day plate motions and plate boundary deformation. *Geophysical Journal International*, *154*(1), 8–34. doi:10.1046/j.1365-246X.2003.01917.x

Kristoffersen, Y., Strand, K., Vorren, T., Harwood, D., & Webb, P. (2000). Pilot shallow drilling on the continental shelf, Dronning Maud Land, Antarctica. *Journal of Antarctic Science*, *4*(4), 463–470. doi:10.1017/S0954102000000547

Krogg, E. J. (1988). The garnet-clinopyroxene Fe-Mg geothermometer-reinterpretation of existing experimental data. *Contributions to Mineralogy and Petrology*, *99*(1), 44–48. doi:10.1007/BF00399364

Kulkarni, P. G. (2006). *Study of dielectric properties of different soil texture at C-Band microwave frequency* [Doctoral thesis, Dr Babasaheb Ambedkar Marathwada UniversityAurangabad]. Shodhganga: a reservoir of Indian theses at INFLIBNET.

Kumar, R. (1986). Note on the P-T Conditions of metamorphism of Schirmacher Range, East Antarctica. Sci. Rep. In *3rd Ind. Sci. Exp. Antarctica*. DOD, Govt. of Ind. Pub. Technical Publication.

Kumar, A., Charan, S. N., Gopalan, K., & Mac Dougal, J. D. (1998). A long-lived enriched mantle source for two Proterozoic carbonatite complexes from Tamil Nadu, southern India. *Geochimica et Cosmochimica Acta*, *62*(3), 515–524. doi:10.1016/S0016-7037(97)00341-4

Kurtadikar, M. L., Popalghat, S. K., & Mehrotra, S. C. (2013). *Laboratory Validation of Dielectric Properties of Earth Resources* [Paper Presentation]. International Expert Meeting on Microwave Remote Sensing, Ahmadabad, India.

Kurtadikar, M. L., Popalghat, S. K., Kulkarni, P. G., & Khare, N. (2009). Indian. *Journal of Geosciences (Prague)*, *63*(2), 241.

Kutschera, W., Paul, M., Smithers, R.K., Stephenson, E. J., Yntema, J. L., Alburger, D. E., Cumming, J. B., & Harbottle, G. (1980). Measurement of the ^{32}Si half-life via accelerator mass spectrometry. *Physics Review Letters, 45*(8), 592.

Kutzbach, J. E. (1985). Modelling of palaeoclimates. *Advances in Geophysics*, *28A*, 159–196. doi:10.1016/S0065-2687(08)60223-X

Lal, D., Nijampurkar, V. N., Rajagopalan, G., & Somayajulu, B. L. K. (1979). Annual Fallout of 32Si, 2l0Pb, 22Na, 35S, and 7Be in rains in India. Proceedings Indian Academy of Science, 88(1), 29.

Lal, M. (1986). Sedimentology of the glacial sand and lake terraces sediments from Schirmacher Oasis and sea bed sediment of Princess Astrid Coast, Queen Maud land Antarctica. Sci. Rep. In *3rd Ind. Sci. Exp. Antarctica.* DOD, Govt. of India Publ. Technical Publication.

Lal, D. (1987). ^{10}Be in polar ice: Data reflect changes in cosmic ray flux or polar meteorology. *Geophysical Research Letters*, *14*(8), 785–788. doi:10.1029/GL014i008p00785

Lal, D., & Peters, B. (1967). Cosmic ray produced radioactivity on the Earth. In *Handbuch der Physik, 46* (pp. 551–612). Springer-Verlag. doi:10.1007/978-3-642-46079-1_7

Lal, R. K. (1993). Internally consistent recalibrations of mineral equilibria for geothermobarometry involving garnet-orthopyroxene-plagioclase-quartz assemblages and their application to the South Indian granulites. *Journal of Meteorology and Geology*, *11*(6), 855–866. doi:10.1111/j.1525-1314.1993.tb00195.x

Lamb, W. M., Brown, P. E., & Valley, J. W. (1991). Fluid inclusions in Adirondack granulites: Implications for retrograde P-T path. *Contributions to Mineralogy and Petrology*, *107*(4), 472–483. doi:10.1007/BF00310681

Larson, K. M., Burgmann, R., Bilham, R., & Freymueller, J. T. (1999). Kinematics of the India-Eurasia collision zone from GPS measurements. *Journal of Geophysical Research*, *104*(B1), 1077–1093. doi:10.1029/1998JB900043

Lee, H. Y., & Ganguly, J. (1988). Equilibrium compositions of coexisting garnet and orthopyroxene: Experimental determination in the system FeO-MgO-Al_2O_3-SiO_2 and applications. *Journal of Petrology*, *29*(1), 93–111. doi:10.1093/petrology/29.1.93

Lieutenant-Commander, J. B., & Dixon, R. N. (1965). *Hydrographic Surveying in the Antarctic.* https://journals.lib.unb.ca › ihr › article › download

Lindholm, R. C. (1987). *A Practical Approach to Sedimentology*. Allen and Unwin Publ. doi:10.1007/978-94-011-7683-5

Liptey, G. (1973). *Atlas of Thermoanalytical Curves*. Akademiai Kiado.

Malaimani, E.C., & Ravi Kumar, N. (2006). *Recent results on Indian plate kinematics with the longer time span of GPS measurements in the global network solution*. EOS Special Volume, 53 B-0893.

Malaimani, E. C., Campbell, J., Görres, B., Kotthoff, H., & Smaritschnik, S. (2000). Indian plate kinematics studies by GPS-geodesy. *Earth, Planets, and Space*, *52*(10), 741–745. doi:10.1186/BF03352275

Malaimani, E. C., Ravikumar, N., Akilan, A., & Abilash, K. (2007). GPS-Geodesy with GNSS Receivers for revalidation of Indian Plate Kinematics with recent plate velocities. *Proceedings of Conference on Global Navigation Satellite System and Its Applications.*

Malaimani, E.C., & Ravi Kumar, N., Padhy, S., Rao, S.V.R.R., Navin Chander, G.B., Srinivas, G.S., & Akilan, A. (2008). Continuous monitoring of seismicity by Indian permanent seismological observatory at Maitri. *Indian Journal of Geo-Marine Sciences*, *37*, 396–403.

Manabe, S., & Broccoli, A. J. (1985). The influence of continental ice sheets on the climate of the ice age. *Journal of Geophysical Research*, *90*(D1), 2167–2190. doi:10.1029/JD090iD01p02167

Mange, M. A., & Maurer, H. F. W. (1992). *Heavy Minerals in Colour*. Chapman and Hall. doi:10.1007/978-94-011-2308-2

Mathur, A. K., Asthana, R., & Ravindra, R. (2006). Arcellaceans (thecamoebians) from core sediments of Priyadarshini Lake, Schirmacher Oasis, East Antarctica. *Current Science*, *90*(12), 1603–1605.

Mathur, A. K., Asthana, R., & Ravindra, R. (2006). Arcellaceans (thecamoebians) from core sediments of Priyadarshini Lake, Schirmacher Oasis, Eastern Antarctica. *Current Science*, *90*(12).

Matzler, C., & Wagmuller, U. (2000). Dielectric properties of fresh-water ice at microwave frequencies. *Journal of Applied Physics*, *20*, 1623.

Ministry of Earth Sciences Technical Publication. (2008) *Dielectric Properties of Antarctic Ices and Soils at 5 GHz Frequency* (Technical Publication No. 20).

Mital, G. S., & Verma, S. K. (1991). *Helicopter-borne magnetic survey around the Indian station in Antarctica.* Session GAM 5.14, XX General Assembly IUGG, Vienna.

Moiola, R. J., & Weiser, D. (1968). Textural parameters: An evaluation. *Journal of Sedimentary Petrology*, *38*(1), 45–53.

Molar, P., & Lyon-Caen, H. (1989). Fault plane solutions of earthquakes and active tectonics of the Tibetan plateau and its margins. *Geophysical Journal International*, *99*(1), 123–153. doi:10.1111/j.1365-246X.1989.tb02020.x

Molnar, A. (1990). review of the seismicity and the rates of active undertrusting and deformation at the Himalaya. *J. of Himalayan Geology*, *1990*, 131–154.

Moran, M. L., Greenfield, R. J., & Arcone, S. A. (2003). Modelling GPR radiation and reflection characteristics for a complex temperate glacier bed. *Geophysics*, *68*(2), 559–565. doi:10.1190/1.1567225

Morra, V., Secchi, F. A. G., Melluso, L., & Franciosi, L. (1997). High-Mg subduction-related Tertiary basalts in Sardinia, Italy. *Lithosphere*, *40*, 69–71.

Murugkar, A. G. (2002). *Dielectric Properties of Seawater* [Doctoral thesis, Dr Babasaheb Ambedkar Marathwada UniversityAurangabad]. Shodhganga: a reservoir of Indian theses at INFLIBNET.

Nairn, R. (2010). *The Challenge of Hydrographic Surveying and Charting the Antarctic.* FIG Congress 2010 Facing the Challenges – Building the Capacity, Sydney, Australia. https://www.icsm.gov.au/sites/default/files/CongressPaper4531.pdf

Navarkhele, V. V. (1996). *Study of dielectric properties of liquids of geophysical interest* [Unpublished doctoral thesis]. Dr Babasaheb Ambedkar Marathwada University, Aurangabad.

Navarkhele, V., Agrawal, R., & Kurtadikar, M. L. (1998). Dielectric properties of electrolytic solutions. *Pramana*, *51*(3-4), 511–518. doi:10.1007/BF02828943

Nayaka, S., & Upreti, D. K. (2005). Schirmacher Oasis, East Antarctica, a lichenologically interesting region. *Current Science*, *89*(7), 1069–1071.

Newton, R. C., & Perkins, D. III. (1982). Thermodynamics calibration of geobarometers based on the assemblage garnet-plagioclase-orthopyroxene (clinopyroxene)-quartz. *American Journal of Mineralogy*, *67*, 203–222.

Ni, J., & Barazangi, M. (1984). Seismotectonics of the Himalayan collision zone: Geometry of underthrusting Indian plate benath the Himalaya. *Journal of Geophysical Research*, *1989*(B2), 1147–1163. doi:10.1029/JB089iB02p01147

Nijampurkar, V. N., & Rao, D. K. (1993a). Ice dynamics and climatic studies on Himalayan glaciers based on stable and radioactive isotopes. *Snow and Glacier Hydrology, 218*, 355–369.

Nijampurkar, V. N., Bhandari, N., Vohra C P & Krishnan, V. (1982). Radiometric chronology of Nehnar glacier, Kashmir. *Journal of Glaciology, 28*(98), 91.

Nijampurkar, V. N., Bhandari, N., Bhattacharya, S. K., Rao, D. K., Sengupta, D., Raina, V. K., & Kaul, M. K. (1985). *Isotopic and T.L. studies of Antarctic ice Samples.* Scientific Report of 2nd Indian Expedition to Antarctica, Technical *Publication* 2.

Nijampurkar, V. N., & Rao, D. K. (1993b). Polar Fallout of radionuclides ^{32}Si, ^{7}Be and ^{210}Pb and past accumulation rate of ice at Indian station Dakshin Gangotri, east Antarctica. *Journal of Environmental Radioactivity, 21*(2), 107–117. doi:10.1016/0265-931X(93)90048-C

NIO-Annual Report. (1988-89). *Antarctic Lake Studies for Palaeoclimatic Interpretations.* Indian Institute of Technology, Kanpur.

Nishiizumi, K., Arnold, J. R., Elmore, D., Ma, K., Newman, D., & Gove, H. E. (1983). ^{36}CI and ^{53}Mn in Antarctic meteorites and ^{10}Be- ^{36}CI dating of Antarctic ice. *Earth and Planetary Science Letters, 62*(3), 407–417. doi:10.1016/0012-821X(83)90011-0

O' Gorman, J. V., & Walker, P. L. Jr. (1973). The thermal behaviour of mineral fractions separated from selected American coals. *Fuel, 52*(1), 71–79. doi:10.1016/0016-2361(73)90016-1

Ohta, H., Maruyama, S., Takahashi, E., Watanabe, Y., & Kato, Y. (1996). Field occurrence, geochemistry, and petrogenesis of the Archean Mid-Oceanic Ridge Basalts (AMORBs) of the Cleavenville area, Pilbara Craton, Western Australia. *Lithosphere, 37*, 199–221.

Ohzono, M., Tabei, T., Doi, K., Shibuya, K., & Sagiya, T. (2006). Crustal movement of Antarctica and Syowa Station based on GPS measurements. *Earth Planets and Space.* https://www.researchgate.net/publication/44258528_Crustal_movement_of_Antarctica_and_Syowa_Station_based_on_GPS_measurements doi:10.1186/BF03351984

Orheim, O., Gjessing, Y., Lunde, T., Repp, K., Wold, B., Clausen, H. B., & Liestol, O. (1986). Oxygen isotopes and accumulation rates at Riiser-Larsenisen, Antarctica. *Norsk. Polarinstitutt. Skr., 187*, 33–37.

Paech, H. J., & Stackebrandt, W. (1995). Geology. In P. Borman & D. Fritzsche (Eds.), *The Schirmacher Oasis, Queen Maud Land, East Antarctica, and its surroundings. Gotha-Perthes.*

Pal, R. K., & Shandilya, S. K. (2008). *Large Scale Mapping in Schirmacher Oasis in Antarctica.* XXII Indian Scientific Expedition to Antarctica – Technical Publication No. 20. http://14.139.119.23:8080/dspace/bitstream/123456789/859/1/cover-pages.pdf

Palanisamy, M. (2007). Snedra ulna (Nitzsch) Ehrenberg: A new generic record in Schirmacher Oasis, East Antarctica. *Current Science, 92*(2), 179–181.

Parthasarathy, G., Sharma, S. R., Ravindran, T. R., Arora, A. K., & Hussain, S. M. (2003). Structural and thermal studies of graphite from East Antarctica. *Geological Society of India, 61*, 335 343.

Passega, R. (1957). Textures as a characteristic of clastic deposition. *Am. Assoc. Petrol. Geol., 41*, 1952–1984.

Passega, R. (1964). Grain size representation by C-M Pattern as a geological tool. *Jour. Sed. Pet., 34*(4), 830–847. doi:10.1306/74D711A4-2B21-11D7-8648000102C1865D

Passega, R., & Byramjee, R. (1969). Grain size image of clastic deposits. *Sedimentology*, •••, 13.

Pattison, D. R. M., & Newton, R. C. (1989). Reversed experimental calibration of the garnet clinopyroxene Fe-Mg exchange thermometer. *Contributions to Mineralogy and Petrology*, *101*(1), 87–103. doi:10.1007/BF00387203

Pearce, J. A., Harris, N. B. W., & Tindle, A. G. (1984). Trace element discrimination diagrams for the tectonic interpretation of granitic rocks. *Journal of Petrology*, *25*(4), 956–983. doi:10.1093/petrology/25.4.956

Perchuk, L. L., & Lavrent'eva, I. V. (1983). Experimental investigation of exchange equilibria in the system cordierite-garnet-biotite. In S. K. Saxena (Ed.), *Kinetics and equilibrium in Mineral reactions* (pp. 199–239). Springer-Verlag. doi:10.1007/978-1-4612-5587-1_7

Perchuk, L. L., Podlesskii, K. K., & Aranovich, L. Y. A. (1991). Thermodynamics of some framework silicates and their equilibria: application to geothermometry. In L. L. Perchuk (Ed.), *Progress in metamorphic and magmatic petrology* (pp. 131–164). Cambridge University Press. doi:10.1017/CBO9780511564444.009

Perkins, D. III, & Chiper, A. (1985). Garnet-orthopyroxene-plagioclase-quartz barometry: Refinement and application to the English river subprovince and the Minnesota river valley. *Contributions to Mineralogy and Petrology*, *89*(1), 69–80. doi:10.1007/BF01177592

Petschick, R., Khun, G., & Gingele, F. (1996). Clay minerals distribution in surface sediments of the South Atlantic: Sources, transport, and relation to oceanography. *Marine Geology*, *130*, 203–229.

Petterson, G., Renberg, I., Geladi, P., Lindberg, A., & Lindgren, F. (1993). Spatial uniformity of sediment accumulation in varved lake sediments in Northern Sweden. *Journal of Paleolimnology*, *9*(3), 195–208. doi:10.1007/BF00677213

Pettijohn, F. J., Potter, P. E., & Siever, R. (1973). *Sand and Sandstone*. Springer-Verlag. doi:10.1007/978-1-4615-9974-6

Philibert, J. (1963). *X-ray optics and X-ray Microanalysis*. Academic Press.

Picciotto, E., Cameron, R., Crozaz, G., Deutsch, S., & Wilgain, S. (1968). Determination of the rate of snow accumulation at the pole of relative inaccessibility, Eastern Antarctica: A comparison of glaciological and isotopic methods. *Journal of Glaciology*, *7*(50), 273.

Picciotto, E., Crozaz, G., & De Breuk, W. (1964). Rate of accumulation of snow at the south pole as determined by radioactive measurements. *Nature*, *203*(4943), 393–394. doi:10.1038/203393a0

Powell, M., & Powell, R. (1977). Plagioclase-alkali feldspar geothermometry revisited. *Mineral Magnetics*, *41*(318), 253–256. doi:10.1180/minmag.1977.041.318.13

Price, J. C. (1985). Ideal site mixing in solid solution with an implication to two-feldspar geothermometry. *American Journal of Mineralogy*, *70*, 696–701.

Provost, A. (1990). An improved diagram for isotope data. *Chemical Geology*, *80*, 85–89.

Raisbeck, G. M., & Yiou, F. (1985). ^{10}Be in polar ice and atmospheres. *Annals of Glaciology*, *7*, 138–140. doi:10.3189/S0260305500006054

Rama Rao, P., Ramana Rao, A. V., & Poornachandra Rao, G. V. S. (1995). Geological, geochemical, geochronological, and palaeomagnetic studies on the rocks from parts of East Antarctica. In *Sci. Rep., 8th Ind. Sci. Exp.* Antarctica. DOD, Govt. of India Publ. Technical Publication.

Rama Rao, P., Ramana Rao, A. V., & Poornachandra Rao, G. V. S. (1999). Geological, geochemical and geochronological and palaeomagnetic studies on the rocks from parts of East Antarctica. *Scientific Report. Department of Ocean Development Technical Publication*, *8*, 97–106.

Ramesh, R., Bhattacharya, S. K., & Pant, G. B. (1989). Climatic significance of δD variations in tropical tree species from India. *Nature*, *337*(6203), 149–150. doi:10.1038/337149a0

Rameshwar Rao, D., Sharma, R., & Gururajan, N. S. (1996). *Mineral Chemistry, Fluid Inclusion Studies and P-T path of Gneiss-Granulite Rocks of Schirmacher Region, East Antarctica.* Twelfth Indian Expedition to Antarctica, Scientific Report, Department of Ocean Development, Technical Publication.

Rameshwar Rao, D., Sharma, R., & Gururajan, N. S. (1997). Mafic granulites of the Schirmacher region, East Antarctica: Fluid inclusion and geothermobarometric studies focussing on the Proterozoic evolution of the crust. *Transactions of the Royal Society of Edinburgh*, *88*(1), 1–17. doi:10.1017/S0263593300002285

Rameshwar Rao, D., Sharma, R., & Gururajan, N. S. (1998). Geothermobarometry and fluid inclusion studies of leucogneiss from Schirmacher region, East Antarctica. *Journal of the Geological Society of India*, *51*, 595–607.

Rameshwar, R. (1995). BGT - the macros driven spreadsheet program for biotite-garnet thermometry. *Computers & Geosciences*, *21*(4), 593–604. doi:10.1016/0098-3004(94)00098-F

Rameshwar, R. (1996). AMPH - a program for calculating formulas and for assigning names to the amphibole group of minerals. *Computers & Geosciences*, *22*(8), 931–933. doi:10.1016/S0098-3004(96)00018-0

Rameshwar, R. (2000). Metamorphic evolution of charnockites and felsic gneisses from the Schirmacher region, East Antarctica. *Gondwana Research*, *3*(1), 79–89. doi:10.1016/S1342-937X(05)70059-9

Rameshwar, R. (2000). Origin of high magnesian metatholeiites of the Schirmacher region, East Antarctica: Constraints from trace elements and Nd-Sr isotopic systematics. *Gondwana Research*, *3*(1), 91–104. doi:10.1016/S1342-937X(05)70060-5

Rameshwar, R. D., & Subba Rao, T. V. (1994). AKP: A two-feldspar geothermometry using Lotus 1-2-3 software. *Himalayan Geology*, *5*, 93–101.

Rameshwar, R. D., Vijian, A. R., & Prabhu, B. N. (1995). Geochemical and Rb-Sr isotope studies of the felsic rocks from the Schirmacher region, East Antarctica. *Geosciences Journal, 20*, 41–59.

Rameshwar, Rao, D. (2006). The evolutionary history of the Schirmacher region, East Antarctica. *Himalayan Geology, 27*, 81–94.

Ravi Kumar, N., Malaimani, E.C., Prem Kishore, L., Chaitanya, T., Akilan, A., Srinivas, G.S., Navin Chander, G.B., & Rao, S.V.R.R. (2004). *Significance and importance of collocating permanent GPS and seismic observatory at Maitri, Antarctica.* Terra Nostra, Schriften der Alfred- Wegner- Stiftung 2004/4.

Ravi Kumar, N., Malaimani, E.C., Rao, S.V.R.R., Akilan, A., & Abilash, K. (2007). Geodetic tying of Antarctica and India by continuous GPS measurements for strain accumulation studies. *Proceedings of Conference on Global Navigation Satellite System and Its Applications.*

Ravi Kumar, N., & Malaimani, E. C. (2008). 10 years of Continuous GPS measurements for Geodetic tying of Antarctica and India for geodynamical and strain accumulation studies in the south of Indian Peninsula. *Journal of Indian Geophysical Union, 12*, 141–148.

Ravi Kumar, N., Malaimani, E. C., & Akilan, A. (2006). Geodynamical Processes between Antarctica and India as revealed by very long base lines between the continents estimated from continuous and long-term GPS measurements. *EOS Special Volume G, 53*, B-0906.

Ravi Kumar, N., Malaimani, E. C., & Akilan, A. (2008b). Strain accumulation studies between Antarctica and India through geodetic tying of two continents from continuous and long-term GPS measurements. *Indian Journal of Geo-Marine Sciences, 37*, 404–411.

Ravich, M.G. & Kamenev, E. N. (1975). *Crystalline basement of the Antarctic Platform.* Academic Press.

Ravichandran, V. (2000). *Scientific Activities Report of Survey of India during XVII IAE 1997-98.* XVII Indian Scientific Expedition to Antarctica – Technical Publication No. 15, 145-154 published by Department of Ocean Development, Government of India. http://14.139.119.23:8080/dspace/bitstream/123456789/499/3/17TH.pdf

Ravikant, V. (2005). Metamorphism of Ultramafic and mafic enclaves within granulites, Schirmacher Oasis, East Antarctica. *Journal of the Geological Society of India, 65*, 279–290.

Ravikant, V., Bhaskar Rao, Y. J., & Gopalan, K. (2004). Schirmacher Oasis as an extension of the Neoproterozoic East African Orogen into Antarctica: New Sm-Nd Isochron Age Constraints. *The Journal of Geology, 112*(5), 607–616. doi:10.1086/422669

Ravikant, V., & Kundu, A. (1998). Reaction textures of retrograde pressure-temperature deformation paths from granulites of Schirmacher Hills, East Antarctica. *Journal of the Geological Society of India, 51*, 305–314.

Ravindra, R. (2001). Geomorphology of Schirmacher Oasis, East Antarctica. In *Proc. Symp. On snow, Ice and Glaciers.* Geological Survey of India, Spl. Publication. No. 53, 379-390.

Ravindra, R. (2001). Geomorphology of Schirmacher Oasis, *East Antarctica. Proceedings. Symposium on Snow, Ice and Glaciers, Geological Survey of India*, *53*, 379-390.

Ravindra, R., Chaturvedi, A., & Beg, M. J. (2000). Meltwater lakes of Schirmacher Oasis – Their genetic Aspects and classification. In D. Sahoo & P. C. Pandey (Eds.), *Advances in Marine and Antarctic Sciences* (pp. 301–314). APH Publishing Corporation.

Ravindra, R., Chaturvedi, A., & Beg, M. J. (2002). Melt water lakes of Schirmacher Oasis - their genetic aspects and classification. In D. Sahoo & P. C. Pandey (Eds.), *Advances in Marine and Antarctic Science* (pp. 301–313). APH Publishing Corporation.

Rawat, S. S., & Mehta, S. K. (2000). *A report on surveying in Antarctica during 16th Indian Scientific Expedition to Antarctica*. Scientific Report XVI Indian Expedition to Antarctica – Technical Publication No. 14, 335-340 published by Department of Ocean Development, Government of India. http://14.139.119.23:8080/dspace/bitstream/123456789/513/3/INTRODUCTION.pdf

Reinick, H. E., & Singh, I. B. (1980). *Depositional Sedimentary Environments*. Springer-Verlag. doi:10.1007/978-3-642-81498-3

Richardson, C., Aarhalt, E., Hamran, S. E., Halmlund, P., & Isaksson, E. (1997). Spatial distribution of snow in west Dronning Maud Land, East Antarctica, mapped by a ground-based snow radar. *Journal of Geophysical Research*, *120*(No. B9), 20243–20359.

Ringwood, A. E. (1974). Petrological evolution of island arc systems. *Journal of the Geological Society*, *130*(3), 183–204. doi:10.1144/gsjgs.130.3.0183

Robert, C., & Maillot, H. (1990). Palaeoenvironments in the Weddel Sea area and Antarctic climates, as deduced from clay mineral associations and geochemical data ODP Leg 11. Proceedings. ODP, Science Research, 113, 51-70.

Robert, C., & Kennett, J. P. (1994). An Antarctic humid episode of Palaeocene- Eocene boundary: Clay mineral evidence. *Geology*, *22*(3), 211–214. doi:10.1130/0091-7613(1994)022<0211:AS HEAT>2.3.CO;2

Rogers, J. J. W. (1996). A history of the continents in the past three billion years. *The Journal of Geology*, *104*(1), 91–107. doi:10.1086/629803

Santosh, M. (1986). Carbonic metamorphism of charnockites in the southwestern Indian Shield: A fluid inclusion study. *Lithosphere*, *19*, 1–10.

Sarin, M. M., Bhushan, R., Rengarajan, R., & Yadav, D. N. (1992). Simultaneous determination of ^{238}U series nuclides in the water of Arabian Sea and Bay of Bengal. *Indian Journal of Geo-Marine Sciences*, *21*, 121–127.

Saunders, A. D., Tarney, J., & Weaver, S. D. (1980). Transverse geochemical variations across the Antarctic Peninsula: Implications for the genesis of calc-alkaline magmas. *Earth and Planetary Science Letters*, *46*(3), 344–360. doi:10.1016/0012-821X(80)90050-3

Scott, L., & Van Zinderen Bakker, E. M. Sr. (1985). Exotic pollen and long-distance wind dispersal at a sub-Antarctic Island. *Grana*, *24*(1), 45–54. doi:10.1080/00173138509427422

Seeber, G. (1993). *Text book on Sattlite Geodesy*. Walter de Gruyter & Co.

Seleshi, Y., Demaree, G. R., & Delleur, J. W. (1994). Sunspot numbers as a possible indicator of annual rainfall at Addis Ababa, Ethiopia. *International Journal of Climatology*, *14*(8), 911–923. doi:10.1002/joc.3370140807

Sengupta, S. (1986). *Geology of Schirmacher Range (Dakshin Gangotri), East Antarctica*. Department of Ocean Development, Technical Publication No. 3, 187-217.

Sengupta, S. (1986). Geology of Schirmacher range (Dakshin Gangotri), East Antarctica. Sci. Rep. 3rd Ind. In *Sci. Exp. Antarctica*. DOD, Govt. of Ind. Pub. Technical Publication.

Sengupta, S. (1994). *Mylonites of the Schirmacher Hills, East Antarctica*. Department of Ocean Development, Technical Publication No. 6, 173- 187.

Sengupta, S. M. (1996). Introduction to Sedimentology. Oxford & IBH Publishing Co-Pvt. Ltd.

Sengupta, P. K., Dasgupta, S., Bhattacharaya, P. K., & Hariya, Y. (1989). Mixing behaviour in quaternary garnet solid solution and an extended Ellis and Green garnet-clinopyroxene geothermometer. *Contributions to Mineralogy and Petrology*, *103*(2), 223–227. doi:10.1007/BF00378508

Sengupta, S. (1986). *Geology of Schirmacher Range (Dakshin Gangotri), East Antarctica*. Department of Ocean Development. *Technical Publication*, *3*, 187–217.

Sengupta, S. (1988). History of successive deformation in relation to metamorphic-migmatitic events in the Schirmacher Hills, Queen Maud Land, East Antarctica. *Journal of the Geological Society of India*, *32*, 295–314.

Sengupta, S. (1988). History of successive deformation in relation to metamorphism-migmatitic events in the Schirmacher Hills, Queen Maud Land, East Antarctica. *Journal of the Geological Society of India*, *32*, 295–31.

Sengupta, S. (1988a). History of successive deformations in relation to metamorphism-migmatitic events in the Schirmacher Hills, Queen Maud Land East Antarctica. *Journal of the Geological Society of India*, *32*, 295–319.

Sengupta, S. (1988b). Precambrian rocks of the Schirmacher Range, East Antarctica, Z. Geol. *Wiss. Berlin*, *16*, 647–660.

Sengupta, S. (1993). Tectono-thermal history recorded in mafic dykes and enclaves of the gneissic basement in the Schirmacher Hills, East Antarctica. *Precambrian Research*, *63*(3-4), 273–292. doi:10.1016/0301-9268(93)90037-3

Sengupta, S. (1997). Contrasting fabrics in deformed dykes and host rocks: natural examples and a simplified model. In S. Sengupta (Ed.), *Evolution of Geological Structures in micro to macro-scales* (pp. 293–319). Chapman and Hall. doi:10.1007/978-94-011-5870-1_18

Sengupta, S., & Bose, S. (1997). Development of successive generations of shear zones in the Precambrian Basement of Schirmacher Hills, East Antarctica. *Gondwana Research, 1*(1), 35–45. doi:10.1016/S1342-937X(05)70004-6

Sen, S. K., & Bhattacharya, A. (1984). An orthopyroxene-garnet thermometer and its application to the Madras charnockites. *Contributions to Mineralogy and Petrology, 88*(1-2), 64–71. doi:10.1007/BF00371412

Setti, M., Marinoni, L., & López-Galindo, A. (2001). Crystal chemistry of smectite in sediments of CRP-3 drill core (Victoria Land basin, Antarctica): Preliminary results. *Terra Antarctica, 8*(4), 543–550.

Sharma, C., Bera, S. K., & Upreti, D. K. (2002). Modern pollen-spore rain in Schirmacher Oasis, East Antarctica. *Current Science, 82*(1), 88–91.

Sharma, C., Chauhan, M. S., & Sionha, R. (2007). Studies on Holocene climatic changes from Priyadarshini Lake sediments, Schirmacher Oasis East Antarctica: The Palynological Evidence. *Geological Society of India, 69*, 92–96.

Sharma, R., & Rameshwar, R. D. (2006). Fluid evolution in the granulite-amphibolite metamorphism of the Schirmacher region, East Antarctica. *Indian Journal of Geochemistry, 21*, 103–117.

Shen, Z.-K., Zhao, C., Yin, A., Li, Y., Jackson, D. D., Fang, P., & Dong, D. (2000). Contemporary crustal deformation in east Asia constrained by Global Positioning measurements. *Journal of Geophysical Research, 105*(B3), 5721–5734. doi:10.1029/1999JB900391

Shrivastava, P. K., Roy, S. K., Srivastava, H. B., & Dharwadkar, A. (2019). Estimation of Paleo-ice Sheet Thickness and Evolution of Landforms in Schirmacher Oasis and Adjoining Area, cDML, East Antarctica. *Journal of the Geological Society of India, 93*(6), 638–644. doi:10.100712594-019-1242-5

Siegert, M. J., Hindmarsh, R. C. A., & Hamilton, G. S. (2003). Evidence for a large surface ablation zone in Central East Antarctica during last ice age. *Quaternary Research, 59*(1), 114–121. doi:10.1016/S0033-5894(02)00014-5

Simmonds, I. (1981). The effect of sea ice on a general circulation model of the Southern Hemisphere. *International Association of Hydrological Sciences Publication, 131*, 193–206.

Singer, A. (1984). The palaeoclimatic interpretation of clay minerals in sediments: A review. *Earth-Science Reviews, 21*(4), 251–293. doi:10.1016/0012-8252(84)90055-2

Singh, M., & Tariyal, J. S. (2012). *Report on the work done by Survey of India team during XXIV Indian Scientific Expedition to Antarctica*. XXIV Indian Scientific Expedition to Antarctica – Technical Publication No. 22, 199-206 published by National Centre for Antarctic and Ocean Research/Ministry of Earth Sciences, Government of India. http://14.139.119.23:8080/dspace/bitstream/123456789/941/1/RRContent.pdf

Singh, R. (1998). *Hydrographic Survey of Approaches to Indian Bay Antarctic Region. Fourteenth Indian Expedition to Antarctica, Scientific Report, 1998.* Department of Ocean Development, Technical Publication No. 12. http://14.139.119.23:8080/dspace/bitstream/123456789/420/3/ARTICLE+13.pdf

Singh, R. K. (1986). Geology of Dakshin Gangotri hill range, Antarctica. Sci. Rep. In *3rd Ind. Sci. Exp. Antarctica*. DOD, Govt. of Ind. Publ. Technical Publication.

Singh, S. M., Puja, G., & Bhat, D. J. (2006). Psychrophilic fungi from Schirmacher Oasis, East Antarctica. *Current Science*, *90*(10), 1388–1392.

Sinha, R., & Chatterji, A. (2000). Thermal structure, sedimentology, and hydrogeochemistry of lake Priyadarshini, Schirmacher Oasis, Antarctica. In *Sci. Rep., Ind. Sci. Exp. Antarctica*. DOD, Govt. of India Publ., Technical Publication.

Sinha, R., Sharma, C., & Chauhan, M.S. (2000b). Article. *Paleobotanist*, *49*, 1-8.

Sinha, R., & Chatterjee, A. (2000). Article. *Journal of the Geological Society of India*, *56*, 39–45.

Sinha, R., & Chatterjee, A. (2000). Lacustrine sedimentology in the Schirmacher Range Area, East Antarctica. *Journal of the Geological Society of India*, *56*(3), 9–45.

Sinha, R., Chatterjee, A., Panda, A. K., & Mitra, A. (1999). Article. *Current Science*, *76*, 680–683.

Sinisilo, A., Grinsted, A., Moore, J. C., Karkas, E., & Petterson, R. M. (2003). Snow-accumulation studies in Antarctica with ground-penetrating radar using 50, 100 and 800 MHz antenna frequencies. *Annals of Glaciology*, *37*, 194–198. doi:10.3189/172756403781815825

Smith, R. I. L. (1991). Exotic sporomorphs as indicators of potential immigrant colonists in Antarctica. In M Hjeimroos, S. Nilson, & G.E. Ghazaly (Eds.), *Proc. 4th International Conference of Aerobiology* (pp. 313-324). Grana.

Srivastava, A. K. (2007). Detailed sedimentological studies of the glaciogenic sediments of Schirmacher Oasis, Antarctica. In *Sci. Rep., 25th Ind. Sci. Exp*. Antarctica. DOD, Govt. of India Publ., Technical Publication.

Srivastava, A. K., Ingle, P. S., & Khare, N. (2010). Textural characteristics, distribution pattern, and provenance of heavy minerals in glacial sediments, Schirmacher Oasis, East Antarctica. *Journal of the Geological Society of India*, *73*, 393–402. doi:10.100712594-009-0019-7

Srivastava, A. K., Ingle, P. S., & Khare, N. (2018). Controlling factor for nature, Pattern, and accumulation of the glacial sediments of Schirmacher Oasis, East Antarctica: Comments on paleoclimatic condition. *Polar Science*, *18*, 113–122. doi:10.1016/j.polar.2018.05.004

Srivastava, A. K., Ingle, P. S., Lunge, H. S., & Khare, N. (2012). Grain-size characteristics of deposits derived from different glacigenic environments of the Schirmacher Oasis, East Antarctica. *Geologos, 18*(4), 251–266. doi:10.2478/v10118-012-0014-0

Srivastava, A. K., & Khare, N. (2009). Granulometric analysis of glacial sediments, Schirmacher Oasis, East Antarctica. *Journal of the Geological Society of India, 73*(5), 609–620. doi:10.100712594-009-0047-3

Srivastava, A. K., Khare, N., & Ingle, P. S. (2011). Characterisation of clay minerals in the sediments of Schirmacher Oasis, East Antarctica: Their origin and climatological implications. *Current Science, 100*(3), 363–372.

Stackebrandt, W., Kampf, H., & Wetzel, H. U. (1988). The geological setting of the Schirmacher Oasis, Queen Maud Land, East Antarctica. *Zeitschrift für Geologische Wissenschaften, 7*, 661–665.

Stål, T., Reading, A. M., Halpin, A., Phipps, S. J., & Whittaker, M. (2020). *The Antarctic Crust and Upper Mantle: A Flexible 3D Model and Software Framework for Interdisciplinary Research -Technology and Code Article*. Frontier Earth Science. doi:10.3389/feart.2020.577502

Steig, E. J., Polissar, P. J., Stuiver, M., Grootes, P. M., & Finkel, R. C. (1996). Large amplitude solar modulation cycles of ^{10}Be in Antarctica: Implication for atmospheric mixing processes and interpretation of the ice core record. *Geophysical Research Letters, 23*(5), 523–526. doi:10.1029/96GL00255

Sterne, R. S. M., & Bodnar, R. J. (1989). Synthetic fluid inclusion VII. Re-equilibration of fluid inclusions in the quartz during laboratory simulated burial and uplift. *Journal of Meteorology and Geology, 7*, 465–484.

Stormer, J. C. (1975). Practical two-feldspar geothermometers. *American Journal of Mineralogy, 60*, 667–674.

Stuiver, M., Denton, G. H., Hughes, T. J., & Fastook, J. L. (1981). History of the Marine Ice Sheet in West Antarctica during the last glaciation: A working hypothesis. In G. H. Denton & T. J. Hughes (Eds.), *The last great Ice Sheets* (pp. 319–439). Wiley & Sons.

Stuiver, M., Grootes, P. M., & Braziunas, T. F. (1995). The GISP2 δ^{18}O climate record of the past 16,500 years and the role of sun, ocean, and volcanoes. *Quaternary Research, 44*(3), 341–354. doi:10.1006/qres.1995.1079

Subrahmanya, K. R. (1996). Active intraplate deformation in south India. *Tectonophysics, 262*(1-4), 231–241. doi:10.1016/0040-1951(96)00005-4

Sundararajan, N., & Rao, M. (2005). A note on the petrophysical properties and geological interpretation in Schirmacher Oasis East Antarctica. *Geological. Society of India, 65*, 497–503.

Sun, S. S., & Mc Donough, W. F. (1989). Chemical and isotopic systematics of oceanic basalts: implications for mantle composition and processes. In A. D. Saunders & M. J. Norry (Eds.), *Magmatism in the ocean basins. Journal of Geological Society of London Special Publications*. doi:10.1144/GSL.SP.1989.042.01.19

Survey of India. (n.d.). *Annual Report 2011-12*. https://surveyofindia.gov.in/documents/annual-reports/annual-report-2011-12.pdf

Synal, H. A., Beer, J., Bonani, G., Sute, M., & Wolfli, W. (1990). Atmospheric transport of bomb produced ^{36}Cl. *Nuclear Instruments and Methods*, *B52*(3-4), 483–488. doi:10.1016/0168-583X(90)90462-4

Thomas, R. J. (1989). The petrogenesis of the Mzumbe Gneiss suite is a tonalite-trondhjemite orthogneiss suite from the southern part of the Natal structural and metamorphic province. *South African Journal of Geology*, *92*, 322–338.

Thompson, A. B. (1976). The mineral reaction in pelitic rocks. Calculations of some P-T-x (Fe-Mg) phase relations. *American Journal of Mineralogy*, *276*(4), 425–454. doi:10.2475/ajs.276.4.425

Thorpe, R. S. (1982). *Andesites: orogenic andesites and related rocks*. Wiley.

Tingey, R. J. (1991). The Geology of Antarctica. Clarendon Press.

Travassos, J. M., & Simoes, J. C. (2004). High-resolution radar mapping of internal layers of a subpolar ice cap, King George Island, Antarctica, *Pesquisa Antarctica Brasileira, 4*, 57-65.

Van der Knaap, W. O., & Jacqueline, F. N. (2019). A recent pollen diagram from Antarctica (King George Island, South Shetland Islands). *The Holocene*, *3*(2), 169–173. doi:10.1177/095968369300300209

Verma, S. K. (1988). Preliminary evidences substantiating the paleorelationship of Dronning Maud Land, Antarctica and Mozambique region, Africa. *Proc. Workshop on Antarctic Studies.*

Verma, S. K., & Mital, G. S. (1988). Geophysical investigations on the ice-shelf around Dakshin Gangotri, Antarctica. *Proc. Workshop on Antarctic Studies.*

Verma, S. K., & Mital, G. S. (1988). Geophysical studies in Antarctica with particular emphasis on the work in Dronning Maud Land. *Proc. Workshop on Antarctic Studies.*

Verma, S. K., & Mital, G. S. (1989). Helicopter-borne Magnetic Survey Between the Schirmacher Oasis and the Wohlthat Mountains, Dronning Maud Land, Antarctica. Detailed Report Submitted to the Dept. of Ocean Development.

Verma, S. K., Mital, G. S., & Dayal, A. M. (1987c). *K-Ar dating of some rocks from the Schirmarcher Oasis, Dronning Maud Land, East Antarctica*. Tech. Pub. no. 4 on Fourth Indian Scientific Expedition to Antarctica, Scientific Report, DOD.

Verma, S. K., Mital, G. S., Dixit, M. M., Reddy, K. N. S., Venkatarayudu, M., & Pantulu, K. P. (1987b). *Seismic Investigation on the ice-shelf near Dakshin Gangotri Antarctica*. Tech. Pub. no.4 on Fourth Indian Scientific expedition to Antarctica, Scientific Report, DOD.

Verma, S. K., Mital, G. S., Rambabu, H. V., Venkatarayudu, M., & Srinatha Reddy, K. N. (1987a). *Magnetic Survey over the ice-shelf around the Indian permanent station in Antarctica*. Tech. Pub. no. 4 on Fourth Indian Expedition to Antarctica, Scientific Report, DOD.

Verma, S. K., Mital, G. S., Rao, G. V., Rangarajan, R., Reddy, K. S. N., & Venkatarayudu, M. (1987d). *Radioactivity measurements on some rock and water samples from Dakshin Gangotri, Antarctica*. Tech. Pub. no. 4 on 4th Indian Scientific expedition to Antarctica, Scientific Report, DOD.

Verma, S. K., Ram Babu, H. V., & Mital, G. S. (1988). Structure of the continental margin between Wohlthat range and Sor Rondane mountains, Antarctica. *Proc. Workshop on Antarctic Studies*.

Verma, S. K., Rambabu, H. V., & Mital, G. S. (1999). *Sub-glacial imaging in Schirmacher Oasis-Wohlthat Mountains region in Antarctica employing heli-magnetic and surface gravity surveys*. IUGG 99 Workshop, Birmingham, UK.

Verma, S. K., Singh, B., Malaimani, E.C., Tiwari, V.M., Prem Kishore, L., & Rao, M.B.S.V. (2004). *India's geoscientific achievements and future programs in Antarctica*. International Scientific Conference on Antarctica and the Southern Ocean in the Global System (XXVII SCAR) and the Annual Meeting of Council of Managers of National Antarctic Programs (COMNAP XVI), Bremen, Germany.

Verma, S. K. (1993). Document highlighting the ongoing and future programs for the Antarctic research by NGRI. DOD.

Verma, S. K., & Mital, G. S. (1994). Geophysical Surveys Around the Schirmacher Oasis, Antarctica. *Twentieth Annual Convention and Seminar of AEG on Exploration Geophysics at Dehra Dun*.

Visher, G. S. (1969). Grain size distributions and depositional processes. *Journal of Sedimentary Petrology*, *39*(3), 1074–1106.

Von Hippel, A. R. (1954). *Dielectrics and Waves*. Wiley.

Voughan, D. G., Corr, H. F. J., & Doake, C. S. M. (1999). Distortion of isochronous layers in ice revealed by ground-penetrating radar. *Nature*, *398*(6725), 323–326. doi:10.1038/18653

Wells, P. R. A. (1979). Chemical and thermal evolution of Archean Sialic crust, Southern West Greenland. *Journal of Petrology*, *20*(2), 187–226. doi:10.1093/petrology/20.2.187

Whitney, J. A., & Stormer, J. C. Jr. (1977). The distribution of $NaAlSi_3O_8$ between coexisting microcline and plagioclase and its effects on geothermometric calculations. *American Journal of Mineralogy*, *62*, 687–691.

Wikipedia. (n.d.). Antarctic Treaty System. In *Wikipedia*. Retrieved from https://en.wikipedia.org/wiki/Antarctic_Treaty_System#cite_note-3

Wilkins, S., Boulaire, Y., & Landct, F. (2013). *Different Approach to Surveying Antarctica -Operation Rock Bottom*. https://www.hydro-international.com/content/article/operation-rock-bottom

Windom, H. L. (1976). Lithogenous material in marine sediments. In Chemical Oceanography (pp. 103-113). Academic Press.

Wynne-Edwards, C. J. C. (1960). The Geographical Journal. *The Royal Geographical Society, 126*(3), 310-315. . doi:10.2307/1793634

Wynn–Williams, D. D. (1991). Aerobiology and colonisation in Antarctica-the BIOTAS program. *Grana, 30*(2), 380–393. doi:10.1080/00173139109431994

About the Contributors

Neloy Khare is presently Adviser/Scientist G to the Government of India at MoES has a very distinctive acumen not only of administration but also of quality science and research in his areas of expertise covering large spectrum of geographically distinct locations like Antarctic, Arctic, Southern Ocean, Bay of Bengal, Arabian Sea, Indian Ocean etc. Dr. Khare has almost 30 years of experience in the field of paleoclimate research using paleobiology Paleontology) / teaching / science management /administration / coordination for scientific programmes (including Indian Polar Programme) etc. Having completed his doctorate (Ph.D) on tropical marine region and Doctor of Science (D.Sc) on Southern High latitude marine regions towards environmental / climatic implications using various proxies including foraminifera (micro-fossil), have made significant contributions in the field of Paleoclimatology of Southern high latitude regions (Antarctic and Southern Ocean) using Micropaleontology as a tool. These studies coupled with palaeoclimatic reconstructions from tropical regions have helped understand causal linkages and teleconnections between the processes taking place in Southern high latitudes with that of climate variability occurring in tropical regions. Dr. Khare has been conferred Honorary Professor and Adjunct Professor by many Indian Universities. He has a very impressive list of publications to his credit including 7 Special Issues of National / International Scientific Journals as Guest Editor). His special issues include, Indian Journal of Marine Sciences (two issues), Indian Journal of Geosciences (one issue), Polar Science (Elsevier), Journal of Asian Earth Science (Elsevier), Quaternary International (Elsevier) and Frontiers in Marine Science as Managing / Guest Editor. He has Authored/ edited many books published by renowned publishing Houses with the International repute like Elsevier, Springer, Taylor and Francis (CRC Press), IGI Global etc., The government of India and many professional bodies have bestowed him with many prestigious Awards for his humble scientific contributions to Past climate changes/Oceanography / Polar Science and Southern Oceanography. The most coveted award is Rajiv Gandhi National Award -2013 conferred by Honourable President of India. Others include ISCA Young Scientist Award, Boyscast Fellowship, CIES French Fellowship, Krishnan Gold Medal, Best Scientist Award,

Eminent Scientist Award, ISCA Platinum Jubilee Lecture, IGU Fellowship, besides many. Dr. Khare has made tremendous efforts to popularize ocean science and polar science across the country by delivering many Invited lectures, radio talks and publishing popular science articles. Dr. Khare has sailed in the Arctic Ocean as a part of "Science PUB" in 2008 during the International Polar Year campaign for scientific exploration in the Arctic Ocean and became the first Indian to sail in the Arctic Ocean.

* * *

K. Abilash has been a scientist at National Geophysical Research Institute, Hyderabad.

Rajesh Asthana has participated in several Indian Scientific Expeditions to Antarctica. He also led the Indian Antarctic team and voyages. His contribution to limnological, glaciological and climate change studies resulted in several national and international publications.

Akilan Athiswamy Nadar is a scientist at NGRI. Working in GPS project. He is involved in GPS data processing and analysis for the past 5 years. He has visited Maitri once as summer member and once as a winter member.

S. K. Bera is a Palynologist by profession and served as a Scientist at Birbal Sahani Institute of Paleobotany, Lucknow.

R. L. Dhar has been a scientist at National Geophysical Research Institute, Hyderabad.

Amit Dharwadkar's contributions in Antarctica include the mapping of Dronning Maud Land and Larsemann Hills. His work in central Dronning Maud Land has added to a better understanding of Gondwanaland evolution. His contributions have been incorporated and resulted in the publication of two maps. The work in various glaciological investigations and climate change has yielded valuable insights in Polar Ice recession patterns and mass-balance characteristics. He has been instrumental in delineating the ice thickness and the Land-Ice-Sea interface using GPR which has tremendous implications on climate change and subsurface topography. He is presently actively engaged in unravelling the geological evolution of Himalaya, the third pole.

D. V. Divakara Rao has been a scientist at National Geophysical Research Institute, Hyderabad.

U. Gowrishankar has been a scientist at National Geophysical Research Institute, Hyderabad.

H. K. Gupta has been a scientist at National Geophysical Research Institute, Hyderabad.

S. S. H. Jafri has been a scientist at National Geophysical Research Institute, Hyderabad.

S. C. Jain has been a scientist at National Geophysical Research Institute, Hyderabad.

R.A. Jani has been the Faculty at the Physical Research Laboratory, Ahmedabad.

M. S. Joshi has been a scientist at National Geophysical Research Institute, Hyderabad.

L. Prem Kishore has been a scientist at National Geophysical Research Institute, Hyderabad.

M. L. Kurtadikar served as a Reader at Department of Physics, JES College, Jalana.

T. Madoom Hussain has been a scientist at National Geophysical Research Institute, Hyderabad.

E. C. Malaimani has been a scientist at National Geophysical Research Institute, Hyderabad.

G. S. Mittal has been a scientist at National Geophysical Research Institute, Hyderabad.

Y.V.V.B.S.N. Murthy has been a scientist at National Geophysical Research Institute, Hyderabad.

B. Nagarajan is presently serving as Faculty in the Indian Institute of Technology, Kanpur, India. Earlier, he served in the Survey of India in various capacities.

A. Nageshwara Rao has been a scientist at National Geophysical Research Institute, Hyderabad.

G. B. Navinchander has been a scientist at National Geophysical Research Institute, Hyderabad.

S. Padhy has been a scientist at National Geophysical Research Institute, Hyderabad.

P. Rama Rao has been a scientist at National Geophysical Research Institute, Hyderabad.

A. V. Ramana Rao has been a scientist at National Geophysical Research Institute, Hyderabad.

D.K. Rao is an Isotope Glaciologist who participated in many scientific expeditions to Indian Himalayan Glaciers. He is a retired scientist from the Physical Research laboratory in Ahmedabad, Gujarat.

D. Rameshwar Rao had his Honor's Graduation degree in Geology, and M.Tech in Applied Geology. His doctoral work was on the gneiss-charnockite transition rocks of Dharmapuri area, Tamil Nadu, India. His areas of interest are petrology, geochemistry and geochronology of igneous and metamorphic rocks. He retired as senior scientist from Wadia Institute of Himalayan Geology, Dehradun, India in November 2018.

I.P. Raju has been a scientist at National Geophysical Research Institute, Hyderabad.

S.V.R.R. Rao has been a scientist at National Geophysical Research Institute, Hyderabad.

N. Ravikumar has been a scientist at National Geophysical Research Institute, Hyderabad.

K. N. S. Reddy has been a scientist at National Geophysical Research Institute, Hyderabad.

V. N. Sarma has been a scientist at National Geophysical Research Institute, Hyderabad.

Sudipta Sengupta is an eminent structural geologist who was educated in Jadavpur University, Imperial college and Uppsala University. She served as a faculty member of Jadavpur University from 1982 to 2016. Sengupta published numerous papers pertaining to structural geology in important journals and is considered as an expert on the subject. She was one of the first Indian women to take part in the Indian Antarctic Expedition in 1983-84 and 1989-90 and conducted pioneering geological studies in the Schirmacher Hills of East Antarctica. She has received many awards including the SS Bhatnagar Prize, DN Wadia Medal, National Mineral Award, Antarctic Award etc. She is a Fellow of the Indian National Science Academy and West Bengal Academy of Science and Technology.

Prakash Shrivastava's contributions through his 5 geo-scientific expeditions including multi-national and multi-institutional Integrated Ocean Drilling Programme, where he represented GSI as well as India amongst scientists from 10 countries. The work on IODP Wilkes Land Expedition 318, indicate high to low grade metamorphic provenance and exhibit various glacial and interglacial phases. Sedimentological studies have elaborated the microenvironment of Schirmacher Oasis and Larsemann Hills. Arctic glaciological work has explained the refreeze layers in snow and mass balance. The geological studies have culminated in the publication of geological map of the Muhlig-Hoffman Mountains by GSI which has been accepted by SCAR.

Rajib Kumar Sinharay is a geophysicist having total 23 years of experinces in industry, research and academia. A scientific member of XXII Indian Scientific Expedition to Antarctica, 2002-2003. Presently working as an Associate Professor in the School of Petroleum, Dr. Viswanath Karad MIT-World Peace University, Pune, India.

A. K. Srivastava served as a Professor of Geology at the SGB Amravati University, Amravati, Maharashtra.

P. K. Srivastava served as a Commander and posted at the National Hydrographic Office, Dehradun. Currently serving Ministry of Earth Sciences, Government of India, New Delhi.

D. V. Subba Rao has been a scientist at National Geophysical Research Institute, Hyderabad.

V. M. Tiwari has been a scientist at National Geophysical Research Institute, Hyderabad.

Md. Ghous Uzzaman has been a scientist at National Geophysical Research Institute, Hyderabad

M. Venkatrayudu has been a scientist at National Geophysical Research Institute, Hyderabad.

S. K. Verma worked as a scientist at NGRI and visited Maitri once as summer member. As a Geophysicist worked on Gravity, Magnetics and Resistivity and Diamond exploration.

B. Vijayagopal has been a scientist at National Geophysical Research Institute, Hyderabad.

M. B. Vyageshwar Rao has been a scientist at National Geophysical Research Institute, Hyderabad.

Index

A

amphibolites facies 58, 81
Antarctica hydrography 260

B

bedrock 2, 98, 101, 110-111, 121-123, 127, 129, 132, 134-135, 213

C

C-band microwave 227-229, 234, 237-238
coastline 4, 160-165, 169, 181, 188, 191, 208, 211
control work 215-216, 222
crustal deformation 99, 112, 116-118, 139-141, 143, 153, 156, 158

D

Dakshin Gangotri 26, 30-31, 59, 62, 83, 96, 98, 101, 103, 119, 121-123, 125-127, 130, 135, 177-179, 181, 186, 194, 202, 205-207, 216-217, 240-242, 245-247, 249, 258
deglaciation 84, 211, 266
desmids 85
dielectric Cell 228, 230, 232, 234
dielectric properties 227-229, 234-239
ductile shear zones 70, 75, 81

E

E-bend 232
EM 98, 102, 124-125, 132
enclaves 28, 34, 51-52, 56, 65-66, 68-75, 77, 81-83, 110

F

firn 123, 130-131, 208, 210, 257
fissure zone 97-98, 102
fluid inclusions 33, 49, 53-54, 57, 59, 63, 66
foolhardy 176

G

Geodynamical Processes 100, 138-140, 152-153, 158
geological studies 1, 196-197, 199-200
Geophysical 38, 60, 95, 97-101, 106, 112, 117-119, 124, 132, 136, 138-139, 157-158, 194, 216, 223, 227, 234, 239, 256-259, 263
geothermal gradients 54
glacial sediments 1-3, 5, 7, 12, 23-24, 31
glaciers 96, 121-123, 128, 130, 132, 134-135, 195, 197, 208, 214, 217, 241, 258
glaciological studies 202, 204, 223
GPS 99-100, 112-113, 115-118, 125, 139-143, 145-146, 148, 152-159, 164-165, 167-168, 173-175, 177-179, 181-182, 184, 186-188, 190-191, 208, 215-223, 225
GPS-Geodesy 118, 139-141, 155-156, 158-159
gravity 86, 98-101, 106-108, 112-115, 117, 120, 170, 216-217, 220, 224
Ground Penetrating Radar 121, 136, 194,

211, 227

H

Hand plane 232
heliborne 98

I

ice core 122, 135-136, 194, 196, 205, 208-211, 240-244, 247-251, 254-257, 259
ice shelf 26, 85, 98, 101, 130, 160-164, 168-171, 173-178, 180-183, 185-189, 191, 202, 205, 207-208, 211
Indian Plate 139, 141, 143, 145, 147, 152-159, 220
isobaric cooling path 56
isostatic rebound 99, 112, 117

L

Larix 84, 87, 90-92, 94
leucogneisses 100, 112
limnological 194, 205, 211-212, 214, 263

M

mafic dykes 66, 68, 70, 73, 75, 80-81, 83
mapping 70, 82, 105, 122, 127, 134-136, 161, 194-195, 197-198, 200-202, 204, 211, 215-219, 222-226, 261
Marine Biomass 260, 265
Meteorological 92, 130, 161-163, 170, 176, 178, 182-183, 192, 196, 209, 216
model ages 48, 262
moraine matrix 84-85, 87

O

Ocean Lithosphere 100, 139-141, 153
oceanographic 160-163, 183, 263
Oceanography 29, 31, 106, 108, 118, 263
OXYGEN ISOTOPE 208, 267

P

palaeoclimatic 23, 26, 31, 88, 92-93, 210, 266-267
paleomagnetic 97-98, 100, 111
palynodebris 84, 86-87
past climate 240
permafrost 122, 134, 207-208
plate convergence 99, 112, 117
Polar Fallout of Radio nuclides 240
polythermal 122, 128-130
post-orogenic 112
Precambrian 2-3, 61, 66, 68-71, 83, 85, 108, 110, 118, 201, 261
Predominant 35, 160, 166
primary production 265
propagation 124, 232, 237
Proterozoic 32-33, 55, 59, 63, 65, 112

Q

Queen Maud Land 29-30, 66, 68, 82-83, 121, 123, 264

R

recession 194, 205-207

S

Schirmacher Hills 30, 65-66, 68-73, 79-83
Schirmacher Oasis 1-5, 23-25, 27-31, 62, 65-66, 82, 84-86, 90, 93, 95-96, 98, 100-102, 104, 106-112, 116, 118-119, 121-123, 127, 133, 136, 176, 194, 196, 198, 201-202, 204-209, 211-217, 219, 221-226, 260-261, 267
Sea Level Measurements 215
Sedimentology 1-2, 5, 24-25, 29-30, 96
Seismological Observatory 139, 146, 158
Seismotectonic 138-139
semi-diurnal variation 169

T

thermal stratification 266
topography 2-4, 98, 102, 105, 117, 121, 127, 129, 132, 169, 176, 181, 188, 191
Two-stage crustal evolution 32

U

Ulmus 84, 87, 94

W

waveguide 228, 232

Recommended Reference Books

IGI Global's reference books can now be purchased from three unique pricing formats: Print Only, E-Book Only, or Print + E-Book.
Shipping fees may apply.

www.igi-global.com

ISBN: 9781799822493
EISBN: 9781799822516
© 2021; 684 pp.
List Price: US$ **195**

ISBN: 9781799844082
EISBN: 9781799844099
© 2021; 279 pp.
List Price: US$ **195**

ISBN: 9781799884262
EISBN: 9781799884286
© 2021; 461 pp.
List Price: US$ **295**

ISBN: 9781799872504
EISBN: 9781799872528
© 2021; 297 pp.
List Price: US$ **195**

ISBN: 9781799875123
EISBN: 9781799875192
© 2021; 416 pp.
List Price: US$ **295**

ISBN: 9781799867098
EISBN: 9781799867111
© 2021; 405 pp.
List Price: US$ **195**

Do you want to stay current on the latest research trends, product announcements, news, and special offers?
Join IGI Global's mailing list to receive customized recommendations, exclusive discounts, and more.
Sign up at: **www.igi-global.com/newsletters.**

Publisher of Timely, Peer-Reviewed Inclusive Research Since 1988

IGI Global
PUBLISHER of TIMELY KNOWLEDGE

www.igi-global.com Sign up at www.igi-global.com/newsletters facebook.com/igiglobal twitter.com/igiglobal